D0164068

QUALITY ASSURANCE
METHODS AND
TECHNOLOGIES

Kenneth L. Arnold
Michael Holler

GLENCOE
McGraw-Hill
New York, New York Columbus, Ohio Mission Hills, California Peoria, Illinois

Library of Congress Cataloging-in-Publication Data

Arnold, Kenneth L., 1957-

Quality assurance : methods and technologies / Kenneth L. Arnold, Michael Holler.

p. cm.

Includes bibliographical references and index.

ISBN 0-02-802333-1

1. Quality assurance. I. Holler, Michael. II. Title.

TS156.6.A76 1994

658.5'62—dc20 94-20426

CIP

Cover photo: Electroglas wafer probe/station, Lawrence Manning/Westlight. Part One photo: Glencoe. Part Two photo: Doug Martin. Part Three photos: Glencoe, courtesy of Billy L. Reiter, Director of Training, Staveley Schools for Non-Destructive Testing (a division of Conam Inspection Inc.); Columbus, Ohio. Preliminary assistance courtesy of the American Society for Non-Destructive Testing, Inc.

Quality Assurance: Methods and Technologies

Send all inquiries to:
GLENCOE DIVISION
McGraw-Hill
936 Eastwind Drive
Westerville, OH 43081

ISBN 0-02-802333-1

Printed in the United States of America.

1 2 3 4 5 6 7 8 9 10 99 98 97 96 95

DEDICATION

This book is dedicated to those professionals who first strive for quality that is 99.99% acceptable and live for their dream of zero defects. It is with this concept in mind that we present this book. We hope that this material will help quality professionals gain a solid understanding of this science.

About the Authors

Kenneth L. Arnold has been in the quality assurance field for over 15 years. He holds an A.A.S. in Aviation Maintenance from Spartan School of Aeronautics, a B.S. in Technical Education from Oklahoma State U niversity, and an M.A. in Business Administration from Norwich University. He is certified by the American Society for Quality Control as a Certified Quality Engineer (CQE) and a Certified Quality Auditor (CQA).

He joined the staff of Spartan School of Aeronautics as an instructor in the aircraft Airframe and Powerplant Mechanics Department in 1979. In 1984 he transferred into the Quality Control, Non-Destructive Testing Department to develop many of the courses for this new program. In 1988 he accepted the position of Director of Student Records and in 1989, Director of Quality Control, Non-Destructive Testing, and Aircraft Powerplant departments.

He joined Dowell Schlumberger as quality engineer for the Tulsa Equipment Manufacturing division in 1990. Between 1990 and 1993, he functioned as quality control supervisor, prior to promotion to quality assurance manager. From this position he led Dowell Schlumberger's Tulsa manufacturing plant to ISO-9001 certification. He actively participated in developing the program for certification at the Tulsa Chemical Manufacturing plant for Dowell Schlumberger. This facility has attained certification to ISO-9002. He has worked as an internal consultant to numerous Schlumberger divisions that have attained, or are currently pursuing, ISO-9001 or -9002 certification. He has also provided seminars to groups and corporations that are working toward certification and is actively working as a consultant for these groups.

Currently he is co-owner and operations manager of Tulsa Equipment Manufacturing Co. TEM was founded in October 1993 and was certified to ISO-9002 in April 1994.

Michael Holler has been in the quality field for 16 years and in quality management for 12 years. He holds an A.A.S. in Aviation Maintenance from Spartan School of Aeronautics and an Associate of Arts from Otero Junior College in La Junta, Colorado.

Since 1982, he has served as Director of International Student Advisors and Director of the Nondestructive Testing Education Department at Spartan School of Aeronautics. Under his leadership, the nine-month technician course was expanded into an associate degree program in quality control. During the curriculum development for the new program, he realized the need for a book that incoporated many philosophies and technologies, and the dream of this book was born.

In 1990 he became the quality manager at Paragon Films, Inc., a start-up company in Broken Arrow, Oklahoma. He has supervised the development of the quality system for this company, which is engaged in continuous operation plastic extrusion. Paragon embraces the concept of Total Quality Management with a commitment to ISO-9001 certification.

PREFACE

During our twenty-plus years in technical education, we have found a need for a textbook that would give a well-rounded background in the science of quality. We noticed that books on quality control seemed to deal with only one philosophy or with statistics. In many cases, the reader is presumed to have prior knowledge of quality concepts. Rarely is the reader exposed to the fact that "there is no one right way" to achieve quality. We have set out to compile as many tools as possible in one book. In so doing, we hope to give a broader base from which to grow.

This book was written on a level that assumes that the reader has no prior knowledge of quality control. It is not designed as an in-depth study into any one area of quality, but is an introduction into the many facets of quality used in industry today. We hope that with the knowledge gained in this book, students can understand a quality organization and possibly offer alternatives for improvement. Extensive references are cited in each chapter that give readers an avenue to deepen their knowledge in a specific area.

The book is divided into three sections. Part I gives an overview of some of the popular philosophies in quality. It gives a brief history of quality through time and in different cultures. Basic tools are explained and the need for them is discussed.

Part II contains special inspection techniques. Several nondestructive examination methods are discussed on an entry level. The need for inspection is discussed, with proper uses and common abuses. Part III contains the statistics commonly used in quality control. It begins by putting the reader at ease with the thought of statistics, and skillfully takes the reader through probability distributions. Great pains are taken to explain how and why certain statistical functions are used in specific circumstances.

We wrote this book to allow maximum flexibility in material presentation at the instructor's option. Any of the book's three distinct sections could be taught first. Additionally, within each section, many sequences could be used. Instructors are encouraged to view the chapters in this book as discrete resources upon which to draw. The intention was to allow the instructor to be as creative as desired when teaching from this book.

Studies of personnel in the quality profession and those studying to enter it show an average education of between twelve and thirteen years. We have written this book on a grade level that makes it as easy to understand as possible. The quality profession is an exciting field, and we hope that this book will transmit some of that excitement—in other words, we want to take out the mystery but leave the magic of quality.

K.L.A.

M.H.

Acknowledgments

I wish to acknowledge the assistance of the following people in helping me complete this work. I thank my daughters Beth, Maggie, and Ginny for allowing me the time needed to complete this book. They experienced many hours when I was locked away writing and editing the manuscript. Without their love and support, completing this book would have been impossible.

KEN ARNOLD

I wish to express my heartfelt thanks to my wife Suzi and my children Greg, Kevin, John, and Angela, who "did without Daddy" for so many hours as I worked diligently at my computer. Their support and encouragement made this book a reality.

MIKE HOLLER

The authors are grateful to Damodar A. Ramanuj, Mid-State Technical College, Wisconsin Rapids, Wisconsin; Samuel L. Redding, Tulsa Junior College, Tulsa, Oklahoma; Thomas Colbath, Northridge Campus, Austin Community College, Austin, Texas; Ruby Ivens, Lansing Community College, Lansing, Michigan; for their detailed, technical, and practical suggestions.

We both express our deepest appreciation to Linda I. Laird and Judith Barton for their help in editing and correcting this book. Their hours of work have helped improve the quality of this work.

CONTENTS

QUALITY ASSURANCE
METHODS AND
TECHNOLOGIES

PART I: QUALITY ASSURANCE

Quality is not a new subject, as is believed by many companies. Many of the techniques that are used today were used hundreds or even thousands of years ago with good results. In the last 30 years, the issue of product quality has grown to the point that in many industries it has become the dominant issue of our time.

This change can be attributed to changes in the attitude of the customer, unprecedented global competition for markets that have traditionally been U.S. strongholds, and the need to improve profits and return on investment for every product produced. These reasons, although not all-inclusive, seem to be among the driving forces that have allowed quality assurance to emerge as a full partner in guiding many companies toward corporate goals.

This change in focus has increased the demand for business professionals who have more than a casual knowledge of the quality sciences. Today's demand is for business professionals who are well schooled in the philosophies and techniques used to determine, control, and improve the quality of the products and services provided by a company.

Part I (Chapters 1–6) provides a solid foundation in quality concepts. This section does not slant toward any one philosophy or

PART OBJECTIVES

In this section, you will be introduced to the basics of the following:

- Quality in history
- What quality is today
- Typical job responsibilities
- Traditional manufacturing organizations
- Effect of culture on quality
- How to change the culture
- Need for quality in today's business environment
- Prevalent quality philosophies
- Methods used to control quality

method; rather, it includes the common philosophies and approaches taken by the leaders in quality. Part I provides exposure to many ideas, philosophies, and methods used in quality assurance and quality control today. We make no judgments on the philosophies or methods. With the diversity of industries today, each of the philosophies and methods has a place where it will work effectively. Decision makers must have a good understanding of the broad scope of the quality issue in order to choose the best approach for their industry. This textbook was written to provide that broad introduction.

CHAPTERS IN THIS PART

QUALITY IN HISTORY

1

OBJECTIVES

After completing Chapter 1, the reader should be able to:

- Identify practices used in quality control today that were used thousands of years ago.
- Identify the first recorded attempt at mass production in the U.S.
- Discuss the approach to management outlined by Frederick Taylor.
- Discuss the effects of the Taylor method in a modern industry.
- Discuss how to change the culture in a modern industry.

There is tremendous emphasis on quality in today's marketplace. The number one selling point for a product is the quality image it projects. One quick look around the supermarket will prove the point. The phrases you will see most often are "NEW AND IMPROVED" or "HIGH QUALITY."

Marketing people know the power of a quality image. Slogans such as "Quality Is Job One"[1] or "The Quality Goes in Before the Name Goes On"[2] are used continually to sell products. Even the lack of demand for large-appliance repair personnel is used to portray the quality of Maytag products.

Newspapers are full of stories about liability lawsuits filed against companies and individuals because of a lack of quality. Product liability insurance is a fundamental need for companies today. Medical bills are skyrocketing partly because of medical liability insurance premiums. There are *no* single-engine, two-seat, trainer aircraft built in the U.S. Liability insurance for the airplane is one-half the cost of the airplane itself ($64,000 per airplane produced in 1985).[3] Not many people want to pay $128,000 for a $64,000 airplane!

If a company does not produce a quality product or service in today's market, that business will not last long. But the emphasis on quality did not develop overnight. Quality has been a serious business for all of recorded history.

1–1 QUALITY IN EGYPT

Quality is directly related to the ability to measure. If you can't measure it, you can't control it. Therefore, all measuring instruments must be "calibrated" to a standard. In ancient Egypt, the length of the Pharaoh's forearm was the unit of measure called the "cubit." A standard, called the cubit stick, was made to match the length of his arm. The cubit stick was the standard against which all other cubit sticks in the nation were measured. Every builder was required to have his or her cubit stick checked and calibrated once every full moon. Failure to comply with this regulation would cost the builder's life. In other words, if a cubit stick was the wrong length, the builder was killed.[4] Quality was very important in Egypt. Maybe that is why the pyramids are still standing.

1–2 BIBLICAL VIEW OF QUALITY

The biblical work ethic (which the U.S. generally professes to embrace) calls for quality. In several places throughout the Bible, quality (in the form of honesty) is commended. Leviticus 19:35 - 36 says, "You shall do no wrong in judgment, in measurement of weight, or capacity. You shall have just balances, just weights, a just ephah, and a just hin. . . ." This passage presents its admonition as an order from God.

To have just weights and measures, there had to be standards available to calibrate those weights and measures.

1–3 QUALITY IN ROME

In the Roman empire, quality was a life-and-death matter. If a building constructed by a Roman engineer were to collapse, and someone died, the engineer would be executed.

Many of the aquifers and bridges built by those engineers are still standing. It is hard to believe that a bridge thousands of years old could still be used. Perhaps the explanation lies in the fact that when the keystone that supported the bridge was laid, the engineer stood under the bridge. If the bridge collapsed, the engineer was the one who died.

This may seem somewhat barbaric, but things have not changed much. As an aircraft mechanic, one of the authors of this text repaired many airplanes, some for people who would not fly the plane unless the mechanic accompanied them on their first flight. The mechanic's flying in the plane is similar to the engineer's standing under the bridge—the effect is the same, only the technology has changed.

1–4 WAR FOUGHT OVER QUALITY

In 1790, France was ripped apart and the metric system was born partly because of quality. The problem with the French was not that they had no standard, but that they had too many standards. Each province had a different standard for its unit of measure. Those in power could buy and sell using whichever standard they found profitable. Such abuses in measurement were among the causes of the French Revolution.[5]

Imagine buying a two-liter bottle of soda only to find that it was smaller than the two-liter bottle purchased yesterday in another town. Continual manipulation of standards can precipitate great strife.

1–5 FRANKLIN EXPEDITION LOST BECAUSE OF POOR QUALITY

In 1845, an expedition searching for the Northwest Passage ended in tragedy. The entire 130-person crew became sick with an unknown disease and died. For over a century mystery surrounded this expedition.

Recently, however, new evidence indicates that the crew died from lead poisoning in their food.[6] Autopsies performed on three bodies that had been buried below the permafrost line proved that the expedition's lead-soldered cans of preserved meats, soups, and vegetables caused the death of the entire party.

Goldner Provisioners had been selected as the expedition's supplier over the objections of the ship's second officer. Although Goldner had never before provisioned a naval ship, it won the contract solely on the basis of its low bid. Poor quality in this case resulted in tragedy.

1–6 QUALITY AND THE INDUSTRIAL REVOLUTION

Before the industrial revolution, products were made one at a time. A craftsperson would make his or her craft product from beginning to end, handfitting every piece.

The quality of the product was judged by how well the pieces fit together. If the product ever broke, a person had to take it back to the craftworker to have another part hand crafted for that particular product.

The first attempt at mass production came in 1798, when Eli Whitney won a contract from the U.S. government for 10,000 muskets. Whitney began by mass producing enough parts to assemble a partial order of 700 muskets. However, the error rate in the parts was so high that only 14 muskets could be assembled from that batch.[7] Although the number that could be correctly assembled was less than exceptional, it did prove that mass production was possible and so, Whitney is credited with the concept of interchangeable parts.

Industry began to build its products piece by piece, instead of building a single item from beginning to end. Henry Ford started the first assembly line, and the responsibility for quality shifted from the craftworker to the supervisor. As the number of workers grew, the supervisor could no longer inspect the work of each worker, and the inspection department was born.

Armies of inspectors were employed to police the quality of the assembly line workers. As productivity increased, even a large number of inspectors could not keep up with the massive amount of products being put out by manufacturing plants. At that point, techniques employing statistics were used to control the quality of the operation.

1–7 THE EFFECT OF CULTURE ON QUALITY

Tremendous changes have taken place since 1900. Changes in manufacturing processes and culture have affected the course of quality control. It is evident how the changes in manufacturing processes could affect quality. What may not be as clear is how culture can affect quality. Therefore, some time must be spent on the cultural aspect of quality, because culture still plays an important role in quality today.

Around the turn of the century, there was a large influx of immigrants into the U.S. Prior to 1860, most immigrants coming to the U.S. were Protestant, English-speaking Anglo-Saxons. Most became successful in the new country.

After 1860, the religious and ethnic backgrounds of the immigrants shifted. From 1860 to 1920, 30 million immigrants came to America from Italy, Poland, Russia, and the Balkans. Most of these people settled in cities and tended to cluster together. Because of their poor English skills, these new Americans were looked upon by some as stupid. As they began to work in factories, they had a difficult time adjusting.

Their ignorance of technology and modern manufacturing methods put them at a disadvantage.

In 1881, Frederick Taylor introduced the time motion studies that earned him the title of "Father of Scientific Management."[8] Wishing to relieve the strain on industry, Taylor attempted to simplify the manufacturing process and thereby overcome the immigrants' lack of English communication and mechanical skills. This also reduced the cost of manufacturing by eliminating wasted time and effort.

The "Taylor Method," as it became known, was an overwhelming success. At that time in the history of the U.S. it was exactly what was needed. In effect, the Taylor Method put all authority into the hands of management. The work force was paid to work, not to think. It caused a polarization of management and labor. However, a polarization already existed for ethnic and religious reasons. Management began to think of labor as stupid and lazy. Labor started thinking of management as unfair and uncaring.

Labor unions began to fight for worker's rights. The adversarial and sometimes violent relationship between labor and management continued through the depression of the 1930s and into World War II. Although the literacy level of workers improved, management still viewed them as ignorant.

During World War II, the manufacturing industry was thrown into a production frenzy. Demanding schedules were inflicted on contractors for the war effort. For a short time, labor and management worked as a team. Productivity and quality were very high.

After the war, the American manufacturing industry was the only one in the world left intact. As such, postwar U.S. businesses supplied over 50 percent of the world's needs through exports and held 95 percent of the domestic market. Demand far exceeded supply.[9] There was not a wide variety of things to buy, and people bought what they could get. This situation enabled manufacturers to dictate product requirements to customers over the next 25 years. To a large extent, American industry pictured its work force as uninformed, apathetic, and illiterate.

Foreign manufacturing industries finally recovered. Supply began to catch up with demand. In contrast to U.S. businesses, overseas companies began to ask customers what they wanted in the products they bought. Products were made to the customers' specifications, using new equipment and with positive input from the work force. The foreign strategy worked to their advantage.

In 1973, an unofficial oil embargo against the U.S. sent the price of oil to new heights. Before 1973, U.S. cars were large and fuel inefficient. Honda met the needs of customers by marketing small, fuel-efficient cars. Today, Hondas are among the most widely owned cars in the U.S. The American automobile industry was forced to compete on Honda's level or die.

In the early 1980s, the bottom fell out of the steel industry on a global basis. Cultural impact on quality can be evaluated by investigating how an American company and a Japanese company dealt with this situation. The steel market disappeared almost overnight. As a result, the American company laid off thousands of workers to cut costs. By contrast, the Japanese company did not lay off any of its workers. Instead of a layoff, everyone in the Japanese company took a pay cut—not just the workers, but everyone from the CEO to the janitors. The Japanese company also cut back its production, but faced the problem of what to do with all the idle workers. The decision was made to put them to work renovating the company facilities. The workers began to paint, rebuild, and clean the entire factory.

As time passed, the market did not improve. In fact, it worsened. The American company experienced a second layoff. The Japanese company experienced a second wage cut and continued to work on the factory. Both companies had virtually stopped all steel production at this point.

More time passed and all but a few American steel workers had been laid off, whereas the Japanese factory was upgraded and modernized. Still the market was depressed. The Japanese company again faced the situation of what to do with all the workers. The decision was made to build another blast furnace. So, using the steel that was stockpiled, the Japanese company began to build the furnace.

Finally, the market began to pick up. Demand increased as the world began to build again. The American company could not meet demands because its workers had been retrained in other fields and were not available. Therefore, a new work force needed to be trained. The Japanese company had its entire work force in place, a beautifully renovated factory, and the second largest blast furnace in the world.[10] The American steel industry still has not overcome this disaster, and America is still struggling to catch up.

One might be tempted to condemn the management of the American company for laying off its workers, but look at the situation realistically. The problem lies in the way the two work forces think about their work. In America, labor and management are on adversarial terms. Neither labor nor management feels much responsibility toward the other. In America, many unions will not allow the workers' pay to be cut, so layoffs are the only way to cut costs.

On the other hand, Japanese culture fosters a great respect on the part of the workers for their company. Also, the management of the companies feels a responsibility for the well-being of its workers much like a parent/child relationship.

Investors, management, and labor all create the culture of a company, just as on a national level it takes all of us to create a culture. If the steel market were to disappear again, the result would be the same. The American company has not changed its culture.

1–8 HOW TO CHANGE A CULTURE

Quality control professionals are limited in their ability to alter the culture of a manufacturing organization. Some change may come through the workers' efforts; however, lasting changes in this area come only from the top down.

It may be difficult to effect a change, but a quality professional should know what variables will affect the culture of an organization. The culture is made up of individuals on the corporate level. People act to fulfill needs, either perceived or real. Armed with this knowledge, a person will better predict the possible success of the changes required in a quality improvement effort.

Abraham Maslow defined the needs of people in Western culture. Maslow's hierarchy of needs (Figure 1-1, next page)lists the basic needs of people in the Western industrial nations on a priority basis.[11]

People will meet their real or perceived needs progressively from the bottom to the top. The result is the same—if a perceived lower need is not met, a person's behavior will be the same as if the need is real. This is the behavior with which the quality technician must deal.

Each person decides what will satisfy each of these needs in his or her life. Only after the individual feels the lower need satisfied can the next-level need be addressed. If an individual is to contribute to the quality improvement organization, he or she must be operating on the upper levels of Maslow's hierarchy.

For instance, if someone does not feel that his or her basic physical needs are being met, his or her energy will be expended in trying to satisfy those needs (food and shelter) and not the needs of a quality improvement plan. People have to become personally involved with

Figure 1–1 Maslow's Hierarchy

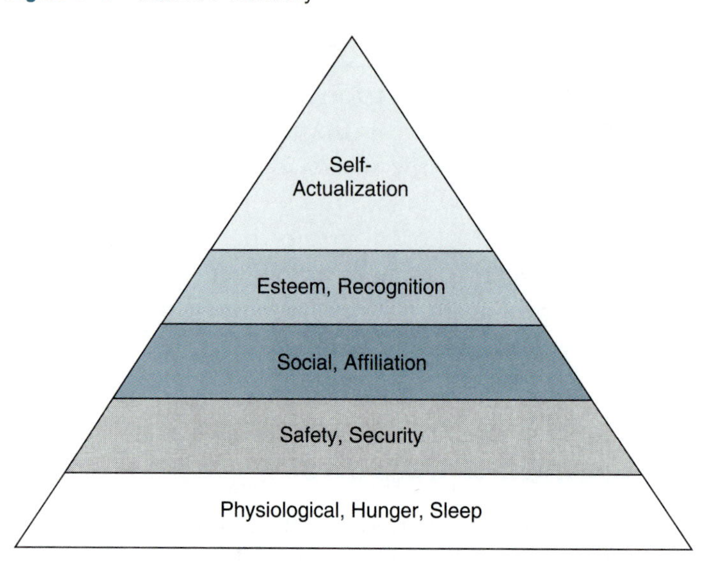

quality for it to improve. They can become involved only when it becomes a priority. Involvement will become a priority only when people feel good about themselves.

You should realize at this point that people's self-image is often connected to the work they do. If a person perceives that his or her work is accepted by others, then he or she will feel good (esteem). Only when a person's identity is secure can that person begin to expand into areas that have not yet been mastered. Self-actualization is the point where a person reaches out from a secure base to explore new areas. The possibility of failure is accepted because the person feels secure in previously displayed abilities.

A quality professional should not be surprised if an improvement effort is not well accepted when there is a morale problem in the company. The basic needs of the workers must be met or the workers' efforts will be spent in trying to meet their own needs.

SUMMARY

Although quality, or the lack of it, has been highly publicized in recent years, it has been an important part of almost every culture. Quality was integrated into people's work. It was taken for granted; rather than being a separate part of doing business, it was the fulcrum upon which business balanced. In some cultures the quality of one's work was literally a matter of life and death. Failure to perform the necessary tasks to ensure quality could cost workers their lives.

Quality became a separate, stand-alone entity only after the beginning of the industrial revolution. It brought with it mass production, interchangeable parts, and the loss of people's identification with the products they made. Great armies of workers began making pieces and parts of the whole. In the United States many immigrants made up this work force. Language barriers and lack of education created great problems in communication. The labor force began to be viewed as a stupid and ignorant group. Deep rifts were created between labor and management. Labor unions gained in popularity.

Frederick Taylor and others began to study the manufacturing process and to view the labor force as a part of the process. The process was broken into small parts that could be communicated easily to the immigrant work force. The labor force was discouraged from thinking about ways to improve the process—that was left to the educated and literate. The resulting culture has remained long after the illiterate immigrant work force became educated. Many studies have been done to try to remotivate workers. To a great extent, the quality movement has led in this fight to put improvement back into the hands of the workers.

KEY TERMS

Calibrate The act of adjusting a measuring instrument to read accurately to a known standard.

Self-actualization The state of mind where a person reaches out from a secure base to explore new areas.

Standard Test pieces of known size used to adjust the reading of a measuring instrument.

Taylor Method The management approach that breaks the manufacturing process into small tasks in a set pattern to be altered only by the management personnel.

HISTORY TEST

1. List three practices used in quality control today that were also used thousands of years ago.
2. What product was made in the first recorded attempt at mass production in the U.S.? What was the result of this attempt?
3. Discuss in a short paragraph the approach to management outlined by Frederick Taylor.
4. Discuss in a short paragraph the effects of the Taylor Method in a modern industry.
5. Discuss in a short paragraph how to change the culture in a modern industry.

NOTES

1. Ford Motor Company, nationwide television advertisement.
2. Zenith Corp., nationwide television advertisement.
3. Jan VonFlatern, "Dealing with Product Liability: One Manufacturer's New Approach" (Paper delivered at the A.B.A. National Institute Litigation in Aviation, 27–28 October 1988).
4. Ted Busch, *Fundamentals of Dimensional Metrology,* 3d ed. (New York: Delmar Publishers, Inc., 1966), 22.
5. Busch, *Fundamentals,* 25.
6. Paula Aspell, *Buried in the Ice,* Tinsel Media Production, Ltd., in association with Telefilms Canada and the Canadian Broadcasting Company for Public Broadcasting System's "Nova."
7. DataMyte Corporation, *DataMyte Handbook,* 2d ed. (Minneapolis, MN: DataMyte Corp., 1986), 1–3.
8. Joseph Juran, *The Taylor System and Quality Control* (a Quality Progress Reprint) (Milwaukee, WI: American Society for Quality Control, 1987).
9. W. Edwards Deming (Speech delivered to the Fortieth Quality Congress, 19 May 1986, Anaheim, CA).
10. Deming speech.
11. Wayne Hoy and Cecol Miskel, *Educational Administration,* 2d ed. (New York: Random House, 1982), 146.

QUALITY

2

OBJECTIVES

After completing Chapter 2, the reader should be able to:

- Define quality as it relates to customers.
- Distinguish between quality assurance and quality control job functions.
- Distinguish among inspection, audit, and surveillance.
- Discuss the job responsibilities of a QA/QC manager.
- Discuss the job responsibilities of a quality engineer.
- Discuss the job responsibilities of an inspector.
- Discuss the advantages and disadvantages of the manufacturing organizations discussed in this chapter.

Before you can begin a comprehensive study of quality control, you must possess a command of some basic terms used in the industry. If you are familiar with these terms (see Glossary) as they are used in some companies, you may find that they differ from the descriptions in this textbook. It will help you understand quality control work, however, if you learn the terms as described in this section. This is not to say that the other definitions are wrong, but the learning process will be much easier if everyone speaks the same language.

2–1 WHAT IS QUALITY?

If 10 different people were asked to define quality, there would probably be 10 different answers. Most of the definitions would deal with absolutes. Phrases such as "the best" would be widely used. When asked what quality they expect, many people say they want the best. This indicates that the general public often perceives quality as an absolute. However, if you watch people's buying habits you will see that what they do is not what they say. If people truly wanted the best of everything, cost would be no object. Most people will buy a sale item or a midpriced item rather than buy the most expensive one. This indicates that people want the best that they can afford.

When you buy something, there are many things to consider—cost, size, performance, economy of operation, warranty, appearance, and reputation of manufacturer. All these things are important and have a bearing on what you picture as "quality." Added together, they give a person a quality image of the product.

The concept of quality as goodness, luxury, or the best is not adequate for a quality professional. For example, a person may think that a Rolls Royce is the best vehicle made. "The best" depends on that person's perspective. This vehicle would *not* be the best in the outback of Australia, the rain forests of South America, or the great sand dunes of the Sahara. The best vehicle for these places would be one with four-wheel drive. The fact that there are not many gas stations in these places would also be important when considering the best vehicle; the Rolls Royce is not designed with gas economy in mind. Therefore, the intended use of the product or service is very important to the quality professional.

There are several ways to define quality that are suitable for the quality professional. The following are accepted definitions of quality:

- Quality is the totality of features and characteristics of a product or service that bear on its ability to satisfy given needs.[1]
- Quality is "fitness for use."[2]
- Quality is "conformance to requirements."[3]
- Quality is the composite of all of the attributes or characteristics including performance of an item or product.[4]
- Quality is the total composite product and service characteristics of marketing, engineering, manufacturing, and maintenance through which the product and service in use will meet the expectations of the customer.[5]
- Quality is the conformance to standards that represent the product's or service's basic characteristic, and are based on customer needs and expectations.[6]

These definitions may seem confusing at first, but read them again with customer requirements in mind. This time they should all sound similar. In every case, *the requirements of the customer are always a consideration in a valid definition of quality.*

To further define product quality, quality can be divided into three major categories:

- Quality of design.[7]
- Quality of conformance to design[8]
- Quality of performance[9]

The quality effort must be involved in each of these three areas to be effective. Let's take a look at quality in each of these areas.

Quality of Design The quality of design is concerned with the specifications for manufacturing. Market research performed by targeting intended buyers is the initial step in design quality. Also important are the intended functions of the product (i.e., luxury, style, and longevity); blueprint tolerances, such as material characteristics and measurement clearances; and manufacturing processes and procedures.

The cost of the item will increase as the quality of design increases.[10] For instance, if the design calls for measurement tolerances to be ±0.0005 in. rather than ±0.005 in., it will cost more to manufacture that part. If a product is made of a material that will last a long time, it will cost more for that material than for a material with a shorter life expectancy. In all areas of design quality, if the design quality increases, the cost to produce will also increase.

Quality of Conformance to Design This area is the one most often associated with quality control. Quality control has its greatest influence on a product's conformance to specification. It deals with whether or not the processes and procedures outlined in the design phase are being used and are effective. Quality control monitors the conformance of material to specification and checks to ensure that tolerances are being held to the specifications called for in the blueprint. Quality of conformance monitors the product and procedures from raw material to shipping the finished product to ensure that all aspects of the product and the production process conform to the written specification.

The cost of the product can actually be decreased as the product quality increases by ensuring that the product conforms to the written specification.[11] Failure to hold specified tolerances can cost a company tremendous amounts of money. These losses come in the form of material costs, rework, shipping, and many other hidden costs that will be covered in more detail later in this textbook. It is not always easy to hold the specified tolerances shown on the blueprint and follow the procedures outlined by a company without quality training. That is why there is such a demand for people with formal quality control training. The quality of the product is the responsibility of every employee, but someone must give them direction. That is the job of management, with the help of the quality control department. As the processes are controlled and the employees trained in the proper methods of assembly, defects are reduced and profits go up.

Quality of Performance Quality of performance is the result of design quality and quality of conformance. If either the design or conformance quality is not up to par, the quality of performance will suffer. For example, if a product is well designed to work under the stresses expected, but the tolerances are not held, the product will probably fail early. Likewise, if a product is not designed to handle the stress to which it will be exposed, no amount of production quality will make it last. Since the performance will suffer if a part is poorly designed or poorly constructed, quality control assurance must be vitally involved in both areas. A company must incorporate a good communication system to give quality control the information necessary to make changes and corrections that will result in increased performance.

So we can see that the term *quality* is not an absolute. It is a diamond with many facets. To further expand the concept of quality management, the terms *quality control* and *quality assurance* should be discussed.

Quality Control and Quality Assurance Often in industry these terms are synonymous. If you are employed in a company in which this is the case, this section will come in handy when reading articles written by people from other companies. If employed by a company that differentiates between the two, this section will give you some background in job duties.

Look at the words *control* and *assurance* before trying to define the complete terms. The word *control* is defined by *The American Heritage Dictionary* as "the authority or ability to regulate, direct or dominate." "Assurance" is defined as "(1) the act of assuring or the state of being assured. (2) a statement or indication that inspires confidence: guarantee."

The word *control* indicates an active role. To control one must take direct and positive action. The word *assurance,* on the other hand, indicates confidence—in other words, an attitude. In the case of quality assurance, the confidence rests on the positive action expected to control the quality of a product.

An analogy is the air traffic control tower at an airport. It is not an air traffic assurance tower. The tower is charged with directing (controlling) traffic, rather than making sure (assuring) traffic exists. A traffic assurance tower would cause a disaster, since there would be no control. With this difference in mind, consider the complete terms *quality control* and *quality assurance.*

> Broadly, quality control has to do with making quality what it should be and quality assurance has to do with making sure that quality is what it should be.[12]

Quality control uses such strategies as inspections and statistical process control to maintain the quality of the product during production at a predetermined level. A quality control department uses vendor audits and vendor surveillance to ensure that incoming product is at an acceptable quality level.

To be effective, control must possess several attributes. If any of these is missing, control will be ineffective. For control of the quality of a product to be effective, everyone involved in the production process must understand the overall approach and the methods used. The approach must follow organization patterns and quickly and accurately report deviations from the acceptable level of quality. The organization must establish methods for corrective action when a deviation is noted, and the overall approach must be economical to the point that the money saved is greater than the amount of money spent. In other words, the system should show a positive return on investment.

Quality assurance uses such techniques as internal audits and surveillance to ensure that the quality organization is accomplishing two goals:

- The organization is following the procedures as they are stated in the quality manual.
- The procedures, when followed, are effective and yield the desired results.

If a company splits the quality effort between quality control and quality assurance, you must study the job responsibilities of each department because each individual company may have different job responsibilities assigned to each area. Generally, you will find that the quality control department is involved in the day-to-day monitoring, evaluation, and adjustment of the processes involved in manufacturing a product. The quality assurance department is usually involved in the actions taken by a company to give confidence that the products meet the targeted quality level. In many companies, the quality assurance department manages the quality, whereas the quality control department performs hands-on evaluation and control of quality.

Quality Assurance Responsibilities In a traditional setting, the quality assurance organization has specific responsibilities beyond the overall duty to set broad goals, provide guidance, and set direction. These specific responsibilities include the following:

- Review qualifications and training of personnel.
- Review quality-related records.
- Develop, document, and implement quality procedures.
- Schedule, conduct, and follow up quality audits.
  - internal (within the company)
  - external (outside the company)

- Be involved in the procurement cycle.
  - preaward audit (prior to signing a contract with a vendor)
  - postaward audit (after the contract is executed by the vendor)
  - review documents (such as the purchase order or the contract)
  - review bidders list
  - develop a list of approved vendors
- Be involved in the design review cycle.
- Schedule and conduct surveillance.
  - vendor (during the execution of a contract)
  - process or program (internally in a company
  - records (traceability for governing bodies)

Quality Control Responsibilities In a traditional setting, the quality control organization has specific responsibilities that center around following the direction given by the quality assurance organization. These activities are directed toward evaluation of the product and control of the processes used to produce the product. Such activities include:

- Establishing hold points for product evaluation
- Qualifying and training personnel
- Reviewing and retaining records
- Analyzing defect data
- Reviewing and following up corrective action reports
- Performing special tests
- Performing inspections
  - receiving inspection
  - in-process inspection
  - final inspection
  - vendor source inspections
- Witnessing tests (e.g., hydrostatic test, nondestructive testing [NDT] or performance testing)
- Reporting on and taking corrective action on nonconforming product
- Conducting surveillance
  - in-process
  - equipment and gauge calibration
  - quality records

Remember, in some companies there is no difference between quality assurance and quality control, whereas in other companies the departments are separated. This is an issue of no great importance provided that the quality effort is company-wide and not pushed by one department only. However, if your company differentiates between the two, you should understand the difference as outlined by the company. No right way or wrong way exists; the right way to organize is whichever way will produce continual improvement in quality and profits.

2–2 INSPECTION, AUDIT, AND SURVEILLANCE

There are three tasks conducted in the quality effort that may cause some confusion to a newcomer in quality. Although similar in some ways, each is unique in the methods used and each has a different purpose. These tasks are inspection, audit, and surveillance. To determine the similarities and differences, let us evaluate the definition of each.

Inspection is a highly defined, close examination of a product or process. The term *inspection* usually refers to the examination process of an object or product.

Quality is not inspected into a product; quality is built into a product. The idea of inspecting quality into a product was widely held through the 1950s and 1960s when companies hired armies of inspectors to check everything as it was completed. This idea is still surprisingly well entrenched in many companies. In a recent visit to a large contractor producing high-tech electronic guidance systems for the U.S. Department of Defense, one of the authors discovered that critical electronic PC boards were 100 percent inspected—by seven different inspectors—before they were accepted and shipped to the military. At incoming inspection (by the U.S. Air Force) 22 percent of the boards were identified as being defective and returned to the contractor. In other words, 700 percent inspection (100 percent inspection performed seven times) still allowed 22 percent bad product to be shipped to the customer. If the customer, too, is inspecting all boards (100 percent inspection), how many defective boards are being overlooked at this level and assembled into the guidance systems?

Inspection, if used properly, can provide a large amount of information about the performance of a process. Inspection should be used as a tool to gather information, not as the ultimate method to ensure a quality product. Specific information about inspection will be covered in Chapter 7.

Quality Audit Quality audit is an inspection of an organization's adherence to the established quality standards. Like the inspection of a product, the audit procedure must be very well defined. The quality manual is used to set up the audit procedure. The purpose of the audit is to ensure that the quality control procedures are in place and that they are being followed. Therefore, the audit contains a checklist of steps in the quality procedures that are vital to the effectiveness of overall product quality.

The quality audit should be designed to answer three basic questions about the organization being audited:

- Does the organization have a quality system? This is usually evidenced by a quality manual, operating manual, or quality procedures.
- Is the quality system being followed? An audit is conducted to determine whether the procedures are being adhered to on an ongoing, consistent basis.
- Is the system effective? Are the results of following the procedures consistent and positive?

The audit may be done internally (by people from the company itself) or externally (by people from outside the company). In either case, the results of the audit are given to upper-level management for further action, because the auditor does not carry the authority to make corrections in the procedures. The auditor should have the ability to follow up on findings and recommendations.

The focus of an audit can be as broad as a system-wide audit of corporate operations. Or it can have a narrower scope, to include only a process used to accomplish a particular goal. The focus might be limited to a specific product. No matter how narrow the scope of the audit, the same three questions need to be answered. The specifics of auditing will be covered in more detail in Chapter 5.

Surveillance Surveillance is a loose inspection process that uses some of the techniques of the audit and some of the inspection. The procedures used in surveillance are much less concise than for the inspection or audit. Surveillance is an objective evaluation to determine how well the quality procedures are being followed in day-to-day production, along with determining how well the procedures, when followed, maintain the quality of the product. Surveillance answers the questions, Is the process of performing as planned, and is the product of acceptable quality?

2–3 TYPES OF NONCONFORMANCE

Quality control's function, defined earlier, is to find what is right as well as what is wrong with a part. These findings may be on the product level, process level, or departmental level. Although it is important to know what is right about a part, finding problems with a part is the objective of most inspections. As such, nonconformances or defects are generally placed in one of three categories based on the overall effect on the product's usefulness. Three categories that are universally accepted are critical, major, and minor.

The quality professional will usually find these three categories used in conjunction with the product, rather than the process used to produce the product. For this reason, the definitions used here come from MIL-STD 105-D.

Critical defect: "A critical defect is a defect that judgment and experience indicate is likely to result in hazardous or unsafe conditions for individuals using, maintaining, or depending upon the product. . . ."[13]

Major defect: "A major defect is a defect, other than critical, that is likely to result in failure, or to reduce materially the usability of the unit of product for its intended purpose."[14]

Minor defect: "A minor defect is a defect that is not likely to reduce materially the usability of the unit of product for its intended purpose, or is a departure from established standards having little bearing on the effective use or operation of the unit."[15]

It must be stressed here that these three categories are not the only possible classifications. A company may break these categories down into smaller sections, if it is useful to do so. These three categories may also be used to isolate nonconformances in procedures or in processes. However, these three categories are the basic standards used in the quality effort to give relative importance to nonconformances.

2–4 JOB RESPONSIBILITIES

The science of controlling quality is as diverse as companies themselves and as the variety of processes within a company. Within this discipline, there are many different jobs requiring many different skills. To explore some of the skills needed and some sample career opportunities in the quality field, let us consider some typical job categories and responsibilities. This list is not exhaustive in its scope or all-inclusive.

Rather, it gives the typical responsibilities within each area. Typically, quality assurance responsibilities are divided into three general categories: quality management, quality engineering, and inspection. Within each of these areas are many job responsibilities, which will be discussed in the next sections.

QA/QC Manager In addition to supervision, the manager of a QA/QC organization has the responsibility of providing upper management with accurate and timely feedback on production. He or she should provide a scorecard on how well production meets the quality objective. The QA/QC manager should set quality goals, provide direction, develop programs, and monitor the progress of the departments involved in meeting the objectives.

Manufacturing fabricates a product at the lowest possible cost in order to offer the best return on the stockholders' investments. This drive for economy must include quality, but must always meet the expectations or objectives of the company. Remember, the company objective is to manufacture a product at the least cost for the customer's "fitness for use."

The QA/QC manager, then, must monitor production to meet these cost and use objectives. This job may include vendor performance evaluations and working with engineering, production, and shipping to investigate and solve customer complaints. More important, a manager has the responsibility of looking at all information provided by the quality organization and drawing up a recommendation for meeting the customer's use needs at the lowest cost. The manager has to weigh every decision constantly in light of two criteria: (1) Will this decision adversely affect the product's quality in the eyes of the customer? and (2) Will this decision adversely affect the ability of the company to make a profit on the product? The manager must constantly find the balance between the customer's needs and the company's needs.

Quality Engineer The quality engineer's job is to analyze quality-related issues, solve problems, train employees, and implement quality-improvement projects. These duties require the quality engineer to have an in-depth, working knowledge of all aspects of quality assurance, statistical problem-solving techniques, and basic philosophies in quality control. The quality engineer must be able to develop a program that will take a company in the direction mapped by the quality manager. This direction must also meet the quality goals set by upper management in a timely and cost-effective manner. The quality engineer must assemble, from drawings and from the manufacturing unit, the quality plan answering the customer's specifications. The quality plan assures management of a comprehensive quality program to meet the customer's requirements without manufacturing delays. A quality plan should include:

- Written procedures
- Inspection methods
- Inspection hold points
- Customer hold points
- Sampling plans
- Definition of defect types (e.g., critical, major, minor, and incidental) to determine acceptance standards
- Description of quality or marketplace "fitness for use"

Typical quality engineer responsibilities include:

- Writing a quality plan for each product.
- Reviewing drawings and purchase orders for correct quality requirements.
- Discussing quality objectives with customers and resolving field complaints.
- Reviewing manufacturing methods to improve quality.
- Reviewing customer rejects.

Good parts, methods, and training all contribute to making a good product. For a quality product to be produced in a consistent and timely manner, each of the pieces of the quality puzzle must be brought together in the right way, at the right time, and in the right amount. To evaluate whether each of the ingredients needed to produce a consistent product to an acceptable quality standard is present, the quality system in place within a company must be inspected. This inspection of the quality system is typically called a quality audit.

The quality engineer is typically part of the audit team to evaluate the internal system in place in one's own company or a vendor's system. In many cases the quality engineer not only plans the audit, but also conducts it.

A quality auditor must make sure, by inspection, that individuals comply with the "quality plan." If the quality plan is followed, the quality objectives will be met. This inspection (audit) of individuals and the system in engineering, purchasing, and planning to ensure compliance is part of the quality engineer's responsibility.

Quality audits are normally an evaluation of how well the company is performing to meet quality objectives. Audits are the measure that upper management uses to judge the organization's compliance to company objectives.

Quality audits use company procedures as a guide to interview individuals within departments to determine compliance to quality manuals and procedures. Typical quality engineer audit responsibilities are to:

- Determine whether design procedures are followed.
- Review designs and drawings for a complete engineering design review.
- Check shop drawings to determine whether procedures for revising drawings are followed.
- Ensure that special processes are reviewed.
- Review customer requests and problems that have been conveyed to engineering and manufacturing.
- Determine whether the sales department takes a customer order accurately and provides manufacturing with the customer's requirements.
- Check quality plans to ensure that they are followed.
- Evaluate procedures and policies to determine that, if followed, the product will be produced to an acceptable quality level.

A quality engineer must have knowledge of the test methods, procedures, and abilities of an inspection or QC organization. The quality engineer (QE) usually works with the industrial engineer (IE) to develop the best quality methods for a company. Thus, the QE must know any special processes that are unique to that company.

Procedures are written to define each step that must be performed to ensure that everyone involved with the product is aware of requirements and potential problems. A quality manual is maintained by the QE detailing the procedures that the organization must follow to ensure a consistent product that meets the customer's expectations. This manual defines company policy to achieve a quality objective. It specifies the intended customer and what the customer expects from the product. It also outlines the procedures to be followed to meet the customer's expectations.

2–5 INSPECTION

Although inspection will be discussed in Chapter 7, the role of the inspector needs to be addressed here. Contrary to the apparent philosophy of many companies today, *quality cannot be inspected into a part. Quality must be designed and built into the part.* If this statement is true, then why discuss the role of inspection? The main roles of inspection under this philosophy are to verify and validate quality and to gather data. It is not the traditional "policing" role of sorting the good from the bad.

In the traditional role, quality is achieved *post mortem.* A bad part is produced, the inspector finds it, and a rejection slip (death certificate) is filled out and filed along with many other death certificates. The inspector has accomplished his or her job, but no other steps are taken to prevent the recurrence of the defect. Inspection merely reacts to defective products being produced. In today's business environment, this reactive approach has been fatal to many companies.

If inspection is used, its purpose should be to gather data for a proactive approach to problem solving. Not only should the inspection be used to identify a nonconforming product, but it should also gather the data needed to identify the root cause of the problem and to find and monitor the remedy. This is the quality engineering approach.

Inspection is traditionally used in one or more of the following areas:

- Receiving inspection
- Source inspection
- In-process inspection
- Final inspection
- Special processes (metrology, nondestructive examination [NDE], and welding)

Receiving Inspection A receiving inspection assures management that items received from vendors meet requirements of the purchase order. But if this were all a receiving inspector accomplished, soon he or she would ruin a company. Why? A purchase order is just a way of telling a vendor what materials, drawings, specifications, and quantities of parts a company needs. Sometimes communication within a company breaks down, for whatever reason, and a purchase order provides a vendor with incorrect or outdated information. A receiving inspector may see that parts are correct in every way according to the purchase order, but the parts may still not be usable for correct production. The parts could be made from the wrong drawing revision, using improper material or heat treatment, or embodying other possible mistakes resulting from revisions not reflected by the purchase order.

The receiving inspector must be alert to these errors. The receiving inspector may use a sample of a product lot to reduce inspection costs. A sampling plan or procedure is a statistical method that uses a small sample from a product lot to make reasonable assumptions about the lot quality.

Typical receiving inspector responsibilities include:

- Verifying parts count
- Identifying material
- Ensuring that mechanical dimensions are within tolerance
- Checking physical properties (hardness, heat treatment)
- Checking paint (color, thickness, and coverage)
- Verifying identification (lot number or serial number)
- Checking fit (assembly compatibility)
- Checking for correct threads
- Verifying that correct version of drawing or specification has been used.

Source Inspection Source inspection is the "eyes and ears" of a company in its vendors' shops. It is always easier to correct problems when they are detected in a vendor's shop than later when the parts are needed immediately in the company's manufacturing process. Many companies emphasize this source inspection as a cost-saving method used to improve shipment schedules.

A source inspector must review the vendor's and manufacturer's shops to ensure that both conform to the quality standards. Many times a source inspector can find misunderstandings in prints, purchase orders, quality plans, or requirements. The source inspector can then bring these to the attention of both the vendor's and company's management.

Sometimes vendors cannot perform or build a product as required by the purchase order. Vendors sometimes take on work without the experience or resources necessary to accomplish a task. The source inspector in the vendor's shop must make this evaluation and report to his or her company.

In-Process Inspection In-process inspection ensures quality during fabrication or assembly. With good parts from the receiving inspection, the shop is expected to assemble or fabricate a quality product. In-process inspection is conducted differently depending on the type of manufacturing method the shop uses (job shop or assembly line).

A job shop is a manufacturing method for low-volume (usually high-cost) manufacturing, such as production of an offshore oil drilling rig or a supercomputer. The assembly line is associated with high-volume (usually low-cost) manufacturing, such as automobiles or microchips. Each manufacturing approach has unique procedures and needs for in-process inspection.

Job Shop An in-process inspector in a job shop ensures that the work performed is according to the correct drawing and procedure, and done by qualified individuals. Quality in a job shop concentrates on the process used to make the product. Much of the in-process inspection is centered around how a job was performed and who performed it. If the process was performed by a qualified person following correct procedures, chances are the product will meet quality standards. Much of the actual in-process quality is delegated to the worker performing the work.

Typical in-process inspector responsibilities include:

- Ensuring that the quality plan is followed and all inspections are performed as requested by the customer.
- Verifying that hold points are performed.
- Ensuring that the customer is aware of future hold points and that the parts meet company inspection requirements before the customer's inspection is performed.
- Verifying that correct drawings are used in the shop and obsolete drawings are destroyed.
- Confirming that correct methods or procedures are available at each work station.
- Verifying that individuals performing special processes are qualified to do so.
- Ensuring that equipment and gauges are calibrated.
- Confirming that the shop meets quality plan requirements.
- Verifying that individual craftspeople sign their work, when required.

Assembly Line An assembly line in-process inspector performs either sampling on the line or 100% inspection as a checker. Assembly line inspection concentrates on the product quality instead of the process used to produce the product.

Assembly lines are designed by industrial engineers and quality engineers (planning) to ensure that each step on a manufacturing line is easy, fast, and able to maintain good quality. This means that much time is spent before production even starts to develop assembly methods that produce good quality. The parts may be redesigned to make assembly easier or to prevent handling or assembly damage.

Great care is taken to divide each assembly step into tasks that can be easily performed by each worker on the assembly line. With all this care to make the assembly easy while maintaining good quality, what does the in-process inspector do? On an assembly line, he or she has tasks much different from those of an in-process inspector in a job shop.

The assembly line in-process inspector is a roving inspector responsible for monitoring the line for problems that may affect quality. Typical in-process inspection responsibilities include:

- Inspecting sample parts from the line.
- Evaluating product performance during assembly.
- Verifying that individuals are using correct assembly techniques.
- Evaluating the general appearance of product.
- Evaluating repairs of in-process damage to parts.
- Ensuring that design changes are incorporated on the assembly line.
- Sampling packaging for completeness.
- Verifying calibration of test equipment.

Final Inspection (Job Shop) The last inspection in a job shop is assembling the proof that a product was made using the correct process. Often, a job shop final-product test is impossible. The best that a job shop can do is to certify that a quality plan has been followed 100 percent. This ensures that each manufacturing step is followed and that the correct drawings and procedures are adhered to by qualified individuals.

The final inspection reviews plan-and-process sheets to assure the company that all steps were correctly performed. The final inspection also determines whether hold points and customer requirements were properly met.

Typical responsibilities of the final inspector include:

- Reviewing signed process sheets.
- Verifying materials used.
- Inspecting the final coating.
- Evaluating completeness of quality verification to be provided to the customer.
- Coordinating final tests with the customer.
- Inspecting shipping containers.
- Inspecting spare parts.

Final Inspection (Assembly Line) Assembly line final inspection involves the inspector as "customer" in the plant. This final inspection is sometimes performed after the product has been boxed and is in storage. The final inspector takes the finished product the way a customer would and evaluates "fitness for use."

In many companies, the only opportunity to evaluate customer reaction to a product is through a final inspection of the finished goods. Such an evaluation of everything from appearance to performance is an important part of the feedback to management and sales, since it gives them an idea of how a customer would rate typical problems.

Typical evaluations by the final inspector include:

- Verifying that the shipping container is correctly labeled and instructions are included.
- Ensuring that all spare parts are in the packing box.
- Ensuring that the product is not damaged during packaging.
- Ensuring that a sample from the shipping department works as advertised.
- Ensuring that operating instructions are complete.
- Sampling of finished goods for life tests to determine whether reliability meets advertised quality.

Chief Inspector The responsibility of a chief inspector is to provide leadership in quality and support for the front-line inspectors. The chief inspector ensures that other inspectors are using correct procedures when performing the inspections.

Many times in the quality profession you will hear, "Quality is everybody's job." However, you will find in practice that it is really nobody's job or worry; it is too easy for most workers to allow someone else to worry about how to improve quality in production. A chief inspector carries the quality message to individual supervisors. His or her attitude toward quality should be an inspiration to all plant employees.

Additionally, a chief inspector supports individual shop inspectors with a strong commitment to quality. The chief inspector assumes the responsibility for questionable quality decisions made by a shop inspector. Not all inspection decisions are defined in inspection plans, designs, blueprints, or standards. An inspector must make these decisions on a daily basis, and the chief inspector reviews the inspection and assumes the responsibilities given to him or her by management: to accept, change, or reject findings made by an inspector.

2–6 SPECIAL PROCESS

Many shops require the use of special processes. For the purposes of this textbook, the discussion of special processes will be limited to metrology (the use of precision measurement equipment), nondestructive examination (NDE), and welding. Many other forms of special processes are used in particular industries. Detailed study of all special processes is beyond the scope of this textbook. However, quality professionals should become well versed in all special processes used in the industry in which they work.

Metrology The accuracy of tools and gauges is very important in meeting quality objectives. The metrology laboratory is a specialized facility where trained personnel calibrate and repair inspection tools.

The laboratory has a self-contained program for calibration, issuance, and timely recall of inspection tools. This responsibility requires detailed records of serial numbers of tools and problems encountered. When tools are found to be defective, the metrology lab must be able to trace them to particular parts manufactured on any given day.

Responsibilities of the metrology laboratory include:

- Maintaining records of all tools and gauges.
- Retaining maintenance and repair records to determine possible misuse of tools.
- Calibrating all tools and gauges to factory standards, based on national standards.
- Recalling tools when recalibration is needed.

Nondestructive Testing Nondestructive testing (NDT), also known as nondestructive examination (NDE), is an important special process used in receiving, in-process, and final inspection. An NDT inspector must understand both the NDT method and the quality responsibility at each stage of manufacturing where NDT was performed. This requires the NDT inspector to know quality control as well as NDT. NDT will be discussed in greater detail in Chapters 8, 9, 10, 11, 12, and 13.

The following is a list of NDT methods performed in industry:

- Eddy current testing (ET)
- Magnetic particle testing (MT)
- Penetrant testing (PT)
- Radiographic testing (RT)—i.e., X-ray, gamma ray, and neutron
- Ultrasonic testing (UT)
- Leak testing (LT)

Regulation of the various job and experience requirements is governed by the American Society for Nondestructive Testing (ASNT)'s *Recommended Practice, SNT-TC-1A.* This document outlines four levels of experience for an NDT technician:

Trainee
Level I
Level II
Level III

These levels of training and experience are also used to determine the responsibility a technician has when performing examinations.

Typical NDT inspector responsibilities include:

- Testing incoming materials to specifications.
- Verifying materials prior to special processes.
- Performing final inspection after final test.
- Comparing materials within a lot to determine uniformity.

Welding In a job shop that performs welding, the welding inspector holds an important position. Many companies have a certification program for this position. The American Welding Society (AWS)'s Certified Welding Inspector (CWI) program is becoming very popular. The AWS-CWI is divided into two categories:

- Technician who must work under the supervision of another inspector
- Certified inspector

Many states now require that all welding of public buildings, bridges, and the like be performed with inspection by a CWI.

Typical welding inspector responsibilities include:

- Witnessing welder and procedure qualifications.
- Verifying welder qualifications.
- Inspecting fit-up prior to welding.
- Visual inspection of welds.
- Performing or reviewing NDE of welds.
- Verifying weld and fabrication materials.

2–7 TRADITIONAL MANUFACTURING ORGANIZATIONS

In an effort to operate efficiently, manufacturing management must divide responsibility among suborganizations. This is done both for the efficiency of specialization and to best use each individual's specific abilities. Therefore, quality organizations are placed within the framework of a company to meet management's requirements for control.

Regarding quality, certain basic company structures can be considered traditional in manufacturing organizations. There are advantages and disadvantages to each. The company's production and sales goals should be considered before passing judgment on any structure.

A company that keeps quality on an equal footing with other divisions of management has long been considered to have the most effective form of organization. An example of this type of company structure is shown in Figure 2-1. This allows quality an equal management voice with manufacturing and engineering—the two most common interfaces.

This organizational structure assumes that the quality manager is, in fact, equal to managers of engineering and manufacturing, but this is seldom true. Manufacturing and engineering have both formal training programs and established, experienced programs to train individuals to become effective managers.

Quality control today is often deficient in formal quality training programs, and there are no established routes to learn the fundamentals through experience. For example, a young engineer may spend time drafting, and learning techniques and procedures. He or she will spend a few years as a junior engineer and then be promoted to engineer with registration as a professional. After years of experience as an engineer, he or she may become an engineering supervisor. With more experience, and some luck, a few may become managers of engineering in 10 to 15 years. Continuing into a senior officer position within most companies is unlikely.

Without an existing training program in most quality control departments, it would be naive to think that a quality manager would have equal status with a manager of engineering. Equal status among quality, engineering, sales, and manufacturing seldom exists.

Figure 2–1 Typical Organizational Structure

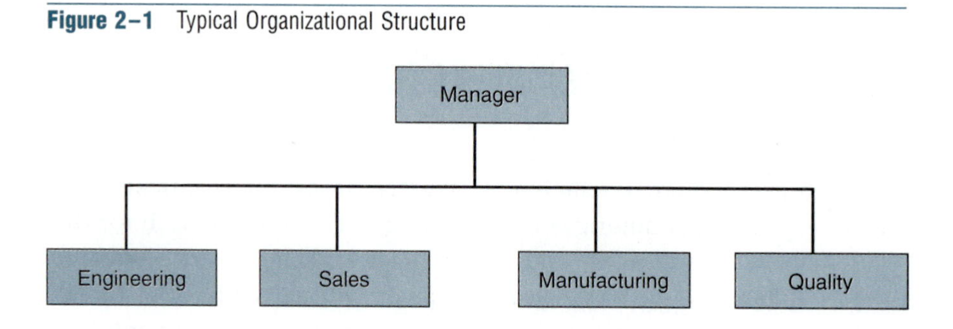

Figure 2–2 Typical Organizational Structure

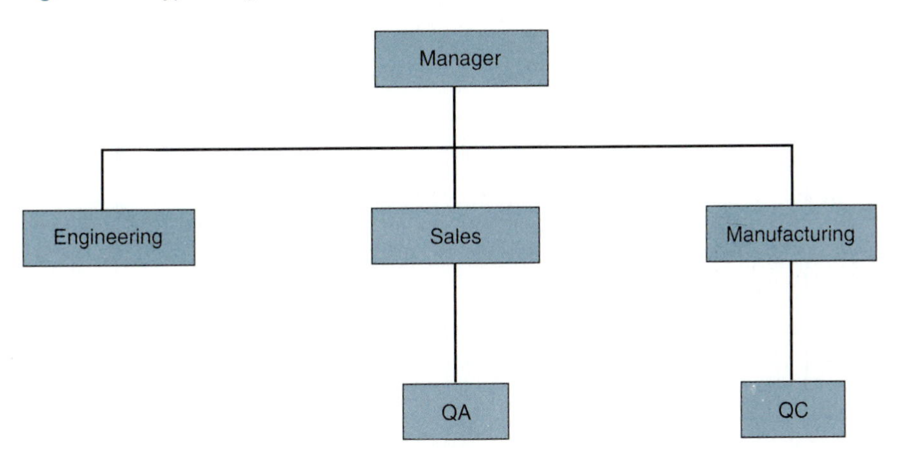

Another traditional organization has two quality departments within its structure (Figure 2-2). An inspection department reports directly to manufacturing, with an additional auditing organization reporting directly to marketing. This allows a predetermined level of quality to be translated into products. For example, if management desires to ship imperfect lots with, perhaps, a 2 percent rejection rate, the inspection provides manufacturing with the data necessary to make decisions that allow this mix.

Before condemning this management position (allowing imperfect parts to be shipped), consider the effect. Basic to all is "fitness for use." One does not expect a cheap product to be perfect. As a consumer, one saves or pays money and hopes the product will perform as expected.

To make all this work, there must be a balance between inspection and manufacturing. This is handled by the quality assurance (QA) department, working directly for the sales department. Quality assurance will act as a "customer" within the organization, maintaining a check on manufacturing to ensure that only an acceptable number of rejects get to the customer. Quality assurance can monitor actual new-product repair rates handled by the service organization to help indicate the effectiveness of manufacturing self-control.

The third traditional organization places quality responsibility within engineering (Figure 2-3). This organization is very effective for process-type industries. Engineering specifies a particular type of manufacturing process. Inspection verifies that the completed product meets that specification. The obvious disadvantage of this configuration is the

Figure 2–3 Typical Organizational Structure

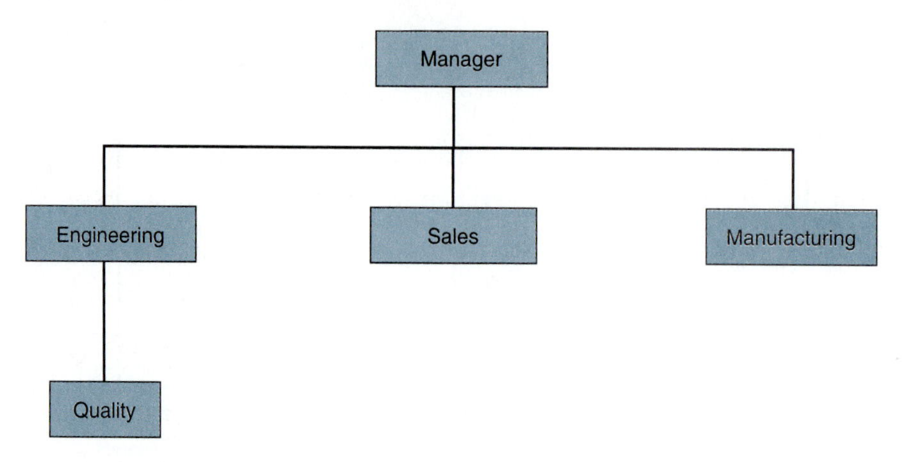

constant requirement for self-control by engineering. This structure sometimes is ineffective when new design drawings or a large number of "running" manufacturing changes require engineering evaluation.

Always remember—*the effectiveness of an organization, not the organization chart, is critical to a productive company.*

Summary

Quality is not a term that possesses a universal, all-encompassing definition in the field. In most cases, the term *quality* is relative to the individual or situation in which it is being applied. For the quality professional, the following definitions are appropriate in most instances.

- The totality of features and characteristics of a product or service that bear on its ability to satisfy given needs
- Fitness for use
- Conformance to requirements
- The composite of all the attributes or characteristics including performance of an item or product
- The total composite product and service characteristics of marketing, engineering, manufacturing, and maintenance through which the product and service in use will meet the expectations of the customer
- The conformance to standards that represent the product's or service's basic characteristic, and which are based on customer needs and expectations

With these accepted definitions in mind, quality can be further divided into three major categories:

- Quality of design
- Quality of conformance to design
- Quality of performance

The analysis of quality is further defined through quality assurance and quality control. Quality assurance involves the management of quality—the act of ensuring or indicating that a program is in compliance and functioning effectively. Quality control—indicating an activist role—uses strategies such as inspections and statistical process control to ensure compliance with product standards.

Within the quality effort are activities such as inspections, audits, and surveillance. A quality inspection is a highly defined, close examination of a product or process. The quality audit is an inspection of an organization's adherence to established quality standards. Quality surveillance is a loose inspection process that uses some of the techniques of both the audit and the inspection.

Through analysis of programs, processes, or products, the quality function identifies instances of nonconformance to stated requirements. Such nonconformances can be placed in one of three categories based on the terminal effect on the program, process, or product's usefulness. The universally accepted categories are:

Critical—a defect that judgment and experience indicate is likely to result in hazardous or unsafe conditions for individuals using, maintaining, or depending upon the product.

Major—a defect, other than critical, that is likely to result in failure, or to reduce materially the usability of the unit of product for its intended purpose.

Minor—a defect that is not likely to reduce materially the usability of the unit of product for its intended purpose, or is a departure from established standards having little bearing on the effective use or operation of the unit.

The science of controlling quality is as diverse as the variety of companies and processes within a company. Inherently, this activity creates many responsibilities. Typical job functions include:

- Quality assurance/quality control manager
- Quality engineer
- Inspectors
  - Receiving
  - Source
  - In-process
  - Final
  - Special processes
- Chief inspector

Many industries use special processes for control and production, and quality professionals must become familiar with them. Three of the more common special processes are:

- Metrology: control and use of measuring tools, gauges, and devices
- Nondestructive testing: inspection methods such as eddy current testing, magnetic particle testing, penetrant testing, radiographic testing, ultrasonic testing, and leak testing
- Welding

Even in a traditional manufacturing organization, quality professionals may occupy one of numerous places in the organizational structure. The organization in which quality is placed on an equal footing with other divisions of management is considered most effective. An alternative structure has two quality departments, with the inspection section reporting directly to manufacturing and an additional auditing sectionreporting directly to marketing. In other operations, the quality function is placed under the responsibility of engineering. The most important factor in determining the placement and structure of the quality department within an organization should be the effectiveness for the company.

KEY TERMS

Audit An examination of records or accounts used in the quality function to determine a company's adherence to established quality standards.
Calibration Comparison of a measurement standard or instrument with another of known, high-level accuracy to detect any variation in accuracy.
Critical defect A defect that could result in hazardous or unsafe conditions for individuals using or maintaining the product.
Defect (discrepancy) A product that does not meet specifications.
First article The first production unit scheduled for acceptance by an inspector or customer.
Fitness for use The ability of a product or service to conform to the end customer's needs.
In-process inspection Inspection performed during manufacturing in an effort to prevent defects from occurring and to inspect characteristics.
Inspection A highly defined, close examination process.
Nonconformance Any condition in which one or more product characteristics do not conform to requirements specified.
Precision The extent to which instruments repeat results with continued measurement.
Quality The degree of conformance of an item to governing criteria.
Quality assurance All those planned and systematic actions necessary to provide adequate confidence that a system or product will perform satisfactorily in service.
Quality control All those actions, relating to the physical characteristics of the material, which provide a means to control quality to predetermined standards.
Surveillance A loose inspection process.

QUALITY TEST

1. Define *quality* as viewed by the customer, and describe how the customer's view will differ from the manufacturer's.
2. Define *quality assurance* and describe how it differs from quality control.
3. Define the term *inspection* and describe how it differs from an audit or surveillance.
4. Discuss in a short paragraph the job responsibilities of a QA/QC manager.
5. Discuss in a short paragraph the job responsibilities of a quality engineer.
6. Discuss in a short paragraph the job responsibilities of an inspector.
7. Discuss in a short paragraph the advantages and disadvantages of the manufacturing organizations discussed in this chapter.

NOTES

1. American Society for Quality Control, *Standard A3,* 1978.
2. Joseph Juran and Frank Gryna, Jr., *Quality Planning and Analysis,* 2d ed. (New York: McGraw-Hill, 1980), 1.
3. Philip Crosby, *Quality Is Free* (New York: McGraw-Hill, 1979), 15.
4. U.S. Department of Defense, *MIL-STD 109,* 1974.
5. Armand V. Feigenbaum, *Total Quality Control*, 3d ed. (New York: McGraw-Hill, 1983), 7.
6. Charles Aubrey, *Quality Management in Financial Service* (Wheaton, IL: Hitchcock Publishing Co., 1985), 7.
7. Joseph Juran, *Juran on Planning for Quality* (New York: The Free Press, 1988), 101.
8. J. Juran, *Quality Control Handbook,* 3d ed. (New York: McGraw-Hill, 1974), 3–4.
9. Juran and Gryna, 2.
10. Juran and Gryna, 2–3.
11. Juran and Gryna, 136.
12. U.S. Department of Defense, *MIL-STD 410-D,* 1974.
13. U.S. Department of Defense, *MIL-STD 105-D,* 1974.
14. *MIL-STD 105-D.*
15. *MIL-STD 105-D.*

NEED FOR QUALITY IN TODAY'S BUSINESS ENVIRONMENT

3

OBJECTIVES

After completing Chapter 3, the reader should be able to:

- Define the reasons for and the effects of the decline of the U.S. market in durable goods.
- Discuss how quality is used to improve sales and profit, and reduce liability exposure.
- Discuss the quality assurance responsibility during engineering, manufacturing, and marketing of a product.
- Discuss the approach used to control quality in a job shop.
- Discuss the approach used to control quality on an assembly line.
- Discuss the approach used to control quality on a continuous operation.

In the 1980s, one thing became clear in business: the U.S. was no longer the world leader in production of consumer goods. The trend of losing market share will continue until U.S. manufacturers decide to ascertain from foreign companies what the industrial world learned from the U.S. 40 years ago. We must produce products at a competitive price, with competitive quality, in mass quantity, and designed to meet consumers' needs. Quality control plays the deciding role in each of these areas. If the quality organization in a company is ineffective, the company will be doomed to a state of customer dissatisfaction and eventual failure.

The erosion of our markets did not happen overnight—it was a slow process that took many years. Unfortunately, this type of decline was often not recognized until the domestic companies were in deep financial trouble.

During the early 1960s, American production in durable goods was at its peak. The erosion, which has taken place over the last 30 years, has left many industries near extinction. A comparison of the levels of products sold on the world market from 1965 to 1989 shows a drastic drop in some markets (Table 3-1).[1] For example, in 1965, 27 American companies produced television sets. Today, only one American company (Zenith) produces televisions.[2]

Many industries are feeling pressure from foreign companies, such as microchip, electronic, steel, computer, textile, clothing, plastics, and aeronautics/space industries. This is the norm rather than the exception. The U.S. was the only country in the industrialized world unable to put a probe into space to study Halley's comet. This was because of major quality design problems in the space shuttle's solid propellant motors.

Table 3–1 U.S. Market Share in Selected Industries

PRODUCT	1965	1989
Machine tools	100%	35%
Cars	76	24
Television sets	90	10
Turntables	90	1

Lack of quality alone is not to blame, but it has played a major role in the downfall of these industries. The prevailing perception among many consumers is that American products are substandard to many of their foreign competitors. Public perception of "the best" has changed to the point that it is having detrimental effects on the U.S. market. The following perceptions were expressed by students in a quality control program.

- The best sports cars are made in Italy.
- The best wines come from France.
- The best passenger cars are made in Japan.
- The best electronics are made in Japan.
- Japan has the best-managed companies.
- The best motorcycles are made in Japan.
- Italian shoes are the best.

Are these perceptions valid? They have been valid in the past, but they may or may not be valid today. Perceived quality superiority of one product over another is as effective at swaying the buying public as is actual quality superiority. If customers believe that one product is better, that product will have better sales.

These widely held perceptions show that the U.S. is losing the world market in the industries that have provided the high standard of living in the U.S. The loss of the world market places the current middle-aged generation in the uncomfortable position of being the first generation in many decades that will pass on a lower standard of living to its children. Regrettably, the decline in market share by domestic manufacturers had been predicted by many people. One of the main factors involved in this decline is foreign competitors' adoption of the philosophy of a few American leaders in the quality control field. For example, the top award for quality improvement in Japan is the Deming Award, named after the late Dr. W. Edwards Deming. The U.S. is now having to relearn these American ideas from our foreign competitors, struggling to keep the U.S. from losing its future to other countries.

Many companies are now striving to develop techniques to improve their prospects in three major areas:

- Marketing
- Profit
- Liability

3–1 MARKETING

The advertising industry has discovered that it pays to exploit quality as a selling tool for many products. This is evident in radio and television advertisements. Many of the products that were traditionally sold on the basis of performance and what is referred to in the industry as "bells and whistles" (options available on products) are now selling "quality."

- Quality is job one. (Ford)
- The quality goes in before the name goes on. (Zenith)
- The serviceman is the loneliest guy in town. (Maytag)
- The best built cars made in America. (Chrysler)
- The finest quality since 1927. (Kraft Foods)
- Fresh from Qualityville (Hiland Dairy)
- A century of reliability and dependability. (AT&T)

The list is endless. But selling a product based on quality requires that the product meet several tests before this tactic will be effective.[3]

- The customer must believe the quality claims.
- The product quality must be as good as or better than that of the competitors.
- The quality advantage must be perceived by the customer as valid and beneficial.
- The quality advantage must be perceived as worth the price.

This discussion of selling products based on quality may seem to have little to do with this chapter. However, selling the product is the most important part of the quality function, because it accounts for a company's ability to make a profit. If a company is not financially healthy in today's market, it will be out of business in short order. Selling the quality of a product becomes vital to every industry.

The general public has received many benefits from improved product quality. The more dramatic examples are the warranties offered by the automobile industry. In the late 1970s, most cars came with a standard 12-month, 12,000-mile warranty. As the quality and reliability of cars improved, companies could increase this warranty without suffering a dramatic change in financial exposure. This has allowed companies to increase warranties by as much as seven times, and this improvement is a major influence in the ability of a company to sell its product. This sales increase has enabled companies not only to stay in business, but also to expand their market share in a stagnant domestic market.

3–2 PROFIT

The area of profitability and cost savings is where amazing stories have been documented through quality improvement.[4] Improving the quality of manufacturing methods, products, and information on a production line has provided tremendous cost savings and improved the quality of the end product.

To demonstrate the potential cost savings and profit enhancement available, suppose a company, ABC Machine and Fabrication, is in the business of making threaded unions. Currently, the threaded unions are made and sold in groups of 100. The raw material is purchased, and 100 threaded unions are manufactured and shipped. The cost to produce each threaded union is a combination of parts, direct labor, and overhead.

The cost breakdown is as follows:

ABC Machine and Fabrication Cost Breakdown

$ 2	Parts and material
4	Direct labor
4	Overhead
$ 10	Delivered cost (the cost of doing business)
12	Invoice price per part (20 percent markup)
$ 2	Profit per part (if all parts are accepted)

Cost and Profit on 100 Threaded Unions

$ 200	Parts
400	Direct labor
400	Overhead
1000	Total cost (to produce 100 parts)
1200	Invoice amount
$ 200	Profit

The total cost of one production run is $1000 and should result in a profit of $200. ABC found that 85 out of 100 threaded unions produced were accepted and 15 were rejected by the customer. If 85 of the threaded unions were good, each good threaded union would actually cost $11.76 to produce ($1000 ÷ 85 = $11.76). The reason for this increase is that the total cost to produce 100 parts ($1000) must be recovered from the revenue produced by the 85 good parts. The cost of parts, labor, and overhead must be recovered from the parts—no matter how many (or how few) of the parts can be sold. If the selling price of the threaded unions were $12 each, this lot of threaded unions would bring $1020 total revenue and yield a $20 profit, as shown below:

$ 1000	Actual cost to produce 85 good parts
1020	Invoice (85 parts)
$ 20	Total profit

The true cost to produce each acceptable part increased to $11.76 from $10, but ABC could invoice for only $1020 (85 parts × $12) instead of $1200 (100 parts × $12). This resulted in the total profit for this order being only $20 instead of the expected $200. A 15 percent reduction in the number of delivered parts resulted in $0.24 profit per part, a return on investment (ROI) of 2 percent, and a profit 90 percent lower than expected.

ABC management saw two alternatives to solve the problem of reduced profits:

- Make more parts on the existing processes to make sure 100 good parts will be delivered.
- Fix the processes to make more of the parts correct the first time.

Both options were explored to determine which would yield the best return.

ABC made a study to determine how many parts should be produced to ensure 100 good parts were made per lot, with the existing process producing 15 percent rejects. ABC discovered that 118 parts would have to be produced to get 100 good ones, when the process used to produce parts is running at a 15 percent reject rate (118 × 0.85 = 100.3 or 100 good parts). A further cost analysis was conducted to find the profitability of this approach. The results of this analysis are as follows:

Total cost of production	118 × $10	=	$ 1180
Cost per accepted part	$1180 ÷ 100	=	$ 11.80
Invoice price	$12 × 100 parts	=	$ 1200
Total part cost			$ 1180
Total profit			$ 20
Profit per delivered part			$ 0.20
ROI			1.67%

ABC determined that producing more parts did not help the situation; in fact, it resulted in a further decay of the profit per part and had the same effect on the ROI. The overall profit was the same $20 as before—but it took an investment of $180 more to make the same money and eroded the ROI.

The second analysis conducted was to determine the impact on profit of making more of the product correctly the first time. ABC ran a comparison of costs and profits if 95 parts had been made right the first time. The comparison is as follows:

85 Good Parts Versus 95 Good Parts		
Part cost	$ 200	$ 200
Direct labor	400	400
Overhead cost	400	400
Total cost	$ 1000	$ 1000
Invoice price	1020	1140
Profit	$ 20	$ 140
Cost per accepted part	$ 11.76	$ 10.52
Profit per part	$ 0.24	$ 1.40
ROI	2%	14%

The same comparison was made for 98 good parts, with the following results:

85 Good Parts Versus 98 Good Parts		
Part cost	$ 200	$ 200
Direct labor	400	400
Overhead cost	400	400
Total cost	$ 1000	$ 1000
Invoice price	1020	1176
Profit	$ 20	$ 176
Cost per accepted part	$ 11.76	$ 10.20
Profit per part	$ 0.24	$ 1.76
ROI	2%	17.6%

If through quality improvement the normal production run of 100 threaded unions would yield 98 good threaded unions, the profit picture would improve dramatically. The cost to run this lot would remain at $1000. The cost per threaded union sold would drop to $10.20 each; with the selling price of $12 each, this lot would yield

$1176. This would result in a $176 profit to the company. The total number of threaded unions sold would increase by only 13, but the profit would increase by $156. Producing more of the parts correct the first time would give a disproportionate increase in profit. *A 13 percent increase in the number of parts accepted (98 instead of 85) results in an 880 percent increase in profits ($176 instead of $20).*

How is this possible? Many approaches are used to determine the profitability of a contract. The definition of profit used here is "all the dollars that are left over after all the bills are paid." Using this definition, how can making 13 percent more product right the first time result in an 880 percent increase in profits? The key is in the break-even point for the contract. The cost to produce the lot of 100 parts is a fixed cost, even if all the parts are good or all the parts are bad. This cost must be fully recovered before profit is made. For the 100 threaded unions, the break-even point is 83 threaded unions. The full $12 selling price of the first 83 threaded unions sold goes to cover the production cost. No profit is made until at least 84 parts are sold. After part number 84, the entire selling price of the parts affects the bottom line as profit. The production cost is recovered on 83 parts, and the profit is made on 17 parts. If a part is rejected, the entire invoice price of that part is deducted from the profit. This is what gives the disproportionate impact on the bottom line when parts are rejected.

As an added benefit, if 98 parts were made correctly the first time instead of 85, the production volume would be increased by 13 percent with no additional overhead. The same amount of raw material was purchased and the same time was spent making 100 threaded unions. No additional workers, machines, or time would be needed to get this 13 percent increase in production. Any company should be interested in increasing production with no increase in production cost.

In the case of 100 threaded unions, this improvement may not seem like much until the numbers are evaluated. What would be the economic effect of a 13 percent increase in a company that produces parts that sell for $50,000 each, if all parts are made right the first time?

Additional profit can be realized by not spending money on producing parts that are defective.[5] In the example of the threaded unions, the assumption was made that the threaded unions that were not produced properly cost only the amount directly spent on production. This is not the case, because many indirect costs are involved, which are greater and well hidden. When attempting to document these costs, several evaluations should be made. "It is normal to obtain only one-third of the total quality costs the first time you try it."[6] This quality cost concept is known as the "iceberg effect." According to this theory, the cost of poor quality can be as much as 20 percent to 40 percent of the sales, which is even higher than the direct costs (parts and labor only) of producing a bad part.[7]

For a company producing 90 percent good parts, the cost of the 10 percent defective product could be staggering. It means that 10 percent of the costs of labor, facility maintenance, floor space, administration, utilities, insurance, and any other costs related to doing business have been spent producing bad parts. These costs are usually sizable enough for a company to seek aggressively for the causes and eliminate them.

3–3 LIABILITY

In today's society, litigation over product liability is common. The number and average size of financial settlements have increased to the point that companies can literally be driven out of business. In many businesses, the cost to the customer of such litigation has prevented products or services from being sold. In the medical profession, for example, doctors who deliver babies are in short supply in some cities because of the cost of malpractice insurance. In the U.S., no aircraft manufacturer produces a two-seat aircraft because the product liability insurance cost doubles the selling price of the airplane.[8] This cost of protection has driven the price of these aircraft above what consumers are willing to pay.

The quality control department should play a significant role in reducing the liability exposure for the company by expanding into every aspect of manufacturing. This involvement should include[9]:

- Engineering
- Manufacturing
- Marketing (sales)
- Warranty

With a quality influence in each of these areas, the liability exposure should be reduced.

Engineering The responsibility of most engineering departments includes designing, redesigning, and evaluating new and existing products. Quality control should have an active part in the following functions:

- Ensuring that safety becomes a formal design parameter.
- Performing design reviews for safety at strategic points of the design formation and approval cycle.

- Ensuring that the design meets all government and industry safety standards.
- Reviewing test data to determine a product's safety and reliability before it is released for production.
- Reviewing all material and material substitutions to ensure that each meets all safety codes.
- Reviewing the tracking method for product traceability (this will allow for a rapid and limited recall if the product is determined to be unsafe after it has been sold on the open market).[10]

If quality control is practiced during the design stage of a product, the end result should be a safer product that can be produced efficiently and have fewer chances of producing a product liability suit.

Manufacturing Manufacturing is where properly followed quality techniques can produce large cost savings. It is also a key area in which product liability exposure can be reduced.

Beyond the cost savings, quality control should be involved in safety

- To ensure that the product that is shipped conforms to the specifications and to government and industry safety standards.
- To provide training for workers to meet safety standards.
- To motivate workers about product safety by explaining why the procedures need to be followed.
- To audit the manufacturing process to determine that the proper checks are made, and that the product is accepted or rejected based on test results alone.
- To provide accurate feedback to upper management as to the state of production and product safety.
- To maintain proper documentation as to the test results on the product.
- To maintain the calibration of the test equipment.
- To plan and perform tests and evaluate test results.
- To determine whether nonconforming product is fit for use.[11]

Marketing (Sales) A sales department may inadvertently give consumers wrong impressions of the use of a product. Quality control should include the following responsibilities:

- Evaluating sales ads for technical and usage accuracy.
- Ensuring that the products are properly packaged and labeled with warnings and remedies.

- Providing safety information to everyone involved with the product.
- Evaluating the user's manual for accuracy and for clear explanation of the safety procedures to follow when using the product.
- Ensuring that the sales force has received proper safety training and is providing this information to the end users.[12]

Warranty The follow-up to problems will provide much-needed information about how the product is performing and how it is being used (and misused) by the general public. Information on how the product is being misused may help engineers to redesign the product to reduce injuries. The quality control department should be involved in the warranty area in the following ways:

- To determine whether a product failure was engineering- or manufacturing-induced
- To establish, in an investigation involving injury or death, whether the product was at fault
- To ensure that service is performed promptly to satisfy the customer, and that the product is as safe as it was when purchased.[13]

The quality control department has a responsibility to reduce the liability exposure of the company. If the quality functions are properly performed, many companies will experience a reduction of injury-causing product failures. Controlling these failures will reduce the number of product liability suits. This can have a dramatic cost-saving effect on the company.

3–4 MANUFACTURING METHODS

The quality methods used to generate savings in the above-mentioned areas—marketing, profit, and liability—are as varied as the types of manufacturing environments in existence. There is no set way to put quality into all manufacturing environments. If it were possible, it would have been done by now. Therefore, the quality professional must be aware of what types of quality procedures work in different manufacturing environments. There are two major types of manufacturing methods, the assembly line and the job shop. They tend to be divided by the type of product manufactured. However, the quality effort should focus on the same criteria for both.

Assembly Line Method In this method, parts of a whole product are mass produced. The manufacture and assembly of pieces are systematically arranged and highly repetitive. The key point is that the procedure *is* highly repetitive. Companies that make televisions, automobiles, telephones, and the like, employ an assembly line.

An assembly line worker may work for years making the same gear for a transmission, and he or she follows the same procedure for every gear made. That gear may then go to an assembler who puts the gear onto a shaft in a precise sequence.

The sequence of events and the procedures are of the utmost importance. They have been well thought out and, if followed on properly maintained equipment, these procedures will produce a serviceable part.

This type of manufacturing process is receptive to statistical analysis. By watching the product as it comes off the assembly line (sampling), it is possible to learn some things about the production process.[14]

No process can produce identical parts every time. All parts vary, but they vary in a predictable manner.[15] If the variation becomes unpredictable, then something has gone wrong in the manufacturing process and the process is said to be out of control. Perhaps the equipment needs repair, the machine setting is wrong, or the base material is substandard. Whatever the reason, something is wrong with the process.

By inspecting the *part,* it is possible to monitor the *process.* At this point, quality control consists of looking at the *average* quality of the product, not the quality of *each* piece that comes off the line. This is much more reliable and less expensive than trying to inspect each part.[16]

The emphasis here is not merely to segregate the bad product, but to control the processes that created it. Therefore, the making of bad parts is eliminated or at least controlled.

There are many other factors that influence the quality effort in the assembly line environment. For instance, communication from one machine worker to another may be hard to maintain on a large assembly line. Also, communication from one department or factory to another may be necessary in very large companies. In such a situation, quality circles may be an advantage (see Chapter 5).[17]

In order to find the most workable system for an assembly line manufacturer, the quality professional must evaluate the environment of the company and compare it with the methods listed in Chapter 6.

Job Shop Method The job shop method produces small numbers of a product. It does not use a highly repetitive process. A worker often works on the product from start to finish.[18] Companies that use a job-shop method include custom machine shops, welding fabrication shops, and repair shops.

With this type of environment, there is not enough volume of product to provide useful statistical data about the process. However, this technique can be used to some extent.

The process itself must be monitored. The quality professional must witness and record the step-by-step sequence of events. Certifying and training assembly personnel to code requirements is a must. Traceability of raw materials must be well defined. The quality unit must inspect the product at certain stages of construction, using certified personnel. If any of these requirements is not fulfilled, the end product may be worthless.

As an example, a boiler would be required to conform to the American Society for Testing and Materials (ASTM) pressure-vessel code. That code demands steel alloys of a specific grade. Therefore, the raw material for the boiler must be traceable back to the foundry that produced it, or it cannot be used. Even if the manufacturer had followed every step of the manufacturing process to the letter, and all inspections had been conducted according to the codes, the boiler would be worthless. If the foundry process could not be verified, that metal could not be used.

By the same standard, if the welding processes were not followed precisely, or the weld process were correct but not done by certified welders, the boiler would be worthless. If inspections were done by uncertified personnel, again, the boiler would be worthless.

It is extremely important that every step of the manufacturing process be closely monitored and recorded in the job shop environment. This demands a great deal of inspection and verification, as compared to an assembly line.

In summary, the quality effort on the assembly line uses characteristics of the product to determine the quality of the process. The job shop quality effort uses verification of the process to ensure the quality of the product.

Continuous Manufacturing Continuous manufacturing processes produce high volumes of product in a continuous operation. The extrusion of aluminum seamless pipe and the production of sustained sheets of plastic film are examples of a continuous manufacturing process. In such instances, the process itself is inspected. These highly automated and computer-controlled processes can produce a consistent product if the process is set up correctly. Inspection techniques in these types of operations are directed toward the process. If the process parameters are correct, the product will be correct. Product inspection in this type of manufacturing can be conducted only following completion of the process. However, continual process monitoring must be conducted to prevent an excessive amount of nonconforming product.

3–5 SUCCESSFUL QUALITY CONTROL PROGRAMS

The need for and success of quality improvement programs are well documented. The reputation of American-made products has taken a beating over the past few years, but the industry is coming back.

Harley-Davidson is a good example of what can happen when a company becomes truly interested in the quality of its product. The name *Harley-Davidson* has been synonymous with American motorcycles for years. However, from the 1960s to the late 1970s, the quality of Harley-Davidsons declined while the quality of foreign motorcycles increased. This caused a loss in market share for Harley-Davidson.

In an attempt to save the company, Harley-Davidson asked the U.S. Congress for—and got—protective legislation until the company could resolve its problems. Much of the management function of the plant was put into the hands of the assembly line workers. Quality circles were implemented, along with a form of just-in-time delivery. (These concepts are discussed in Chapters 4 and 5.)

An old overhead-production line conveyer system was shut down, which drastically reduced the number of defective products. Parts that had been made were stored until needed for motorcycles on this conveyer system. In many cases, parts were not used for as long as six months. If defective parts were not immediately identified, however, the worker producing the parts had no way of knowing how to prevent these defects.

Workers addressed problems of inventory control, machine maintenance, and data generation—all of which had contributed to financial problems, low worker morale, and a poor reputation with consumers.

The company reemerged as a world leader. Harley-Davidson, in an unprecedented move, asked Congress to repeal the protective legislation a year early because they were ready to compete in the global market.[19]

In Bay City, Michigan, a General Motors plant was suffering from the decline in the U.S. auto market overseas. The plant's management had furloughed 40 percent of the workers in an effort to cut expenses. Yet all these efforts seemed to have no positive effect.

A new plant manager asked the workers to help solve the problems, in the belief that the solution would be to make an environment where workers could make high-quality parts efficiently and at low cost. In one year, the rejection rate was reduced by 44 percent, and production increased by 41 percent. These changes saved millions of dollars, increased productivity, and increased the morale of the work force. When workers began to feel ownership over their jobs, absenteeism went down and profits went up.

The benefit of quality programs is not limited to the private sector. The public sector has also started to use this type of management to save taxpayer dollars.[20]

The Alameda Naval Air Station in California overhauls and rebuilds aircraft for the U.S. Navy. This facility was losing $1,000,000 a week and also lost over 50 percent of its work force to private industry in the late 1980s. To solve this problem, several quality control methods were introduced. The regulations and rules were rewritten, reducing the rule book from over 400 pages to a pocket-sized book of less than 20 pages. Management changes included giving to the commanding officer the responsibility and authority to make needed changes to bring the station under financial control. The commanding officer in turn involved the workers to begin improving quality and reducing cost.

The result was that the station went from sixth place out of six naval repair stations to second place in only six months. This rating was based on productivity and repair cost, time, and quality. In 1989, this repair facility accomplished the same amount of work as in 1988 with an overall reduction in operating cost of $40 million.

One dramatic cost savings to the U.S. Navy came when a team attempted to repair two damaged aircraft by splicing the nose of one airplane to the undamaged rear section of another. As a result, the two planes were made into one good plane. The repair cost was $1,500,000, as opposed to the $30 million replacement cost of the airplane.[21]

The city of Phoenix (AZ) has involved its employees in a quality improvement program that has resulted in many improvements in city services. The basic concept was to ask the workers to identify areas where improvement of services could be made. Under the new program, the average response time for an ambulance has dropped to four minutes, which is five times better than the privately operated services in the area. That is unexpected in a city of over one million people.

The Department of City Service continues to stress the quality of service issue with its workers. Several cost savings and quality improvement ideas have been generated. One idea to change the way streets were repaired improved the quality of the streets and resulted in a $50,000 per mile savings. Another team of maintenance workers developed a piece of equipment to keep high-pressure wash hoses from getting tangled and damaged. This idea not only speeded up the work, but reduced damaged equipment and saved the city $72,000 a year. The city of Phoenix has found that teaching and following quality control philosophies have reduced the operating costs of government and have vastly improved the services provided to the citizens of Phoenix.[22]

SUMMARY

The need for quality control in today's businesses is well established. The results of properly implemented quality programs where companies pay attention to quality will be

- Reduced manufacturing cost
- Increased production
- Safer products
- Higher quality reputation[23]

With this recognition of quality a company will realize advantages that should result in greater market share, higher profit, and reduced liability to the company.

In today's business environment, what company would refuse these benefits? Surprisingly, many American companies refuse to implement quality programs. This is partly responsible for a decay in American manufacturing strength. If this trend continues, this generation will leave the next generation a lower standard of living. This could be avoided if more American industries would get involved in the quality revolution that is sweeping the rest of the world. With the proper approach, companies that are on the brink of disaster can turn the situation around in record time and emerge as world leaders in their field.

KEY TERMS

Assembly line manufacturing A manufacturing method in which the assembly pieces are systematically arranged and the operations are highly repetitive.

Continuous manufacturing A manufacturing method that produces a long run of a single product. Examples of this type of operation are the production of plastic stretch film and the production of sewing yarn.

Defect (discrepancy) Any deviation of an item from specified physical, metallurgical, or chemical requirements.

Defective A product that does not meet specifications.

Fitness for use Selling price and the customer's end use.

Iceberg effect A concept used in describing the total cost of poor quality. This concept implies that up to 75 percent of the total cost of poor quality is hidden and not readily identifiable. Even though these costs are not seen, the costs are present.

In-process inspection Inspection performed during manufacturing in an effort to prevent defects from occurring and to inspect characteristics.

Inspection A highly defined, close examination process.

Job shop manufacturing A type of manufacturing that usually involves a small number of parts being produced one at a time. It does not lend itself to production line manufacturing.

Lot A quantity of product manufactured to specific requirements using the same manufacturing methods.

Nonconformance Any condition in which one or more product characteristics do not conform to requirements specified.

Quality The degree of conformance by an item to the governing criteria.

Quality assurance All those planned and systematic actions necessary to provide confidence that a system or product will perform satisfactorily in service.

Quality control All those actions, relating to the physical characteristics of the material, that provide a means to conform quality to predetermined standards.

Random sampling A process in which each element of a population has an equal probability of being sampled.

Sample One or more units of product selected at random from the material or process represented.

Variation Differences in a measured characteristic caused by random chance disturbances and by identifiable disturbances.

NEED FOR QUALITY TEST

1. List three reasons for and the effects of the decline of the U.S. market in durable goods.
2. Discuss in a short paragraph how quality is used to improve sales and profit, and to reduce liability exposure.
3. Discuss in a short paragraph the quality assurance responsibility during engineering, manufacturing, and marketing of a product.
4. Discuss in a short paragraph the approach used to control quality in a job shop.
5. Discuss in a short paragraph the approach used to control quality on an assembly line.
6. Discuss in a short paragraph the approach used to control quality in a continuous operation.

NOTES

1. Tom Peters, *Leadership Alliance* (Video Publishing House, Inc., 1988).
2. A. Weston, *Losing the Future* (NBC Special Report, November 1988).
3. Joseph Juran and Frank Gryna, Jr., *Quality Planning and Analysis,* 2d ed. (New York: McGraw-Hill, 1980), 459.

4. W. Edwards Deming, *Out of the Crisis* (Cambridge, MA: MIT Press, 1985), 371–87.
5. Philip Crosby, *Quality Is Free* (New York: McGraw-Hill, 1979), 11.
6. Crosby, *Quality Is Free,* 103.
7. Joseph Juran, *Juran on Planning for Quality* (New York: Free Press, 1988), 1.
8. Jan VonFlatern, "Dealing with Product Liability: One Manufacturer's New Approach" (Paper delivered at the A.B.A. National Institute Litigation in Aviation, 27–28 October 1988).
9. Juran and Gryna, 502–7.
10. Juran and Gryna, 502–7.
11. Juran and Gryna, 502–7.
12. Juran and Gryna, 502–7.
13. Juran and Gryna, 502–7.
14. Bonnie Small, *Statistical Quality Control Handbook,* 2d ed. (Indianapolis, IN: AT&T Technologies, 1985), 4.
15. Harvey Charbonneau and Gordon Webster, *Industrial Quality Control* (Englewood Cliffs, NJ: Prentice-Hall, 1978), 48.
16. Ellis Ott, *Process Quality Control* (New York: McGraw-Hill, 1975), 171–72.
17. W. S. Rieker, *Quality Control Circles,* 2d ed. (California: Quality Control Circles, Inc., 1977), 30.
18. A. V. Feigenbaum, *Total Quality Control,* 3d ed. (New York: McGraw-Hill, 1983), 852.
19. Tom Peters, *Leadership Alliance* (Video Publishing House, Inc., 1988).
20. Peters, *Leadership Alliance.*
21. Tom Peters, *Excellence in the Public Sector* (WETA, Washington, DC, 1989).
22. Peters, *Excellence in the Public Sector.*
23. John Groocook, *Chair of Quality* (New York: John Wiley and Sons, 1986), 1–8.

PREVALENT QUALITY PHILOSOPHIES

4

OBJECTIVES

After completing Chapter 4, the reader should be able to:

- Discuss the approach to quality taken by W. Edwards Deming.
- Discuss the approach to quality taken by Joseph M. Juran.
- Discuss the approach to quality taken by Philip B. Crosby.
- Discuss the approach to quality taken by Armand Feigenbaum.
- Discuss the use of quality circles to improve quality.

Global competition has increased in recent years; consequently, good quality management techniques have become more critical to the survival of all kinds of industries. Several philosophies have gained wide recognition in the quality industry. This chapter briefly describes some of these philosophies. A more in-depth investigation can be accomplished by reading books that describe each specific philosophy.

Each philosophy is unique in some way, but they all have many areas in common. A company may tend toward one concept or another, but in reality, very few companies use only one philosophy. It will be very helpful for quality professionals to be able to draw from each school of thought when deciding on a course of action in quality improvement.

Every industry is different. Therefore, unique quality problems are created. Although each problem is unique, problems do tend to fall into general categories. The quality professional may find that one technique works better in a given plant than another.

In studying these philosophies, you should look for their similarities and differences. Think back to Chapter 1 and consider how each might work in different types of manufacturing organizations. Try to apply each in a job shop and an assembly line environment.

4–1 W. EDWARDS DEMING

W. Edwards Deming (1901 – 93) is one of the preeminent figures in the quality control profession. Deming worked as a consultant to the U.S. War Department (later the Department of Defense) during World War

II. After the war, Deming tried to interest American industry in the use of statistics to improve productivity and quality. Most of the rest of the world had been devastated by the war, so at that time America was able to sell everything it could produce. Therefore, American industry was not receptive to quality improvement.

The W. Edwards Deming Award for Quality is a national award presented annually in Japan to the company that has demonstrated the greatest quality improvement effort and to the individual who has been responsible for the greatest quality improvement. These are extremely high honors in Japan.

Deming's techniques tend to center around statistical controls on manufacturing processes. He believed that 85 percent of quality problems are generated by management. Deming developed a philosophy to facilitate quality improvement in a company through better management.[1] There are 14 steps to this philosophy aimed at management.[2]

1. Create a consistency of purpose toward the improvement of products and service, with a plan to become competitive and to stay in business. Decide to whom top management is responsible.
2. Adopt a new philosophy. We are in a new economic age. We can no longer live with commonly accepted levels of delays, mistakes, defective materials, and defective workmanship.
3. Cease dependence on mass inspection. Instead, require statistical evidence that quality is built in to eliminate the need for inspection on a mass basis. Purchasing managers have a new job, and must learn it.
4. End the practice of awarding business on the basis of a price tag. Instead, depend on meaningful measures of quality along with price. Eliminate suppliers who cannot qualify with statistical evidence of quality.
5. Find problems. It is management's job to work continually on the system (design, incoming materials, composition of material, maintenance, improvement of machines, training, supervision, and retraining).
6. Institute modern methods of on-the-job training.
7. Institute modern methods to supervise production workers. The responsibility of supervisors must be changed from sheer quantity to quality. Quality improvement will automatically improve productivity. Management must prepare to take immediate action on reports from supervisors concerning barriers such as inherited defects, machines poorly maintained, poor tools, and fuzzy operational definitions.

8. Drive out fear so that everyone may work effectively for the company.
9. Break down barriers between departments. People in research, design, sales, and production must work as a team to foresee production problems that may be encountered with various materials and specifications.
10. Eliminate numerical goals, posters, and slogans for the work force. Ask instead for new levels of productivity without providing methods.
11. Eliminate work standards that prescribe numerical quotas.
12. Remove barriers that stand between the hourly workers and their right to pride of workmanship.
13. Institute a vigorous program of education and retraining.
14. Create a structure in top management that will push the above 13 points daily.

These 14 steps are comprehensive in their coverage of management techniques. They are worth examining in greater detail.

1. Create consistency of purpose toward improvement of products and service, with a plan to become competitive and to stay in business There are two major types of problems in industry. First, there are the daily problems of production (maintenance, budget, forecasts, production schedule, etc.). These problems are a continual source of irritation. Second, there are the problems of tomorrow (new product development, new material development, market trends, and advancements by competitors).

If quality improvement is focused on the daily activity and problems of production, there can be no lasting impact on quality. Production fluctuates daily; therefore, management must begin to look at the big picture. Daily "disasters" demand the manager's attention. The problem comes when the manager falls into the habit of managing by disaster.

On a higher level, managers must stop looking at quarterly dividends as the means of setting objectives. Long-range planning is extremely important. Plans must reach out 3 years, 5 years, 10 years, even 20 or 30 years. Long-range planning should cover such areas as

- New services or products
- New materials required and their costs
- Methods of production and possible changes in equipment
- New skills required
- Training and retraining of personnel
- Training of supervisors
- Cost of production

- Cost of marketing
- Performance improvement
- Satisfaction of the end user

Companies need to allocate funds into these areas and expand their budgets for research, education, equipment, and maintenance. Such allocations require a change in the company's philosophy—which leads to the second step.

2. Adopt a new philosophy In postwar America, industry was overwhelmed by the demand for products, and companies could sell almost anything they produced. If there was a workmanship failure, the customer could take it to be fixed in a repair shop.

Times have changed! That type of business practice will send a company to bankruptcy in a hurry. Unfortunately, many industries have gotten into the habit of giving short shrift to quality, many to the point that they have forgotten how to meet the customer's requirements.

If a company is to survive in today's marketplace, a philosophy of quality must be instilled in every employee. If everyone from the president to the assembly line worker becomes involved, it will have a lasting impact on quality and productivity.

3. Cease dependence on mass inspection In the past, the quality department was called the inspection department, although this type of thinking is on the decline. When a company finds defects by inspection, it is not improving quality—it is only keeping the bad product from reaching the customer. It may not even do a good job of catching all the bad product. Tremendous losses are involved with this kind of detection. In some cases, the cost of inspection can be prohibitive.

The causes of the defects must be removed from the manufacturing process. Inspection can then be used to ensure that the process is working as it should. When using inspection to check the process rather than the product, only samples need to be taken. If these samples begin to show a defective product, the process can be altered before high costs are incurred.

With these types of controls in place, a company can be confident about the outgoing quality of its product. The purchasing companies can buy the product with confidence that it will function properly.

4. End the practice of awarding business on the basis of a price tag When Neil Armstrong awaited launch atop the huge rocket that was about to carry him to the moon, he was quoted as commenting, "It gives me no great comfort to know that the thousands of assemblies which make up this rocket were supplied by the lowest bidder."[3] He was speaking out of personal interest—his very life depended on the functioning of the machine in which he sat. Remember the Franklin expedition in Chapter 2? The food contract, awarded to the lowest bidder, cost the life of every person in the party.

Companies need to realize that their very existence depends on the function of the products and materials they buy. They can no longer allow the price tag to determine who receives their contract. Companies must begin to demand statistical proof of quality from their vendors. The price tag of the product a company uses is not the only money spent on that product. For example, a company buys a circuit board from a vendor, to be used in its own computer. The quality of the computer depends on the quality of the vendor's board. If the vendor's board is of poor quality, the purchasing company may need to repair it, replace it, ship it back, slow its own production, or experience a host of other problems that cost money and therefore reduce productivity.

Purchasing agents must be trained in statistical techniques that will show which vendors ship a consistently acceptable-quality product. Companies must insist that vendors supply statistical data that indicates the quality of the products they make.

Companies must become aware that the processes of their vendors are an extension of their own process. If indeed a company understands this, it will help refine its vendors' processes. That will help reduce everyone's cost and increase profits.

Purchasing agents should not be blamed for the quality of the incoming product. Until recently, many companies have required them to award contracts on price alone, regardless of product quality. Upper management must change the way it thinks about awarding contracts. It must end the practice of awarding business on the basis of price tag alone.

5. Find problems Managers must always be on the lookout for ways to improve process procedures. Manufacturing processes play only a small part in a company's overall performance. Other improvements may come in the areas of design, incoming materials, maintenance, machine improvement, training, customer service, and any other area that has an effect on the final product.

If a process is in statistical control, process improvements cannot be made by looking at the defective parts. The process is working as it was designed to work. Therefore, the process must be studied and analyzed for changes in its structure. This should be a continuing effort on the part of managers.

6. Institute modern methods of training Training can be expensive. It includes:

- instructor's wages
- employee's wages
- machinery downtime
- possible slowdown in production

These are just a few of the related costs. Although it is much more costly for a company not to train its employees, it is harder for a company to track the cost of not training. Training appears to cost money rather than save it. However, this could not be further from the truth. Costs of training are temporary, but production and quality increase when the training is completed. When training is left to trial and error, production goals may never be attained, and the production losses are endless. The costs associated with poor quality arising from lack of training are also never-ending. Rework, scrap material, poor morale, and many other costs must be accepted as a result of the corporate culture.

To a large extent, these costs could be reduced by training. This requires a line item in the budget for such a purpose. In fact, it may require a new department with a budget of its own. Is it becoming clearer what is required in a philosophy change?

7. Institute modern methods to supervise production workers It is important for industrial management to realize that today's workers have a much higher educational level than in Taylor's time (see Chapter 2). Today's assembly workers and machine workers are generally literate. The work force of today, if given a chance, can be of great value to management. Who can determine the problem of the machine better than the operator?

Supervisors need to have the authority to revise processes and repair machines. Workers must also be given the training and tools to effect improvement. Management must find ways to break down the barriers that hinder communication between departments and between labor and management.

8. Drive out fear Labor and management have had an adversarial relationship for many years. This is a cultural attitude prevalent in the U.S. It is not the case in many overseas countries. Japan, for instance, does not embrace this concept. Japanese executives consider that the workers are there for the good of the company and the company is there for the good of the workers. This makes everyone reach for the same goal—the good of the company.

Management must make every effort to remove workers' fear of job loss from the work place. Workers must be encouraged to make suggestions that would improve quality. They must also see some response to the suggestions they make. If improvement is needed from a worker, methods of improvement must be suggested. Intimidation will only breed insecurity and cause workers to cover their mistakes better.

9. Break down barriers between departments Each department has experts in its own area of manufacture. Often, individuals from one department assume that everyone in the plant knows as much about their area as they do. This is very seldom the case. Therefore, one department often unknowingly causes quality problems for another department.

These problems can be reduced by interdepartmental meetings on quality improvement. Representatives from engineering, production, sales, shipping, and research and development should get together on a regular basis. Their focus should be on improving existing processes and coordinating future processes.

10. Eliminate numerical goals, posters, and slogans This type of rhetoric serves only to inflame workers. Placing posters on the walls proclaiming that quality is the most important aspect of the job, without management follow-through, amounts to lying to the workers. It proves that management sees the problems of the company as originating with the workers. Numerical quotas and slogans give the workers nothing to help them improve the quality of their work and merely place unrealistic requirements on the workers.

11. Eliminate quotas Most production problems come not from the workers, but from the system. The employees will work to the limits of the system. To demand more will cause morale problems and job insecurity among the work force. Within a short time, frustration will set in for both labor and management.

If management is to see long-term improvement in quality and production, it must modify the manufacturing process. This may be done by upgrading machinery, training operators, or possibly instituting a whole new process. Demanding zero defects or increased production without giving the workers a means to accomplish the task will only breed frustration.

12. Foster pride in the work being performed Everyone wants to feel good about the job he or she performs. In fact, our self-image is often tied to how we feel about our job. This need is basic in everyone.

It is amazing how industry tries to diminish this need in an individual. The need is disregarded when a company fails to provide the workers with consistent accept-and-reject standards, or when it requires people to work on a machine that is not in proper adjustment. When the company does not allow the workers to do the best job possible, it breeds hostility toward the company.

Often, the product quality advertised is not consistent with what the workers have experienced. Workers may take no pride in the product of such a company, fostering labor/management problems as well as problems of quality and productivity.

13. Institute education and retraining The personnel needs of companies change as demand for the product changes. Often the company's answer to such a change is to lay off some workers and rehire others. In so doing, a company loses all it has invested in the furloughed workers.

Companies can reclaim the experience of the work force by training employees in other areas. This gives the company employees with a good background in more than one manufacturing process. It allows the workers to expand their areas of expertise. In turn, they feel better about themselves and their company.

14. Create management that will push the 13 steps daily The last step to the Deming philosophy is to keep it going. This philosophy will only be as effective as management's commitment to it.

This approach is not a bandage to keep people from looking at a company's quality problems. It is a management philosophy that will eventually correct quality problems and permanently increase productivity.

Therefore, management must keep these steps progressing on a daily basis. Quality improvement never stops.

4–2 JOSEPH M. JURAN

Juran is a well-respected authority in the quality industry and has written many books on the subject. In 1979, Juran established the Juran Institute, which promotes quality in every facet of industry service.

The Juran philosophy is a combination of concepts put together from the ideas of other noted quality professionals and Juran's own ideas. Juran's basic approach is to use whatever tools are needed to solve quality problems one at a time, and then put a system in place to keep them from recurring once the process is fixed. Juran believed there is no hard-and-fast rule that applies to every situation. Rather, there are different rules and concepts that apply to each situation. An overview of Juran's philosophy would be to:

- Obtain quality attitudes
- Obtain quality solutions
- Obtain quality planning and control

These steps each allow for several different methods of achievement. Different methods work for different situations. Juran believed that following these steps will lead to quality in product or service.

According to Juran, the first step to quality is to obtain a quality attitude. A quality attitude is a mind-set that quality is the most important factor in any decision. He believed that everyone within an organization must have a quality attitude, from the line worker to the chief executive officer. To obtain a quality attitude, the worker must believe that quality is desirable and attainable. Without this belief, a quality attitude is impossible. Fortunately, the majority of society holds this belief to some extent.

Juran suggested that there are three types of worker: the inhibitor, the conservative, and the innovator.[4] The *inhibitor's* desire for quality is outweighed by the belief that he or she has certain rights that are more important. For example, a product development manager may block or inhibit a promising proposal because the idea did not begin with, or would not be conducted by, the product development department. The *conservative's* desire for quality is superseded by his or her desire for security. He or she fears that an unproven process may endanger his or her position if the company uses it and it does not work. On the other hand, the *innovator's* desire for quality far outweighs everything else. He or she is not bound by the past nor the fear of losing job security. The innovator will boldly implement a new process. According to Juran, these three types of people must be approached in different ways.

The innovator needs to be approached only with logical reasoning. The conservative must be shown a successful precedent obtained by innovators. The inhibitor may have to be coerced by peer pressure. Once everyone else has changed, the inhibitor will acquiesce because he or she is standing alone. All levels of an organization must participate in the acquisition of quality attitudes. Top management must state a policy toward quality and set the example by living up to the policy. Middle management must make quality attitudes a part of every job including its own. It can also create channels for the open communication of ideas. Breakthrough in quality attitudes is only the first step.

After the attitude has been adjusted, the second step can begin. Problems must be identified and a plan must then be created to correct the specific problems. There are two types of problems according to Juran: sporadic problems, which are dramatic deviations from the status quo, and chronic problems, often thought of as inevitable because they are longstanding and very difficult to solve. Often, chronic problems require a remedy through change. Juran uses the concept of "vital few and trivial many." This concept is called the "Pareto Principle."[5] Under this principle, the vital few account for a large portion of the problems, whereas the trivial many account for only a small part of the total problem. This concept is discussed in more detail later in this study. Juran uses this principle to assign priorities to problem solving.

Once problems have been categorized and prioritized, Juran advocates the use of a "steering arm" and a "diagnostic arm" to set up a plan to solve the problem.[6]

The steering arm consists of people who provide the following[7]:

- Definition of the specific aim of the improvement program
- Ideas on possible causes of the problem
- Authority to experiment
- Authority to implement the solution

The diagnostic arm consists of people who determine the causes, not the remedies, of a problem.[8]

Both of these arms work together to identify the problem, find the cause of the problem, and set up a solution for the problem and/or plan for production. Both work on the basis of Juran's concept of "fitness for use." Standards are set by the customer, who determines when the product is fit for use.

Once a plan has been established, it must be implemented. Implementing a plan always proves to have obstacles and steps of its own. Juran suggested that the first step is overcoming resistance to change. Any proposed change may first be considered a threat to the workers' culture. Acceptance of change involves two steps: unlearning old habits, attitudes, and beliefs; and learning the new. This can be promoted by establishing some means of praise or approval. Juran recommends several ways in which change can be promoted. Planning and implementation should involve those who are affected. It is important to work with the leadership of the culture and to choose the right time to implement change.

Once resistance to change has been reasonably overcome, it's time to take action. The diagnostic and steering arms are composed of individual supervisors, who retain the responsibility for decisions, action, and results. Implementation of the plan by the various departments is explained by Juran in the spiral diagram (see Figure 4-1).

The model is a closed 360° loop in the form of a spiral that shows progression through time. Each individual process and department works with and for another. Each subsystem must meet its individual goals if the plan is to be successful.

Taking action will not in itself implement change. There must be a transition period. The transition involves retooling, exchange of information, training, and following progress to ensure the sequence stays on course. Training can take time, and the more complicated the process, the more time it will take. Implementing a formula to take the place of a rule of thumb takes techniques that reward workers for using it and that make it easy to do. This will increase the probability that the formula will be used.

The information gathered by the diagnostic and steering arms must be communicated to the various subdivisions and departments. Once

Figure 4–1 The Spiral of Progress in Quality

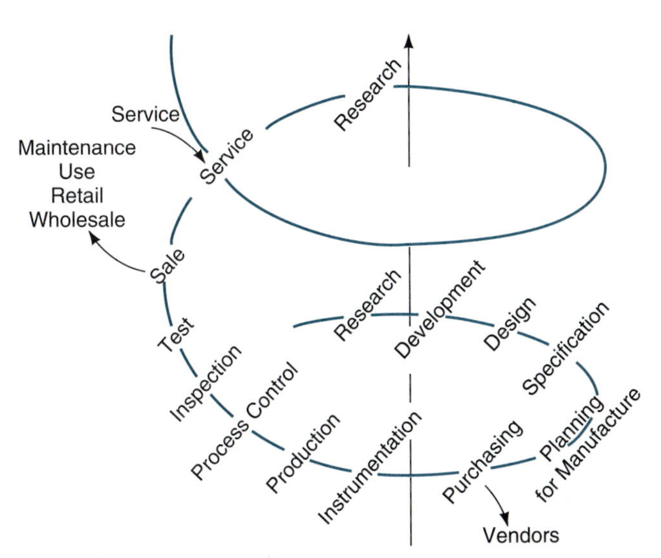

Source: J. M. Juran, *Quality Control Handbook*, 4th ed. (New York: McGraw-Hill Book Company, 1988). Permission granted by McGraw-Hill Book Company.

the transition has taken place, the progress must be followed to the next step required to obtain quality.

Control is the final step in Juran's system to obtain quality in a product or service. Control includes

- Choosing the subject of control
- Using a unit of measure
- Setting and resetting a standard
- Choosing and using a type of sensor.[9]

Control is adherence to standards, staying on course, and preventing change of direction. Identifying a control subject, such as part tolerance, cost, morale, corporate image, or the like, is a must. No progress can be made without this. After a control subject has been picked, a unit of measure must be chosen. Juran advocates the use of numbers to reduce debate over the meaning of the objectives—for example, the question of "how poor is poor quality?"

Even with a unit of measure, a decision cannot be made without a set standard. Again, Juran's concept of fitness for use is applicable. The fitness standard is set by the customer; however, other standards need to be established to meet this end requirement.

Standards must be attainable, economical, applicable, consistent, and all-inclusive. They must be understandable, stable, maintainable, legitimate, and equitable. Standards begin to deteriorate the minute they are established, so formal reviews must be made. Standards also give a goal stating what should be done.

A sensor is needed to determine exactly what is being done. A sensor is a device used to sense specific stimuli and convert them into information. In a company, the most familiar sensors are time clock cards, materials requisitions, inspector's gauges, and sales representatives' call records, to mention a few. Juran notes three types of sensors—those that sense before the fact, those that sense during the fact, and those that sense after the fact. Sensing should be done in accordance with agreed opinion. Without an agreed sensor, the validity of the information will undoubtedly be questioned. With the sensor in place, actual performance can be measured and the tasks of developing quality attitudes, quality plan setting, plan implementation, and quality control can be repeated.

4–3 PHILIP B. CROSBY

Zero Defect (ZD) is a program that originated during the production of the U.S. Army Pershing missile system in the 1960s. It was developed by

Philip B. Crosby, quality manager for Martin Marietta Corporation's Martin Company in Orlando, Florida.

It seemed that the missile program was plagued by one disaster after another, usually because of small instances of human carelessness. It was on December 21, 1961, that Crosby's program was really noted as a success. On that day, Cape Kennedy received its first Pershing missile from the Martin Company with zero discrepancies.[10]

The whole concept of ZD is to stop defects before they happen instead of waiting to find them during an inspection of the part, or worse, by the consumer at a later date. To do this, the definition of a defect and how it came to be must be known. First, the defect (or worker error) is caused by either a lack of knowledge, working with the wrong tools or without the proper facilities, or by the worker's inability to keep his or her attention on the work being done. Crosby's message was to motivate workers toward a mind-set that promotes business ethics and quality.

Crosby proposed that the thinking behind these "worker errors" is cultural. Americans are brought up to think that "to err is human" and "nobody is perfect." All through life workers have been given a certain margin of error. For example, C means passing work in school. Workers carry this thinking into their jobs, where they are allowed a certain percentage of errors, or defects, and can still remain on the job.

However, no matter how mediocre the worker's own performance, each worker tends to expect more than mediocrity from the people around him or her. For example, most people expect their doctor always to be right. Few people would fly on airplanes if they thought that they could walk away safely from only 70 percent of landings. Realizing this, Crosby decided that if he was going to be able to initiate his ZD program, he would have to promote a constant, conscious desire to "do the job right the first time."[11]

Crosby believed the attainability of this goal was directly proportional to the attitudes or desires of individuals. Individual attitude is set, to a large extent, by corporate attitude. To get the worker to respond, direct management action is needed at the highest level. Crosby developed five steps for management to follow:

- Present a challenge.
- Back the challenge with action of its own.
- Establish standards to be met.
- Check results.
- Act in accordance with the results to recognize accomplishments.[12]

There are four absolutes that are an integral part of Crosby's concept:

- The definition of quality is conformance to requirements.
- The system of quality is prevention.
- The performance standard is ZD.
- The measurement of quality is the price of nonconformance.[13]

ZD is a broad-based approach to quality improvement, independent of technology or particular quality techniques. The program is based on the concept that employees perform work at the standard set by management; therefore, employees perform a poor quality of work because of low standards adopted by management. "Management must repent by setting a standard of zero defects for their employees and sponsoring a program that allows the organization to make progress toward the new goal."[14] There are 14 steps to implement a zero defects plan[15]:

- Management commitment
- Quality improvement teams
- Measurements
- Cost of quality
- Quality awareness
- Corrective action
- Zero defects planning
- Supervisor training
- Zero Defects Day
- Goal setting
- Error cause removal
- Recognition
- Quality councils
- Do it over again

Each of the above-listed steps is discussed in the following pages.

1. Management commitment If the program is to be successful, the commitment of the top-level managers is essential. It must be more than lip service; the work force must be able to see a continuing commitment on the part of management. A corporate policy on quality must be developed. Top-level staff meetings should include reports on the quality improvement process, quality costs, and conformance. The chief executive officer and the chief operations officer need to make clear statements to their employees about quality to motivate the workers.

2. Quality improvement teams In this phase, the company brings together representatives from each department to form the team. The purpose of the team is to "improve things around here." This is the starting point for low-level motivation programs. Programs such as quality circles are useful here.

3. Measurements Here the company's quality measurements will be reviewed. What are the company's measurements? Are they working? Questions such as these should be asked and answered. From this, a quality status can be recorded. It will show areas of possible improvement that can be measured, give corrective action, and document actual improvement.

4. Cost of quality The cost of quality is where changing what is wrong will profit the company. There are two areas of cost of quality: the price of nonconformance and the price of conformance. The price of nonconformance is all the costs incurred by doing things wrong. The price of conformance is all the costs incurred by doing things right the first time. The first is the cost of a lack of quality; the second is the cost of doing business. Management must compile the quality costs to evaluate the scope of the financial loss the company is presently experiencing. In most cases, putting a dollar value on the cost of poor quality will motivate upper management toward improvement.

5. Quality awareness This is management's effort to inform the employees about the measurements of the costs of nonquality. This is done through improvement materials such as booklets, films, and posters. Communication inside a corporation is very important. If the company does not have an overall communication system, quality awareness will never be successful.

Some companies have quality slogans such as "Do it right the first time" to help increase quality awareness. Everyone needs to be aware of management's commitment to quality, what the quality policy says, and the cost of doing things wrong. It may be helpful to make the company's quality awareness known to the general public, usually in the form of advertising the quality program. The goal of such a program is to project a quality image to the buying public.

6. Corrective action The purpose of corrective action is to identify and eliminate problems forever. People are encouraged to talk about problems found through inspection, audits, or self-evaluation. Identifying a problem is the first step in corrective action. The idea here is to fix the problem, not fix blame.

7. Zero defects planning For the program to have the most impact, all employees should be informed of the program and its goals on the same day. The day can be devoted to "kicking off" the ZD program. Three or four members of the quality team are selected to investigate how other companies have run their Zero Defects Day.

Speeches could be made by representatives from unions, city government, and customers. Information on the goals and objectives should be given to everyone at the same time. Generally, this is done by department heads who have been trained in the ZD way. Often the company hosts a dinner or coffee to show its commitment.

8. Supervisor training A formal orientation with all levels of management should be conducted prior to implementing all the steps. All managers must understand each step well enough to explain it to their staff. The proof of understanding is the ability to explain it.

Eventually, all the supervisors will be tuned into the program and realize its value for themselves. Then they should concentrate their action to further the program.

9. Zero Defects Day The establishment of ZD as the performance standard of the company should be done in one day; that way, everyone understands it the same way. Supervisors explain the program to their employees and do something different in the facility so everyone will recognize that it marks a "new attitude" day. Making a day of the ZD commitment provides an emphasis and a memory that will be long-lasting.

10. Goal setting Goal setting happens automatically right after measurement. Some quality improvement teams try to do Steps 1 through 14 consecutively, but actually most of them run in parallel (e.g., quality education never stops). The first six steps are all accomplished by management and need to be completed first. When the measurement of quality costs begins, people should immediately start to think about goals. The ultimate goal is ZD, and that is everyone's objective. Goals should be chosen by the group as much as possible, and should be put on a chart for everyone to see. Minor goals should not be listed; the goals should be obtainable but should also make the department work to obtain them.

11. Error cause removal Individuals are asked to describe any problem that keeps them from performing error-free work. This is not a suggestion system; all they have to do is list the problems. The appropriate team will develop the answer. It is important that all problems listed be quickly acknowledged, generally within 24 hours. People will learn that their problems can be heard and addressed. Once employees learn to trust this communication, the program can go on forever.

12. Recognition Very few companies recognize their good performers. Many managers feel that people are being paid to do their jobs well and that should be that. Under this quality program, awards are established to recognize those who meet their goals or perform outstanding acts. The prizes or awards should not be monetary, because the act of recognition is what is important. For this recognition to be motivating, people must be on the social affiliation level or esteem recognition level on Maslow's Hierarchy of Needs, referred to in Chapter 2. Genuine recognition of performance is something people really appreciate. They will continue to support the program, even if they did not personally receive an award.

13. Quality councils Quality professionals and team chairpersons should be brought together regularly to communicate with each other and to determine the necessary action to upgrade and improve the quality program to be initiated. This council is the best source of information on the status of the program.

14. Do it over again The typical program takes 12 to 18 months to install. By that time, turnover and changing situations will have wiped out much of the education effort. Therefore, it is necessary to set up a new team of representatives and begin again. For instance, ZD Day should be marked as an anniversary. The company might give a special lunch for all employees. The point is that the program is never over. Repetition makes the program perpetual. If quality is not ingrained in the organization, it may never be attained.

4–4 ARMAND FEIGENBAUM

Dr. Armand Vallin Feigenbaum is the originator of Total Quality Control (TQC). TQC is an approach to quality and productivity that has profoundly influenced the competition for world markets in the U.S., Japan, and throughout the industrialized world.

For 10 years, Feigenbaum managed worldwide manufacturing operations and quality control for the General Electric Company. He was the founding chairman of the International Academy for Quality and is past president of the American Society for Quality Control, which has presented him its Edwards Medal and Landcaster Award for international contributions to quality and productivity.[16]

TQC is an extensive, companywide, quality improvement program. The Total Quality system gets everyone involved by developing an agreed-on companywide and plantwide work structure documented in effective, integrated technical and managerial procedures. This gives coordination to the actions of the work force, machines, and information of the company in the most practical ways, ensuring customer quality satisfaction and economical costs of quality.

Often, a company's quality control department deals only with production. The quality effort in the production area is extensive; however, this is not the only area in which quality costs are incurred. Feigenbaum realized that quality plays an important role in business. From the time of the initial contact with the customer to installing and servicing the product, the customer's requirements must be met.

The customer comes in contact with more than just the product in the course of business. Often, a problem in shipping, records, invoices, or any of a number of other areas of administration will influence how a customer feels about the product. TQC is a customer-based improvement program. If the customer is not satisfied, there is a lack of quality.

Refer to Feigenbaum's definition of quality in Chapter 2: "The total composite product and service characteristics of marketing, engineering, manufacture, and maintenance through which the product and service in use will meet the expectations of the customer."[17] Feigenbaum is saying that the customer forms an opinion of the quality of the product by more than just the product itself. The customer must be satisfied in all areas of a transaction. The processes and procedures of marketing, engineering, purchasing, manufacturing engineering, manufacturing supervision, shop operation, mechanical inspection, functional testing, shipping, installation, and service must all meet the requirements of the customer.

American businesses can no longer develop procedures and processes in these areas and then expect the customers to change their requirements to fit. Competition in today's worldwide market is too great.

Feigenbaum defined 10 product and service conditions that must be met or considered to satisfy customer requirements[18]:

- Specification of dimensions and operating characteristics
- Life and reliability objectives

- Safety requirements
- Relevant standards
- Engineering, manufacturing, and quality costs
- Production conditions under which the article is manufactured
- Field installation, maintenance, and service objectives
- Energy utilization and material conservation factors
- Environmental and other side effects considerations
- Cost of customer operation and use, and product service

The aim of these conditions is that quality establishes the proper balance between the cost of the product or service, and the customer value it renders (including essential requirements such as safety).

Determining quality and quality costs actually takes place throughout the entire industrial cycle. This is why real quality control cannot be accomplished by individually concentrating on inspection, product design, reject troubleshooting, operator education, supplier control, statistical analysis, and reliability studies, as important as each individual element is. To install this concept, an alteration in the major business management strategy is required.

> The total-customer-satisfaction-oriented concept of quality, together with reasonable costs of quality, must be established as one of the primary business and product planning and implementation goals and performance measurements of the marketing, engineering, production, industrial relations, and service functions of the company. Assuring this customer-satisfaction quality and cost result must be established as a primary business goal of the quality program of the company and of the quality-control function itself—not some narrower technical goal restricted to a limited technical or production-oriented quality result.[19]

4–5 QUALITY CIRCLES

Quality circles rely heavily on the work force to identify and resolve the quality problems that exist within a company. Who better to identify problems than the people working the machines? Quality circles are often used in conjunction with the TQC concept discussed previously. A quality circle is a small group of employees and their supervisors from a work area who voluntarily meet on a regular basis to study QC and productivity improvement techniques. These techniques are used to identify and solve work-related problems, present their solutions, and ensure that they work.

There are 14 criteria used in the formation of quality circles in a company:

- Quality circles are small (they range from 4 to 15 members).
- All members come from the same area (this gives the circle its identity).
- The supervisors come from the same area as the members.
- Supervisors are usually, although not always, the leaders of the circles (but they do not issue orders or make decisions).
- Participation is voluntary (this means everyone in the shop can join or refuse to join, quit or rejoin).
- Circles meet once a week, on company time, with pay.
- Circles meet in special rooms away from work areas.
- Circle members receive special training in the rules of quality circle participation (e.g., the mechanics of running a meeting/management presentation, techniques of group problem solving, brainstorming, flowcharts, Pareto analysis, and cause-and-effect diagrams).
- Circle members, not management, choose the problems on which they will work.
- Circle members collect all information, analyze problems, and develop solutions.
- Technical specialists and management assist circles with information and expertise whenever asked to do so.
- Circles receive advice and guidance from an advisor who attends all circle meetings, but who is not a circle member.
- Management presentations are given to those managers and technical specialists who would normally make the decision on a proposal.
- Circles exist as long as the members wish to meet. They can inactivate themselves or reactivate; they can meet for one month or for years.

Although there are 14 criteria, not all are required for each individual business. The program may range from a small company with a single quality circle requiring only 1 or 2 of the 14 steps, to a large corporation with thousands of quality circles, which may need all 14 criteria in each of their circles.

It should be mentioned here that the requirements of Maslow's Hierarchy of Needs must be met within the company if the quality circles concept is to work. To contribute ideas and implement solutions with others, employees must feel secure about their jobs and the work they do (see Chapter 2).

Careful planning must be done before the circles are started. The circles concept is not a temporary cure; it is an ongoing philosophy of personal, as opposed to personnel, management. A cultural change must accompany the philosophy. If change is not done properly, poor labor/management relations may result.

4–6 CONCLUSIONS

At this point it should be apparent that all of the philosophies have many points in common, and each has points that are in direct contradiction to other philosophies. Distinctions should be drawn among the philosophies to define and separate them.

Deming's approach can be summarized by the use of statistics to allow the operator to obtain detailed information about the process by measuring the product. Deming developed a philosophy that requires a companywide effort to improve the quality system, resulting in a higher quality product. His is an upper-management-down approach.

Juran's philosophy can be summarized as improving quality on a problem-by-problem approach. His technique to improve quality is to evaluate a problem, using some statistics, and to develop a plan to eliminate the cause and keep it from recurring. His approach can be effectively implemented at an intermediate management level, but is most effective when approved and supported by upper management.

Crosby's view can be summarized as motivating the work force and business ethics. Crosby advocates business ethics in his slogans "Do it right the first time" and "If it's not right don't ship it" as the foundation of his approach. Crosby's philosophy primarily concerns the motivation of workers and management. Crosby does not recommend the use of statistical analysis as a means of identifying and solving problems. He believes the workers can, if properly motivated, identify and eliminate 90 percent of the quality problems that occur in a plant.

Feigenbaum coined the term *Total Quality Control.* His approach is to get everyone involved, motivate both workers and management, use statistical analysis, and systematically improve quality. His approach requires upper management's total support and guidance.

Quality circles have evolved through the guidance of many quality professionals. The basic idea behind quality circles is to motivate workers and management to study a problem intensively to find the solutions. The motivation behind the circles is to allow everyone—from janitor to plant manager—to have the same opportunity for input regarding solutions to problems. Everyone is equal when solving a problem in a quality circle. Statistical analysis is used to help identify and solve problems.

Along with the points in common among the philosophies are points that are different. The main points in which the philosophies differ are the methods used to motivate workers, use of statistical analysis, and thoroughness of the initial program. Many authors focus on these differences instead of exploring the proper way to improve quality. Let us address and explore the differences at this point. One word of caution—this discussion of the differences should not be interpreted as a statement that one philosophy is better than another. Each method has had tremendous successes and can be used to improve quality.

Motivation is one area where the philosophies differ most. Deming and Crosby approach motivation from opposite directions, whereas Juran and Feigenbaum are somewhere in the middle.

Deming's system eliminates slogans and banners from the factory. Deming believed this approach insults the workers' intelligence and could be counterproductive if the company does not live up to the slogans that hang in the shop. Deming suggests that true motivation comes when a worker is given total responsibility for quality, along with the authority to make the needed changes to improve quality.

Crosby believes that the use of slogans, signs, education, meetings, motivational speeches, lunches, and recognition is the proper approach. These two philosophies are in direct conflict with each other.

Juran and Feigenbaum are close to each other in the way they advocate motivation. Both these philosophies empower the worker, explain the "why" behind a policy, allow the worker to have input into the resolution of a problem, and treat the worker with dignity.

Each of the approaches to motivation has a good track record in industry. The right approach to motivation is the one that works effectively in a shop.

The use of statistical analysis is a second point where these philosophies differ. At the extremes are Deming and Crosby. Juran and Feigenbaum are again in the middle.

Deming is a reformed statistician and advocates the use of statistical analysis to identify, solve, and control virtually any variance in a process. This requires tremendous amounts of data to be collected and analyzed. It also requires in-depth training of the work force in statistical analysis.

Crosby is at the other extreme in his approach. He advocates no use of statistical analysis, because it tends to overwhelm the work force and the output can be adjusted to give a false impression.

Juran and Feigenbaum advocate the use of statistical analysis when the problem or process under study requires its use. Both believe that if the problem or process can be studied without the use of statistical analysis, statistics will only cloud the issue and could make the solution to the problem harder to find.

The third area of disagreement is in the extent of the systemwide initial improvement program. In this area, Juran is at one extreme, whereas Deming, Crosby, and Feigenbaum take the opposite view.

Juran, though stating that a systemwide approach will yield the greatest improvement, developed his approach to solve problems on a case-by-case basis. This approach can be implemented even when upper management takes a wait-and-see attitude or gives less than total support to the improvement program.

Deming, Crosby, and Feigenbaum developed philosophies that require total support by upper management. If this support is not present, successful implementation of these programs will be at risk.

SUMMARY

Each of these philosophies has had successes and failures. The best approach for many companies is to take the applicable parts of each philosophy and "custom make" a program that better fits the particular organization. If the decision is made to follow one approach to the exclusion of all others, the choice should be carefully made. The only thing worse than not having a quality improvement program in a company is to have a program that fails. If no program is present, the credibility of the quality control department is in question; if a program fails, the credibility of the quality control department is totally destroyed. Pick the philosophy carefully, deciding which one has the best chance of success in your specific situation. If none fits the given situation, do not be afraid to take the best methods from one or more philosophies and apply them to that situation. To have a custom-fit program, you may have to custom build the approach. Do not expect a custom fit with an off-the-shelf philosophy.

KEY TERMS

Chronic problems Problems that are longstanding and hard to solve.
Conservative A worker whose desire for quality is superseded by his or her desire for security.
Fitness for use Definition for quality as proposed by Joseph Juran.
Inhibitor A worker who fears the quality effort will encroach on his or her rights.
Innovator A worker whose desire for quality outweighs other motives.
Management by disaster The tendency of mid-level managers to neglect long-range goals because of continuous crisis situations.
Pareto principle The proposition that 20 percent of the problems account for 80 percent of the effort.
Sporadic problems Problems that are dramatic deviations from the status quo.
Total quality control Armand Feigenbaum's concept concerning quality throughout all levels of a company.
Vital few and trivial many Phrase coined by Joseph Juran denoting the Pareto principle.
Zero defect The program implemented by Philip Crosby while at Martin Marietta Corporation.

PHILOSOPHIES TEST

1. Discuss in a short paragraph the approach to quality taken by W. Edwards Deming.
2. Discuss in a short paragraph the approach to quality taken by Joseph M. Juran.
3. Discuss in a short paragraph the approach to quality taken by Philip B. Crosby.
4. Discuss in a short paragraph the approach to quality taken by Armand Feigenbaum.
5. Discuss in a short paragraph the use of quality circles to improve quality.

NOTES

1. W. Edwards Deming (Speech delivered to the Fortieth Quality Congress, 19 May 1986, Anaheim, CA).
2. W. Edwards Deming, *Quality, Productivity and Competitive Position* (Cambridge, MA: MIT Press, 1982), 16.
3. Deming, Speech.
4. Joseph Juran, *Juran on Planning for Quality* (New York: The Free Press, 1988), 299.
5. Joseph Juran and Frank Grynz, Jr., *Quality Planning and Analysis,* 2d ed. (New York: McGraw-Hill, 1980), 20–21.
6. Joseph Juran, *Managerial Breakthrough* (New York: McGraw-Hill, 1964), 16.
7. Juran, *Breakthrough,* 184.
8. Juran, *Breakthrough,* 184.
9. Juran, *Breakthrough,* 185.
10. J. Halpin, *Zero Defects—A New Dimension in QA* (New York: McGraw-Hill, 1979), 5.
11. Philip Crosby, *Quality Is Free* (New York: McGraw-Hill, 1979), 144.
12. Glenn Hayes, *Quality Assurance: Management and Technology,* 7th ed. (Capistrano Beach, CA: Gallant/Charger Publications, Inc., 1985), 137.
13. Crosby, *Quality Is Free,* 111.
14. Halpin, 8.
15. Crosby, *Quality Without Tears* (New York: McGraw-Hill, 1984), 101–15.
16. A. V. Feigenbaum, *Total Quality Control,* 3d ed. (New York: McGraw-Hill, 1983), 852.
17. Feigenbaum, 7.
18. Feigenbaum, 9.
19. Feigenbaum, 18.

METHODS USED TO CONTROL QUALITY

5

OBJECTIVES

After completing Chapter 5, the reader should be able to:

- Discuss the concept of "cost of quality."
- Discuss the use of a Pareto analysis.
- Discuss the use of a fishbone analysis.
- Discuss the concept behind and use of statistical quality control.
- Discuss the concept of design of experiments.
- Discuss the use of quality circles.
- Discuss the concept of self–quality control.
- Discuss the use of quality audits.
- Discuss the concept of just-in-time delivery.

In Chapter 4 the philosophies of Juran, Deming, Crosby, and Feigenbaum were discussed. The philosophy is the basic belief about the proper management style and course of action that should be taken by a corporation. A distinction should be made at this point between philosophies and methods. To implement any philosophy properly, many methods must be used.

The difference between philosophies and methods can be illustrated by the way an automobile mechanic approaches fixing a car. The mechanic has a philosophy about how to troubleshoot the problem. Using that approach the mechanic diagnoses the root cause, then goes about correcting the problem. The approach taken to identify the root cause is the same as the philosophy about how to improve quality. After identifying the root cause, the mechanic selects the tools needed and fixes the car. In quality control, the methods (like the ones that will be discussed in this chapter) are the tools used by quality professionals to improve quality. These might be statistical process control, design of experiments, or quality audits. To identify and correct a problem, the quality professional selects the proper method to identify, then eliminates, or corrects, the root cause. If a quality professional is to be effective in consistently improving quality, he or she must understand and be able to use a wide variety of techniques to identify and remedy quality problems.

This chapter provides an overview of 12 methods used to identify and correct quality problems. These methods were selected to give a broad foundation in quality concepts. They are:

- Quality cost concepts
- Pareto analysis

- Fishbone analysis
- Statistical quality control
- Design of experiments
- Evolutionary operations
- Quality circles
- Self-quality control
- Quality audits
- Just-in-time delivery
- Computer-integrated manufacturing
- Quality function deployment

5–1 QUALITY COST CONCEPTS

The concept of quality cost has been the starting point for many quality programs. One of the reasons this is so is that the cost of quality has become staggering. "The cost of quality is on the order of ten percent [10%] of the economy."[1] This means that the potential for savings is also enormous. What makes this cost more shocking is that quality costs are only those costs that are related to making an unacceptable product. These costs would disappear if the product were made right the first time. A quality professional must make a thorough accounting of these costs to determine where the majority of a company's revenue is being lost. When this is located, the company can take corrective action to eliminate the financial drain.

Quality costs can be divided into four areas, depending on where these costs occur. These costs incurred are[2]:

- Internal failure
- External failure
- Appraisal
- Prevention

If these costs can be identified with precision, they can be eliminated.

Internal Failure Internal failure costs are those costs that are associated with locating the defective product before it is shipped to the customer. These costs include the following categories[3]:

Scrap The total loss in labor, material, production costs, storage, and inspection of material that is defective and cannot be economically repaired.

Rework The total cost to correct a defective product and make it fit for use, i.e., cost of processing, labor, material, and storage.

Retest The total cost to inspect the product after it has been reworked.

Downtime The time, if any, that production is halted due to defective material. In continuous processing mills, this may be a factor because material from one process is directly fed into the next process. If the defective product is produced in the first process, it is removed and no product is present for the second process to use. This causes downtime for the equipment and labor force. For example, a newspaper press might be halted because of a major printing error.

Yield loss The cost of lower profit that could be eliminated if the production process were improved. A company that makes a soft drink would experience yield loss if it had poor control of the amount of liquid put in the bottle. This cost could become excessive. Suppose 50,000 bottles were filled to 32.5 ounces instead of 32 ounces. This example would lose the cost of production and the profit on 781 bottles of soft drink.

Disposition The cost to determine what to do with a nonconforming product. In many cases, this cost would include the cost of a material review board (MRB). An MRB is a group representing management, engineering, and production that decides the action to be taken on the nonconforming product.[4] The total cost of this board would be included in the disposition.

Facilities These are the total facilities costs of producing a nonconforming product. On average, if a company produced 5 percent defective product, then 5 percent of the costs of the facilities, administration, and production would be used to produce this product.

External Failure External failure costs are those that are associated with shipping a defective product to the customer, including the following[5]:

Complaint adjustment This includes the costs to investigate and make good the complaint.

Returned material This includes the total cost to replace or refund the purchase price of the defective product. It also includes the costs of storing, return shipment, and disposition of the defective product.

Warranty cost This includes all costs associated with servicing and repairing the defective product. For many companies, this cost includes regional repair facilities and repairpersons.

Allowances The cost of discounting a substandard product to a customer, for example, clothing seconds and tire "blems."

Appraisal

Appraisal costs are all costs associated with determining the condition of the product the first time through.[6]

Inspection and testing These include the costs of test equipment, material consumed during testing (e.g., X-ray film, chemicals), labor, facilities, and any product that is consumed during destructive testing.

Maintaining test equipment This includes the costs to calibrate and repair test equipment.

Prevention

Prevention costs[7] are those that are incurred to keep the internal and external failure costs to a minimum.[8]

Quality planning This includes all costs related to planning and implementing the quality plan for a company or a process. These would include the costs of labor, material, and facilities.

Quality data acquisition, analysis, and quality reporting These include all costs associated with gathering, analyzing, and reporting on the quality status to middle and upper management.

Process control This includes the costs of material, labor, and facilities that are dedicated to controlling the output quality of a process used to produce the product.

Improvement projects These include all costs involved with planning, implementation, and maintenance of projects directed toward the improvement of the output quality of a process.

The interrelationship of the four costs can be determined in the module for optimum quality costs (Figure 5 - 1).

If 100 percent of the product being produced were defective, the internal and external failure costs would be the total cost of production. The appraisal and prevention costs would be a zero output. As the product approaches 100 percent good, the internal and external failure costs would be reduced; however, to achieve this improvement, the appraisal and prevention costs would go up. Initially, the internal and external failure costs will be reduced at a greater rate than the appraisal and prevention costs will increase. This would result in a good return on investment. As the product quality gets closer to a 100 percent good product, the return on investment goes down. At a point near 100 percent good product, $1 invested in improving the product yields a $1

Figure 5–1

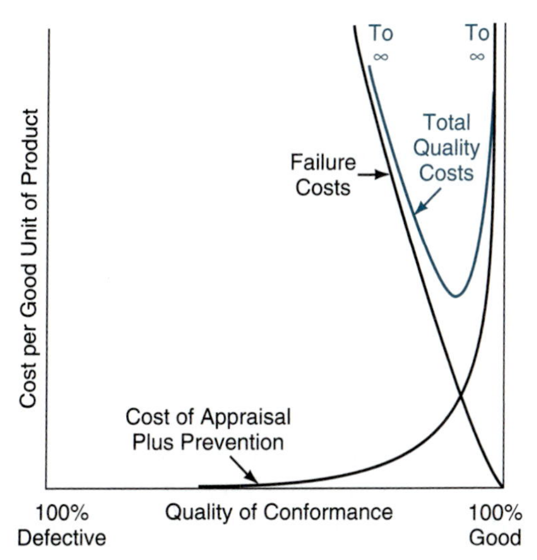

Source: J. M. Juran, *Quality Control Handbook*, 4th ed. (New York: McGraw-Hill Book Company, 1988).

reduction in internal and external failures. Further improvement will yield a negative return on investment.

A quality manager must always keep the interest of the company in mind. The point where a $1 investment yields a $1 savings is the economic state of control. Unless industrial requirements force further improvement, this is the point at which a system should be put in place to maintain the process at this optimum point.

5–2 PARETO ANALYSIS

In the 1870s, Vilfredo Pareto made a study of the uneven distribution of wealth among people in Italy. From this information, he formulated mathematical models to quantify this uneven distribution. Use of this type of analysis was not explored fully or applied universally until Juran found the same uneven distribution in quality control. In May 1975, an article appeared in *Quality Progress* that described the uneven distribution, referring to it as the Pareto principle.[9]

The Pareto principle means that all potential problems do not occur at the same rates. Some will occur many times more often than others. In fact, about 20 percent of all potential problems occur about 80 percent of the time, whereas the remaining 80 percent of the kinds of problems occur only about 20 percent of the time.[10]

For example, if a company making flashlights found that it had to reject bad lights after they were produced for 10 reasons, about 80 percent of the rejections would be for two of the reasons. The other 20 percent of the rejects would be for one of the remaining eight reasons. This example shows the 80/20 rule, also known as the "trivial many and the vital few" theory.[11] In the example here, the "trivial many" reasons for rejecting the lights are the eight that accounted for only 20 percent of the rejects. The "vital few" are the two reasons that accounted for rejecting 80 percent of the product. From these data, graphs can be drawn to complete the Pareto diagram.

The Pareto principle has few limitations on its application. If a study were made of cars in a parking lot, by manufacturer, the results would indicate that about 80 percent of the cars would have been produced by 20 percent of the manufacturers represented in the lot. The Pareto principle, which can be used in many applications other than quality control, can be used to isolate the vital few groups from the trivial many.[12]

After data are gathered, they are shown on a specialized vertical bar graph or column graph. Data are arranged in order, from the group with the highest number to the group with the smallest number (descending order).[13]

On a piece of graph paper, draw a vertical and a horizontal scale. The vertical scale shows the number of occurrences within each group. The horizontal scale shows the groups. Plot, with tic marks, the number of occurrences in each group. The resultant pattern clearly shows which groups make up the vital few.[14] The data in Table 5-1 are from the analysis on the flashlights referred to earlier.

Table 5–1 Flashlight Inspection Analysis

REASONS FOR REJECTION	NUMBER REJECTED
Bad switch	63
Bad ground spring	14
Wrong size light socket	8
Bad light	5
Cracked lens	3
Warped lens	2
Warped reflector	2
Clouded reflector	1
Clouded lens	1
Oversize screw head	1

From data given, a graph can be made that shows which areas produce the greatest result for the effort. To produce the graph, draw the horizontal and vertical scales. The horizontal scale shows reasons for rejection in descending order. The vertical scale shows the percentage of the total number of occurrences that each group of occurrences makes up.

The graph in Figure 5-2 reveals that groups 1 and 2 are the vital few groups with the majority of defects in them. They make up 77 percent of the total number of the defective product. To reduce effectively the total number of rejected flashlights, groups 1 and 2 should be reduced or eliminated. If this can be done, the overall amount of defective product could be reduced substantially. As stated earlier, this method of grouping data will help identify where the most effort should be applied to reduce or eliminate defects—in other words, to give the most return for the money. Groups 1, 2, and 3 should be the three areas where money is spent to correct defects. The rest should be left alone.

One variation on the Pareto analysis would be a scale added on the right to show cumulative sums. This scale would start with 0 on the bottom and end with 100 percent on the top. The scale helps identify the point of diminishing return for the groups. The point of diminishing return is where more money will be spent to fix a problem than will be saved by the problem being fixed. A point should be plotted above each group to show the cumulative sum of all preceding groups. This curve is referred to as a Lorenz curve. The data just discussed (with a cumulative-sum curve) are shown in Figure 5-3.

Figure 5–2 Pareto Diagram

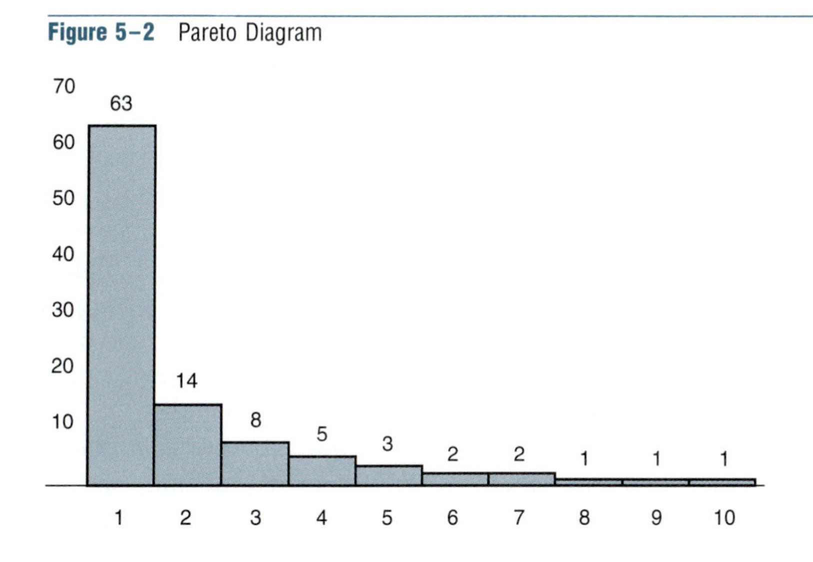

Figure 5–3 Pareto Diagram with Lorenz Curve

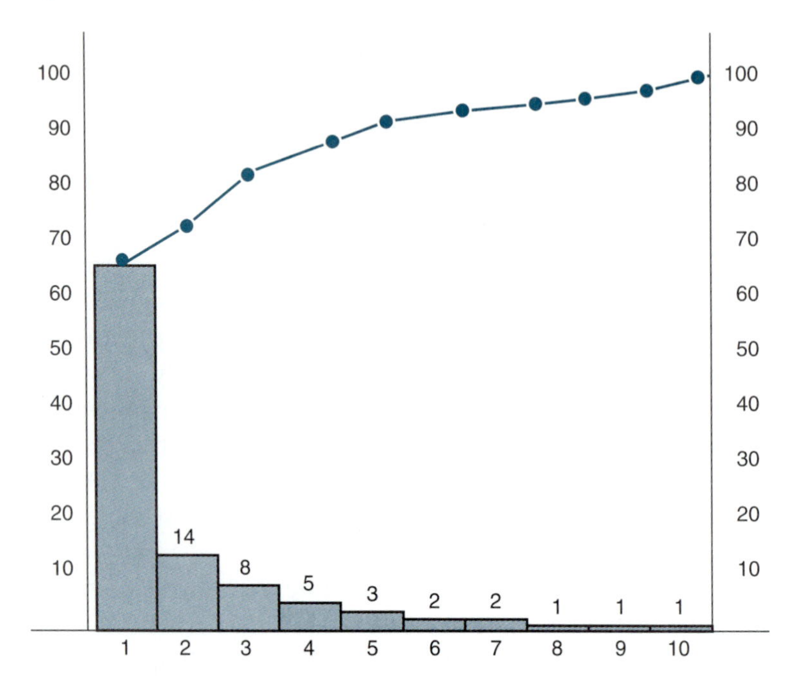

The Lorenz curve (cumulative-sum line) shows the percentage of total data, which helps to find the vital few groups.[15]

A Pareto analysis presented to management usually shows the dollar loss in each group, which helps management decide which projects are most cost effective. This information makes the Pareto analysis a powerful business tool.

5–3 FISHBONE ANALYSIS

The fishbone diagram is also known as a cause-and-effect diagram or an Ishikawa diagram. This diagram examines a process and lists the possible areas where a problem under investigation could occur. The purpose of this method of graphing is to simplify the flow of material through a complicated process. This method graphically depicts the causes that could produce the effect under study. We will discuss the method of constructing the cause-and-effect diagram to help explain the use of this tool.[16]

The centerline of the graph is used to identify the problem under investigation. This problem should be very narrowly defined or the result will be a diagram so complicated it will be useless. In the

Figure 5-4 Fishbone Diagram

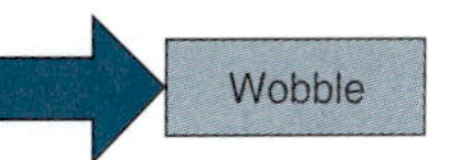

example in Figure 5-4, the final product has an undesirable wobble when operated. This is defined on the centerline.

The major areas that could cause the problem are identified and placed on the centerline as major branches. In the example in Figure 5-5, the factors are workers, materials, inspection, and tools.

Figure 5-5 Fishbone Diagram

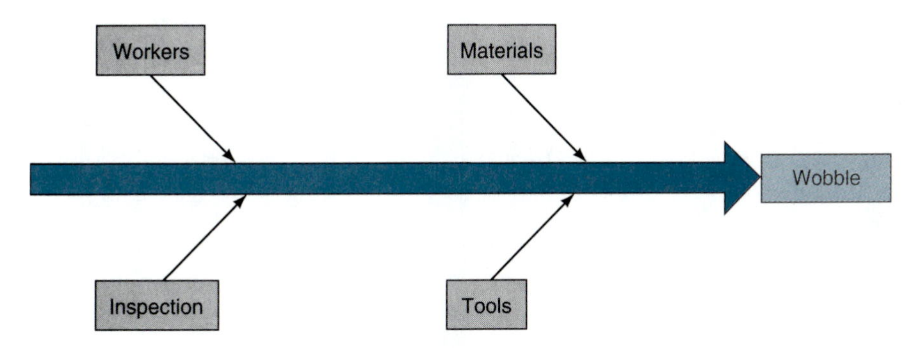

The next step is to evaluate each branch for possible causes (Figure 5-6). Four areas were identified as possible causes for the wobble—for the center axle, G, the bearing size was listed. These "twigs" are attached to the branch. Each branch is evaluated and twigs are added as appropriate.

Figure 5-6 Fishbone Diagram

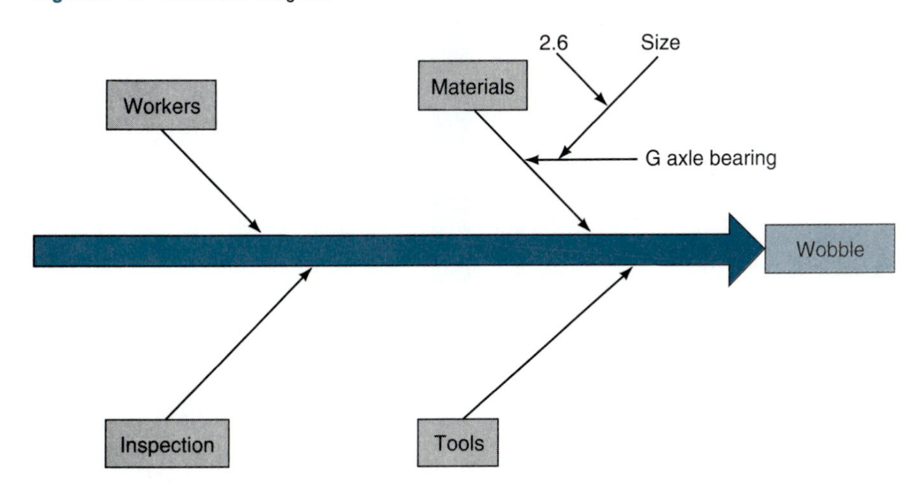

Figure 5–7 Fishbone Diagram

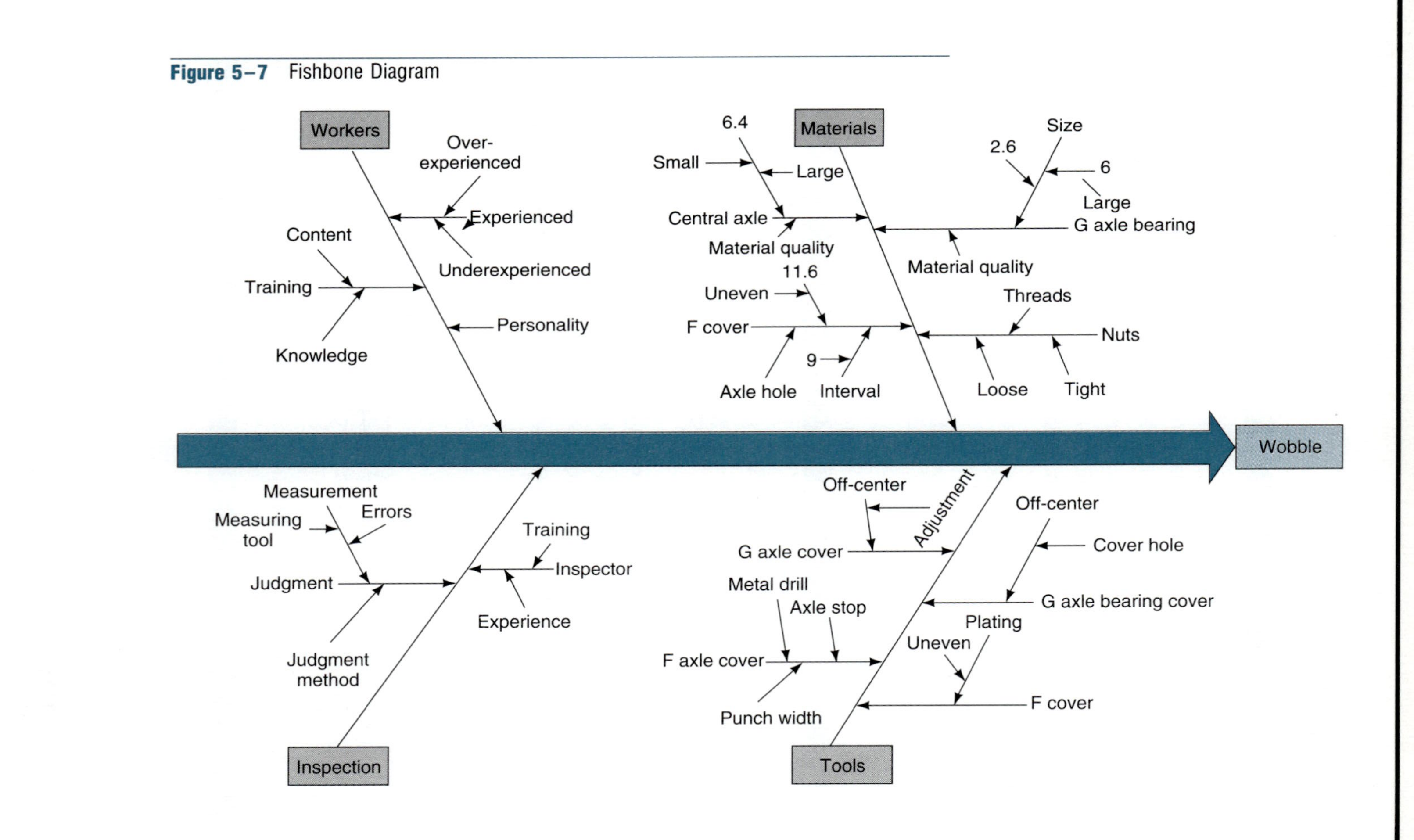

Figure 5–8 Fishbone Diagram

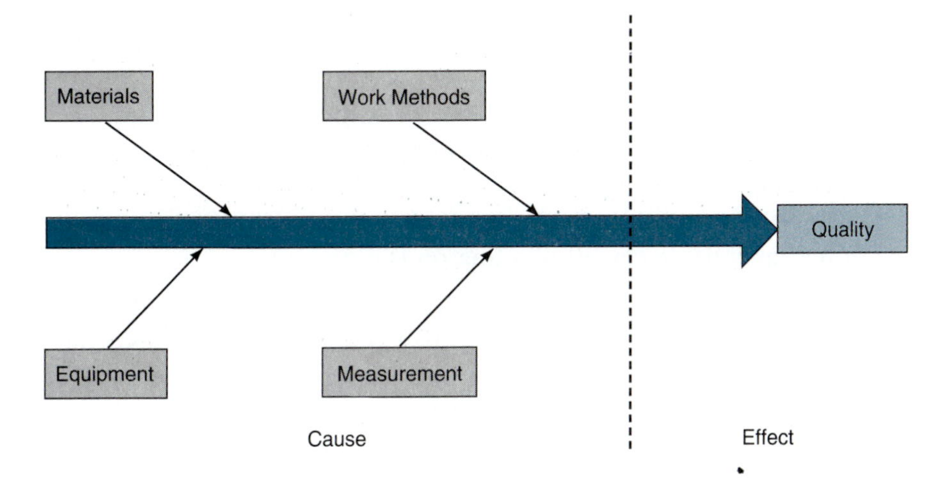

The last step is to evaluate each twig to determine the possible causes under each. In the example in Figure 5-7, under the branch covering the material and the twig covering the nuts, the possible problems are listed. The possible causes here are that the nuts are too tight or not tight enough, or the threads are wrong.[17]

This process of further dissection could continue to include the nut-thread twig, which could itself have two twigs added, and the process would continue[18] (Figure 5-8).

Once the diagram is constructed, the next step is to review the diagram to ensure the causes listed will, in practice, cause the effect under investigation. After you determine that the diagram is complete, each of the possible causes is then investigated to determine which ones are actually causing the effect under study. Several of the causes may be present in the process and have an effect on the product.

One word of caution: Many times a quality professional will get so caught up in the process of charting as to view the chart as an end result. If the chart is used properly, it will zero in on the true source of the cause that is producing the effect under investigation. If the chart is the end result, a lot of time will be spent and the result will be a beautiful chart, but the product will still have the problem.

5–4 STATISTICAL QUALITY CONTROL

Statistical quality control (SQC) is a tool that can generate vast improvement in a production process. Statistical quality control is a method of collecting data from a process to determine whether it is running with natural variation or if variation is being induced by an outside cause.

The three basic rules on which statistical process control is built include the following[19]:

- Everything varies.
- The variation that normally occurs in groups of like items is predictable.
- When the variation in like items becomes unpredictable, an outside force (assignable cause) is acting on the process.

The predictability of variation is the key to SQC. If outside-diameter measurements were taken from 1000 ball bearings, all the sizes would not be the same—some would be larger than, some would be very close to, and some would be smaller than the target size. The predictable part of this size variation is that most of the ball bearings would be grouped closer to the target size and fewer would be grouped above or below target. If the bearings were further away from the target value, there would be fewer good bearings.[20] If the bearings were grouped and counted based on size, the distribution would look like the bell curve shown in Figure 5-9.

When data have been gathered and put on a bar graph, it can be determined that no outside disturbance exists when the shape of the distribution looks like a bell curve. If an outside disturbance existed in the production process, the distribution would have a different shape.[21] The curve may be skewed or bimodal. Both distributions are shown in Figure 5-10.

Both distributions show that the processes producing these distributions have an outside disturbance acting on them. It is important to determine if outside disturbances exist, because this variation can be located and eliminated. When this disturbance is eliminated, the distribution will become normal and form a bell curve. In this way, the company saves money. It produces the product at a lower cost, and

Figure 5–9 Bell Curve

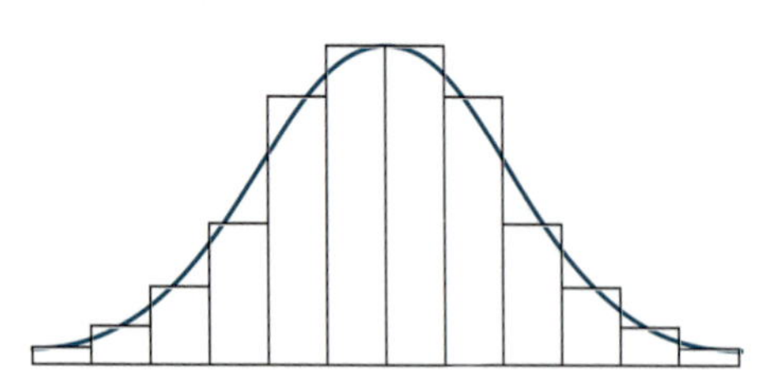

Figure 5–10

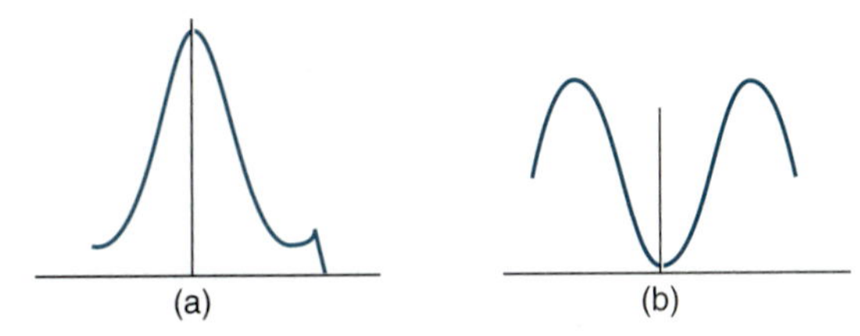

control of the process results in control of the product. If a process results in product distribution other than normal, the process is said to be out of control.[22]

The product produced by a process gives the most information about the process. If the process is out of control, the product distribution will be other than normal. Such a process is no longer predictable. Without a predictable process, management becomes reactive instead of proactive.[23] Reactive management traditionally over-responds to a situation. Managers become like fire fighters, working on one "fire" until they can reduce it to a point where it is smaller than another fire, at which time their attention will shift to the larger fire. If a fire is not completely put out, it will smolder for days, weeks, or months until it has sufficient strength to flare up again.

Management by reaction ensures that the same problems have to be solved over and over. Until a process is evaluated and all outside disturbances are removed from the process, it cannot be effectively operated or managed.

Several methods are used to determine the variation of a process. The most popular method used extensively in today's manufacturing is the control chart. A control chart is used to evaluate whether a product being produced on a production line is exhibiting normal variation or is exhibiting a distribution that would indicate an outside disturbance acting on the process. One of the more popular charting methods is the X-bar and R chart. This chart is actually two charts. The X-bar chart documents averages; the R chart shows ranges. The advantage of this type of chart is that it can be used to determine the variation of a product as it comes off a continuous production line. This means that a company can determine whether an outside disturbance of the process exists without having to produce a large amount of product. Since an outside disturbance of the process results in producing material that has to be rejected, it becomes vital to determine as soon as possible if a disturbance exists.[24]

To construct an X-bar and R chart as a process produces parts, five parts are selected at random. The size under study is measured on each of the parts. These five readings are averaged together to identify the point that is plotted on the X-bar chart. The range between the largest reading and the smallest reading is calculated by subtracting the smallest reading from the largest reading. This range is plotted on the R chart. This allows a determination to be made about the size and spread of the product.

The X-bar chart allows you to plot the average size of a small group of data. The R portion of the chart is used to plot the range of sizes of the same small group of product. These readings are plotted on a chart, and the plotted points are evaluated for signs of other than normal variation.[25] Figure 5-11 is an example of an X-bar and R chart showing the normal variation for a process.

Note that the points on both charts vary in an up-and-down pattern. This shows that a process is operating with normal variation. The unique point of the random pattern on these charts is that, if the data were plotted on a bar graph, the pattern produced would be a bell curve. As stated earlier, if the data plot to produce a normal distribution, no outside variation exists.[26]

A product being produced from a process that is out of control gives a different and unique pattern. Examples of charts showing processes that are out of control are shown in Figure 5-12.

One thing that each of these charts has in common is that the patterns do not show random variation. If each were plotted at the end of the chart, none would produce a normal bell curve. In addition, each of these charts was produced by a process that is out of control. This process can be improved by removing the outside disturbance that is adversely affecting the process.

Figure 5-11 X-Bar and R Chart Normal Variation

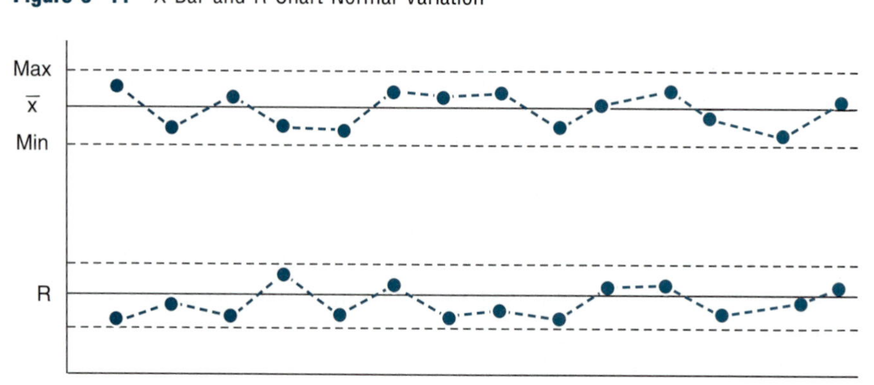

Figure 5–12

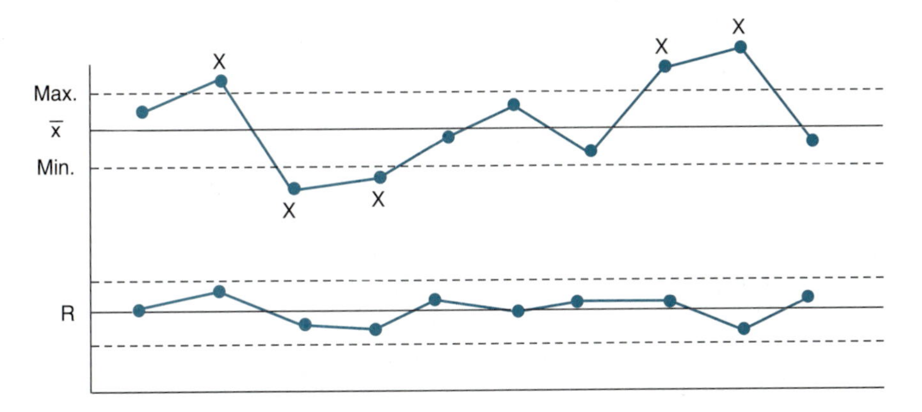

Many other methods are used with statistical process control. It is beyond the scope of this text to study them all. The main point is this: Statistics is a scientific method used to measure the product to make determinations about a process. The statistics are effective in determining when a process can be improved and, once a problem is solved, are used to monitor the process to ensure the problem will not return. This allows the quality professional to solve a problem once and move on to the next problem. There is nothing more wasteful than solving a problem that has been fixed many times before but never monitored to keep it from returning.

Statistical process control allows an operator to control a process much as a driver controls a car. When a car is rocketing down the road, the operator is not controlling it completely. If the operator were in total control of the car, he or she could put the left front tire dead center on the lane stripe and not vary a fraction of an inch left or right. Actually, this is not possible. The road has dips and bumps, the tires have uneven wear, the mechanical linkage between the tires and the steering wheel may be loose, and the driver will overcorrect for these conditions. When a driver is going down the road, he or she makes small corrections to the variation in direction. If the car starts to drift to the right, the driver corrects by turning slightly left; as the car drifts left, the driver corrects right. The process of driving a car involves determining in what direction the car is going, deciding whether this is the desired direction, and correcting the direction if it is not desirable. All drivers make thousands of small corrections to the variation in direction. Processes can be controlled the same way. The product is measured to determine the size; the quality professional determines whether this average size is acceptable and makes a correction to the process if the size is wrong.

The advantage of SQC is that, with constant evaluations, consistent correction can be made. In most processes, variation can be identified and corrective actions taken well before a bad product is made. This is the point of total process control. If consistent monitoring is not performed, the process could have large changes between the times it is checked. This would be like trying to drive a car by looking out the windshield for 15 seconds to correct the car's direction, only to close your eyes for an hour and continue driving blind. It would be crazy to try to drive a car this way, but many companies try to produce a product by evaluating one piece of product every hour. The process may be adjusted based on a single piece of product. To be effective, the process should be continually monitored and adjusted only when the chart shows other than normal variation.

Properly understood and implemented, SQC can have drastic savings not only through the elimination of scrap, but also in allowing the workers to produce a higher-quality product with less effort. SQC allows a process to be improved and ensures that the improvements are maintained.

5–5 DESIGN OF EXPERIMENTS

Design of experiments is a method of process improvement. This method is usually used to increase control of a process once the outside variation has been removed. This technique strictly controls variables in a process, allowing the experimenter to change some of the input variables and measure the result. The experimentation continues by changing many input variables and measuring the results. After all the variables under investigation have been changed according to the experimentation plan, the results should show which input variables have an effect on the product and what settings should be used to produce the best product.

A simple process such as drilling a hole in a piece of metal would be suitable for a formal design of experiment. The first step is to identify the desired result that is going to be measured. For this example, the hole's inside diameter is the measured result. The second step is to make a list of input variables that could affect the result. A list of variables that could affect the hole size includes the following:

- Metal alloy being drilled
- Drill bit type

- Drill bit speed
- Drill bit down-pressure
- Drill bit temperature
- Metal temperature
- Lubricant used
- Drill motor input voltage
- Drill bit angle
- Atmospheric conditions (humidity, air temperature)
- Amount of air circulation
- Security of the part being drilled

Many other variables could be listed that affect the result of this operation, but for the purpose of this example, these 12 items will be used in the experiment. The third step is to evaluate the list to choose the variables that will be held constant and those that will be changed during the experiment.

Variables Held Constant

- Metal alloy being drilled
- Drill bit temperature
- Metal temperature
- Drill motor input voltage
- Drill bit angle
- Atmospheric conditions (humidity, air temperature)
- Amount of air circulation
- Security of the part being drilled

Variables to Be Adjusted

- Drill bit type
- Drill bit speed
- Drill bit down-pressure
- Lubricant used

Each of the variables is assigned two settings. These settings are used to set up the experiment. Experience determines what settings would most likely have the greatest positive results.

Variable	Setting 1	Setting 2
A. Drill bit type	Diamond	Carbide
B. Drill bit speed	600 RPM	1000 RPM
C. Drill bit down-pressure	10 psi	20 psi
D. Lubricant used	None	Cutting milk

To start the experiment, the quality professional devises a plan that will allow each of the variable settings to be tried with every possible combination of the other variables (see Table 5-2).[27]

Table 5–2 Experiment Combinations

1. A1 B1 C1 D1	9. A1 B2 C2 D1
2. A2 B1 C1 D1	10. A1 B2 C1 D2
3. A1 B2 C1 D1	11. A1 B1 C2 D2
4. A1 B1 C2 D1	12. A2 B2 C2 D1
5. A1 B1 C1 D2	13. A2 B1 C2 D2
6. A2 B2 C1 D1	14. A2 B2 C1 D2
7. A2 B1 C2 D1	15. A1 B2 C2 D2
8. A2 B1 C1 D2	16. A2 B2 C2 D2

The experimentation plan requires 16 repetitions of the experiment to cover all combinations. The variables not under investigation must be held as constant as possible. Each of the experiments is conducted and the inside diameter of the hole measured. The results of the 16 experiments can then be evaluated to determine how much effect each variable change has on the hole diameter. Then, the combination that gives the most desirable effect on the hole diameter can be determined.

This type of experimentation should not be conducted on a process that is producing a product to be sold. It requires very close control of all variables, and it produces a product that would be considered defective.

5–6 EVOLUTIONARY OPERATIONS

A second form of structured production experimentation is the evolutionary operation (EVOP). In this form of experimentation, two or three independent variables are selected for experimentation. This form of experimentation differs from formal, designed experiments. EVOPs are conducted during normal production of a product, whereas designed experiments are performed when the process is shut down from normal production.[28]

In the example of drilling a hole, the independent variables that could be used in an EVOP might be the drill speed and the down-pressure. A starting point for each would be selected and a graph made for the operation (Figure 5-13).

A part is drilled at the starting settings and the size is marked on the chart. One of the variables is adjusted, a second hole is drilled, and the

Figure 5–13

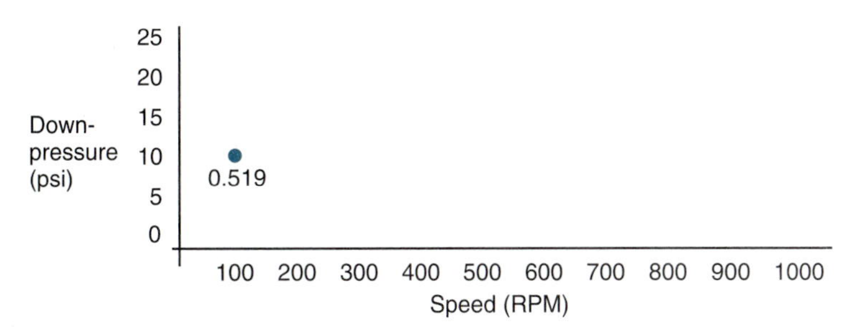

results are recorded (Figure 5 - 14). The first variable is returned to its original setting and the second variable is adjusted and checked. The results are plotted (Figure 5 - 15).

The points are connected to make a triangle, and all three results are evaluated to find the least desirable effect. In this case, the first attempt at 100 RPM and 10 psi is farthest away from the target size. The next trial will be made opposite the least desirable effect. This would be at 200 RPM and 15 psi. The results are plotted and a new triangle is formed (Figure 5 - 16).

The three points that make up the new triangle are evaluated to find the least desirable effect and the settings adjusted to make a new triangle opposite this effect. In this case 0.518 in. is the least desired. The new setting would be 200 RPM and 20 psi. A new triangle would be made, and the trials would continue as long as the results were changing the hole size closer to the target value (Figure 5 - 17).

When the target size is reached, or further adjustment would result in moving away from the target size, the optimum settings have been

Figure 5–14

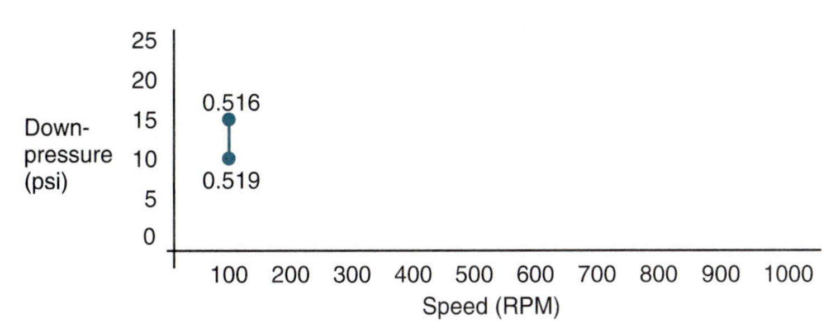

Figure 5–15

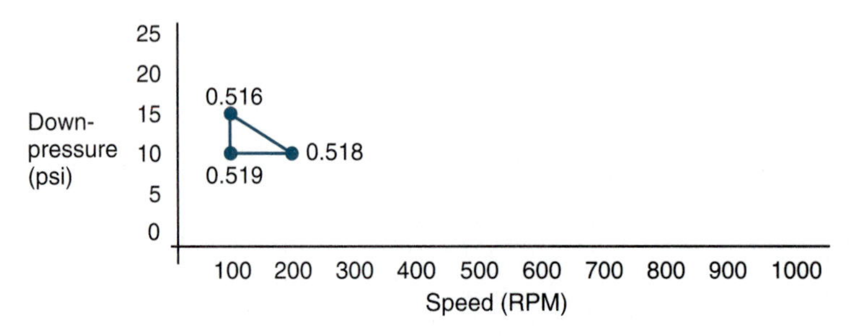

Figure 5–16

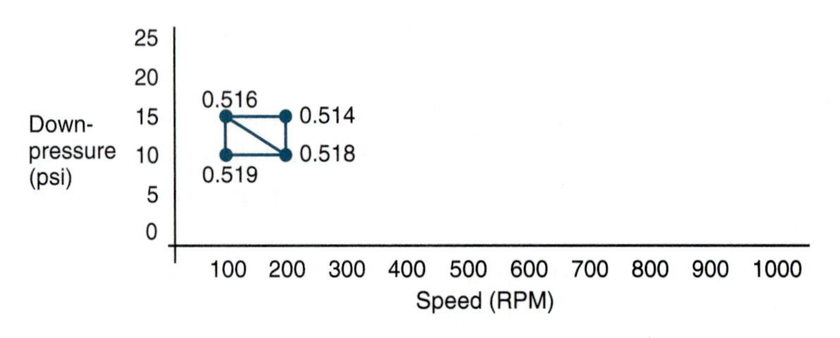

Figure 5–17

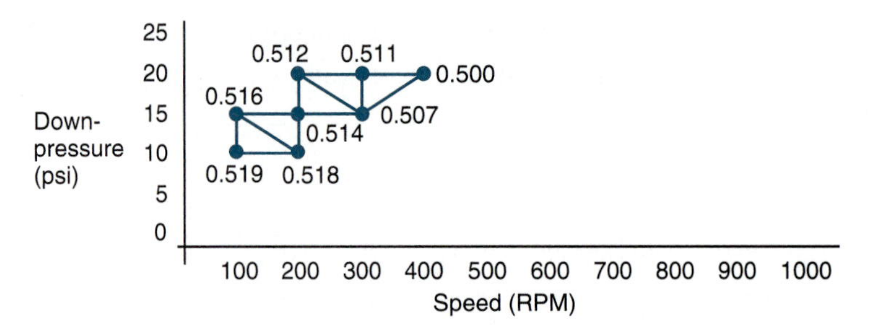

reached. In this way, the best settings can be located for the best result. At this point, new variables could be selected and a new EVOP started.

Evolutionary operations can be used to find the best combinations of settings by making a small adjustment and measuring the output. A quality professional should undertake further study so that he or she understands the entire process before attempting an EVOP on a production process.

5–7 QUALITY CIRCLES

Quality circles have enjoyed both tremendous successes and miserable failures. The difference between the successes and failures seems to be the group dynamics. The focus of quality circles can quickly turn from quality-related issues to complaints that are unrelated to the focus of the group and detrimental to the group's objectives. With a properly trained group proctor and motivated members, the outcome from quality circles can be beneficial to the company and rewarding to each of the group members.

5–8 SELF–QUALITY CONTROL

Self-control is a concept put forward by J. Juran and F. Gryna.[29] Basically, it states that defects are controlled by one of two groups—operators or management. The key to determining whether a defect is management controllable or operator controllable is determining whether the operator is in a state of self-control. Use the following questions:

- Does the operator know what he or she is supposed to do?
- Does the operator know how well he or she is actually doing in a timely manner?
- Does the operator have the ability to take regulatory action?

When a defect is identified, these criteria of controllability are applied. The defect is classified as a management-controllable or operator-controllable defect.

A defect is operator controllable if, and only if, the answers to the three questions of controllability are *yes.* The operator did know what he or she was supposed to do, knew how well he or she was actually doing in a timely manner, and had the ability to take regulatory action. At first glance, it may seem that most defects are operator controllable, but before we reach that decision, let's take a closer look at the criteria.

Criterion No. 1: Does the operator know what he or she is supposed to do? If a person is given a golf club and a ball with instructions to hit the ball, stand well behind him or her. For a person to become somewhat good at golf, he or she needs instructions on how to hold the club, how to swing, and how to correct for a slice or a hook. If an operator is put on a machine and told to produce a part without the proper knowledge of correct operation, watch out!

This knowledge of what one is supposed to do commonly comes from four sources:

- A written specification
- A product standard
- A process standard, which may consist of written instructions or verbal instructions defining the worker's responsibility for deciding the action to take in a given situation
- Formal training in the proper procedure for achieving a desirable end

The product standards must be evaluated for specifications that are vague or in direct conflict with other specifications. A common specification in many industries directs a worker to polish a surface "to a high luster." What is the criterion that "high luster" should be compared against—a "dull luster"? "High luster" is a subjective criterion to the operator and to the inspector. Another commonly used subjective term is "smooth finish." A good specification should define *smooth* so that the operator and the inspector can have a common evaluation point.

Conflicting specifications occur when, if one specification is met, a resulting specification may be out of limits. In the process of making an oil cooler, fins are placed between sheets of flat sheet metal. The fin height is allowed to be 0.003 in. higher than the exact size specified but no lower than the exact size (+0.003 to 0.000). As many as 500 layers of fin, sheet stock, fin, and sheet stock are stacked and the parts welded together. When the welding is complete, the overall height of the completed cooler is measured. This height must be within 0.5 in. of the blueprint height. On the average, if each of the fins were 0.0015 in. taller than the print specification, the finished oil cooler will be 0.75 in. taller than print size. This would result in rejection of the cooler because the stack height was one-quarter of an inch taller than the specification. This conflict occurs when a product is built using parts that are all within specification, but after final assembly, the end result is out of specification. This type of conflicting specification is not the operator's fault and, therefore, is classified as a management-controllable error.

The process standard must also be defined. As the use of exotic materials becomes widespread, process standards must evolve to the point that the step-by-step method, along with cycle times, is specified. Many parts that were made out of steel five years ago are now being made of composite materials, such as carbon fiber, graphite epoxy, and Nomex. If the cycle time for working with the material is not spelled out exactly, the operator will have to experiment until the cycle time is discovered.

Defining responsibility is the process by which an operator is instructed what to do if a condition exists in a process. What is the proper course of action if an operator who is producing a part determines that the raw material is defective? If an operator is not instructed what to do in such a situation, he or she is not responsible for taking the wrong action.

Formal training is a vital link in the operator's knowledge. Many times operators are trained in an on-the-job setting that causes poor work methods to be inherited. If a worker is checked out by Joe, who was checked out by Sam, who was checked out by Mary, who was checked out by Jill, who taught herself how to run a process, many bad habits will be passed down and much job knowledge will be lost. The proper work methods must be taught in a formal setting.

To evaluate whether criterion no. 1 is being met by an operator, a checklist should be produced and an objective evaluation performed. The checklist to be used to determine whether the first criterion has been met should include the following questions:

1. Are written specifications and work methods legible and conveniently accessible to the operator? If written in more than one place, do they all agree?
2. Does the specification define *required* tolerances, as opposed to *suggested* ones?
3. Are visual specifications (color, for example) understood by the operators, and are visual standards displayed in the work area?
4. Are the operators and inspectors using the same specifications? Are the operators using the same instruments to measure the product as it is being produced as the inspectors are using to inspect?
5. Is the operator informed as to the use of the product that he or she is producing?
6. Has the operator been formally trained to understand the product and process standards? Has this training been validated by testing? Is periodic retraining available?
7. Does the operator receive specification or tolerance changes immediately every time?

Many other questions could be asked, depending on the process. If the answer to any of these questions is "no," it is highly questionable that the defects an operator produces are the operator's fault. More likely, the defect is management's fault for not providing the knowledge about what the operator is supposed to do.

Criterion No. 2: Does the operator know how well he or she is presently doing? For a golfer to improve his or her swing, he or she must know immediately where the ball went. It would do no good to blindfold a golfer; allow him or her to hit 100 balls; and report that 25 went straight, 50 went to the right, and 25 went to the left. That method of practice would result in a very frustrated golfer and no improvement in his or her game.

Many operators are required to work in a total vacuum of information about the product they are producing. In many companies, the operator is told after a production lot of a product is made and inspected whether the product was good or bad. Without timely feedback, the bad product is not the operator's fault.

Proper feedback can take several forms:

- "Read at a glance" gauges that show only the important factors over which the operator has control
- Statistical control charts that the operator is trained to evaluate
- Verbal or written comments from the inspector evaluating the product during production
- The operator having the time and equipment to check samples of the product during production

Without timely feedback, the operator cannot know that poor product is being produced and cannot be expected to correct a defect.

The quality professional should compile a checklist to determine whether the second criterion has been met. It should include the following questions:

1. Are precise gauges provided to the operator that give a numerical reading instead of simply sorting the good product from the bad product? Are they regularly calibrated?
2. Is the operator properly trained to evaluate the results of product testing and to correct for an out-of-tolerance condition?
3. Is a system in place to audit the operator to determine compliance to instructions?
4. Are inspection results provided to the operator immediately, while production is still under way?

Criterion No. 3: Does the operator have the ability to regulate the process?

A golfer may know how to swing a club properly and be able to see the flight path of the ball, but if the golfer does not have the knowledge or ability to change his or her swing, no improvement can be expected. Likewise, if an operator knows how to process a part and knows the parts are being wrongly produced, but does not have the knowledge or ability to change the process, no improvement can be expected.

Varying the process usually means adjusting a variable on a machine (e.g., speed or pressure) or adjusting the human component (e.g., lighter blows with a hammer or less torque on a bolt). For effective regulation, the process must be able to produce the part within specifications. It must also respond to changes in a consistent and predictable manner. The operator must be trained in how to regulate the process, and the act of regulation must not be distasteful to the operator. An example of distasteful adjustment would be if the controls were in a hostile environment (e.g., extremely hot or cold or in chemical fumes). Without the ability to regulate the process to eliminate a defective product, the operator is not responsible for a bad product.

The quality professional should compile a checklist to determine whether the third criterion has been met. It should include the following questions:

1. Has the process been measured to ensure that it is able to produce a good product? Is there ongoing evaluation on schedule to ensure that it continues to meet the specifications?
2. Has the operator been trained in how and when to adjust or shut down the process?
3. Is the equipment under a preventive maintenance program?
4. Are the operators aware of the results of the adjustments they make, both negative and positive?
5. Do some operators have a "knack" that could be taught to all operators?

The three criteria should be evaluated to determine whether an operator meets all of them. If any of the three is not met by the operator, then defects are management controllable. What is the result of studies of controllability in industry?

> The box score in most controllability studies indicates that over [80%] are management-controllable and [20%] are operator-controllable. This ratio does not appear to vary greatly from one industry to another, but it does vary considerably from one process to another.[30]

For many managers, the concept that management is responsible for 80 percent of all defects is not well received. This area must be addressed or the bulk of the defects will never be investigated and solved.

5–9 QUALITY AUDITS

Quality audits are used as management tools to determine (1) if the company's quality system is capable of producing quality products, and (2) if the system is actually producing the quality of product of which it is capable. With properly conducted audits, it is possible to determine if the product is good, if the quality system is effective, what quality system problems exist, and what corrective actions should be taken to correct the problems.[31]

If real life in industry went exactly by management planning, only two types of people would work in any industry—those who planned the work and those who conducted the plan. Actually, however, managers must continually make decisions that will allow for continuous improvement to stay competitive with the market. Managers fail in these decisions if they do not document the results and incorporate them into the quality plan. Decisive and directed actions should be taken on these decisions. If managers fail to follow through on the directed action, the problem they are trying to solve will usually become worse. Many managers make further decisions on the worsening problems, which starts a cycle of problem/decision/increased problem/decision, and so on. If documentation and directed action are not completed, many times mismanagement is the result. This mismanagement will almost always result in decay in the quality of the product being produced.

Quality audits are an unbiased evaluation of the various aspects of the quality system. They are conducted by a separate part of the quality control department in an attempt to evaluate the overall effectiveness of the quality plan and to determine whether the plan is being followed. The primary focus of most audits is the quality procedures manual. The decisions that are made about the manual all center around one question: If the procedures are followed, will a quality product be produced? Before the audit team evaluates the implementation of the procedures, it must first decide whether the procedures are adequate for the task. If the procedures are not adequate, it will be useless to conduct the conformity portion of the audit.[32]

All quality audits have one thing in common—they are very unpopular. A sign found hanging in a shop that was recently audited describes the feelings of many workers. It said, "Auditor: One who comes upon the battlefield when the battle is done and attacks the wounded." Nobody likes someone to come in and evaluate the methods that are being used to complete a task. With the pressures production workers face each day, an audit is the last thing most want. Several techniques can be used to overcome much of this resistance.

- Openly communicate the purpose of the audit to the people who are being audited.
- Let the workers know ahead of time what areas will be evaluated.
- Look for *what* is at fault, not *who* is at fault. Workers will be much more willing to help you find the what than the who.
- Work from established guidelines; inform everyone about these guidelines.
- After the audit is complete, let the people involved know what will be in the audit report.
- Allow the people being audited to have input into the content of the audit report.

When conducting the audit, the quality professional should review the goal with all involved. The goal of the quality audit should be to measure the effectiveness of the quality procedures against the quality management objectives. It should evaluate the conformance and the ability to conform to these procedures. The auditor should also continually ask, "What is the value of this procedure to the stated goals of the company?" Procedures should be evaluated for their overall contribution to business, ease of interpretation, and ability to be followed, and the audit should reach a decision about whether each is really necessary.

If the audit is conducted with these goals in mind, the results will be positive attitudes toward implementing the recommendations, an improved quality program, quality procedures that are understandable and useful, and a more satisfied work force. Failure to perform proper quality audits will result in missed opportunities to improve quality and profits.

Quality audits should be used as a tool to keep the entire organization working toward the same goals. They should be structured to assess the strong areas of the organization as well as the weak areas. For most industries, this is one function that could stand vast restructuring and redirection. If this restructuring does not occur, the quality audit will be viewed as a necessary evil to be endured instead of the powerful quality improvement tool that it should be.

5–10 JUST-IN-TIME DELIVERY

Just-in-time (JIT) delivery is a system that is being used by many American manufacturers. This concept takes a radical departure from the normally accepted method of producing and storing parts.

In traditional manufacturing, raw materials are fed into one end of a factory and finished materials are delivered out the other. The processes that take the raw material to the finished product are structured to flow from process to process. As parts are produced at one work station, they are moved to the next process at whatever volume and speed the first process can produce. The product is then stored at the second process until it is needed. If the first process produces a product at a faster rate than the second, the product will start to pile up at the second process.

The problem is that the operator of the first process will not know that the production rate is too high. The cost of storage will go up; with some materials the product may go bad before it is ready for use at the next process.

This type of production is much like delivery of food. A person who wants pizza and soda for one may not order a small pizza and one can of soda to be delivered because it does not sound right to ask for home delivery of so small a quantity. Instead, he or she may order a medium pizza and a six-pack of soda. The leftovers, after all, can be stored. This same mentality in production causes a need for intermediate storage areas for the extra parts produced at the previous process.

JIT delivery works under a unique system. Product transportation from one process to the next is the responsibility of the process that needs the parts rather than the process producing the parts. The process that needs the parts will get only the amount of parts needed to continue production when the parts are needed. At the second process, the parts to make an assembly needed by a third process are kept in the place where the parts are made. When no more room is available, production is stopped. When the third process again needs parts, the operator of the second process can see that parts have been taken and can restart production to replace the parts that were removed. The cycle then starts over again.[33]

This approach is much like shopping at a supermarket. Shoppers buy food in the amounts needed to make fresh meals without overloading food storage areas. When they run out of important items, shoppers return to the store and again purchase just what is needed.

Several advantages have been identified with JIT. It results in great cost savings in the scheduling of work. If one process is much faster than the next process, the faster one will often be shut down. The manager of that area has the opportunity to reassign workers from the faster processes to the slower processes, which will result in overall increased production. Less product is stored in the plant, which frees shop space that can be turned into production space. Less money is tied up in idle parts, liberating dollars for improvement projects.[34] The work can be scheduled more easily and the typical month-end overload can be reduced. Most production lines run with low output at the beginning of the month, while the work is being planned. By the middle of the month, parts are available for the production of the final product. Near the end of the month, a production frenzy is under way to meet the monthly numbers. JIT can reduce this "egg in a snake" effect.

JIT can be tremendously successful in companies, but a manufacturer should exercise caution before attempting it. The result of a poorly planned attempt at JIT has resulted in production shutdowns and even the demise of some companies.[35] This concept should be completely investigated before a company makes the decision to try it.

5–11 COMPUTER INTEGRATED MANUFACTURING

As companies seek to increase profits, reduce cycle time, and enhance customer satisfaction, the computer is taking on primary importance. Computer integrated manufacturing (CIM) is a computer system that is used to take a product from concept through shipping.

Up-front data exchange during design and engineering is enhanced by the use of computer-aided design software. This three-dimensional software integrates with the other computers in the system to exchange information. The system of mainframe and/or personal computers connected to digital and numeric manufacturing equipment offers data feedback during the manufacture of the product. Sophisticated visual inspection equipment that can be programmed for different sequences is included in the system to offer image processing and analysis.

The data supplied by these computers is shared with other computers in the system to control inventories and shipping. This data is likewise shared with accounting for the billing process.

There are eight key functions to the CIM network.[36]

1. Design and drafting: Computer-aided design (CAD) uses a central processing unit (CPU) in the form of a micro-, mini-, or

mainframe computer. It uses storage devices like floppy- or hard-disk drives, high-resolution graphics, and input/output devices such as a keyboard, printer, plotter, and a CRT display.

2. Product scheduling and control: A host mainframe may be used to accomplish master scheduling coupled with material management and requirements planning, from procurement to shipment.
3. Process automation: Programmable controllers give direct numerical control of processes, inspection, and testing. They also use robotics, flexible manufacturing systems, and other automated equipment to perform certain processes and to direct shop floor activities.
4. Process control: Programmable controllers use sensing equipment to report conditions that require operator intervention.
5. Material handling and storage: Minicomputers may be linked into the system to control automated storage and retrieval of finished goods and purchased parts based on picking schedules and requisitions.
6. Maintenance scheduling and control: Host mainframe or minicomputers may be used to supply preventive maintenance schedules and report equipment downtime by cause. Spare part inventory and management reports are also easily drawn from these data.
7. Distribution management: Using the host mainframe, order processing, sales reporting, and invoicing can be planned from the data, along with warehousing and transportation.
8. Finance and accounting: Host mainframe data can generate operating reports, forecast for future demand, and analyze costs. It will also automate billing.

5–12 QUALITY FUNCTION DEPLOYMENT

Quality function deployment (QFD) is a system for translating and integrating identified customer requirements into internal company functional requirements. This is done at each stage of a product's life cycle—from concept through research and development, production, delivery, and service.

QFD is a sophisticated means of applying the requirements of the customer to a matrix that will ensure that the requirements are met by some function within the production process. QFD forces the company to adopt common terminology in all functional areas. When all the departments speak the same language, it is easier to achieve a goal. It provides the team with a better understanding of all functional needs throughout the organization and encourages the documentation of consensus findings through the process. This requires an additional time investment in the product definition stage but results in shorter design and redesign times. This allows for more efficient processes, greater profit margin, and increased customer satisfaction.

The procedure may be complex or simple, depending on the customer's requirements and the stratification of the supplier. For larger and more complex applications, computer software is available. In many cases the requirements can be met by the use of the "house of quality," which is the planning matrix shown in Figure 5-18.

To begin, the requirements of the customer are listed on the left side of the matrix in the area labeled "Customer requirements." They are prioritized in the column to the left. Next, the functional requirements labeled "Design requirements" are entered into the upper box. In the area marked "Relationships," the team indicates where the customer requirements are met by the design requirements. This is most commonly accomplished by the use of symbols indicating whether the design requirement is of strong importance, average importance, or slight importance to the customer requirement.

After the relationships have been marked, the "correlation matrix" is used to note interactions between the design requirements. These interactions may enhance or obstruct customer requirements. The interactions can be marked with a + to show enhancements and – to indicate obstructions that must be addressed. Next, the importance of each design requirement is indicated by prioritizing them in the matrix marked "importance rating." It may be helpful to evaluate the company's performance relative to competitors in the box marked "competitive evaluation."

Note that this process is performed by a team of individuals. One or more brainstorming sessions should be used to accomplish consensus on the matrix. This process may be labeled the *product planning phase* of the operation. Refer to Figure 5-17. Note that the next phase, "parts deployment," uses the design requirements of the planning phase in place of the customer requirements. The importance ratings are used to prioritize the requirements.

The steps of the matrix are repeated until all phases of operation have been addressed. This process may extend as far as field mainte-

Figure 5–18 House of Quality

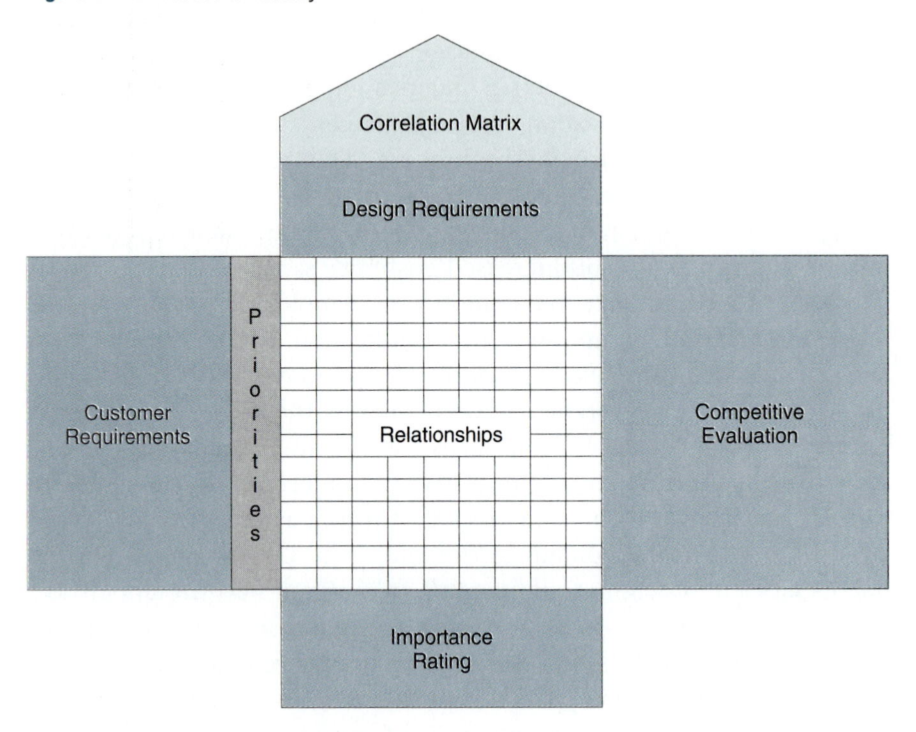

nance, if that is identified as a requirement of the customer—the matrix can be used to whatever degree is required to meet the customer's needs. This method requires dedicated, up-front work but returns great dividends in reduced redesign, rework, and field maintenance.

5–13 PROCESS IMPROVEMENT AND PROBLEM SOLVING

The job responsibilities of a quality professional are diverse. He or she may work on process improvement, problem solving, or any of dozens of other functions in the quality department. The new student of quality may find it hard to keep straight all the tools to be used on those jobs. The methods listed in this text are by no means comprehensive, but they are common and the reader will most likely come in contact with them again.

It is worth spending time to put the methods covered in this chapter and in Chapter 15 into perspective. These methods and others

Figure 5–19 Quality Function Deployment

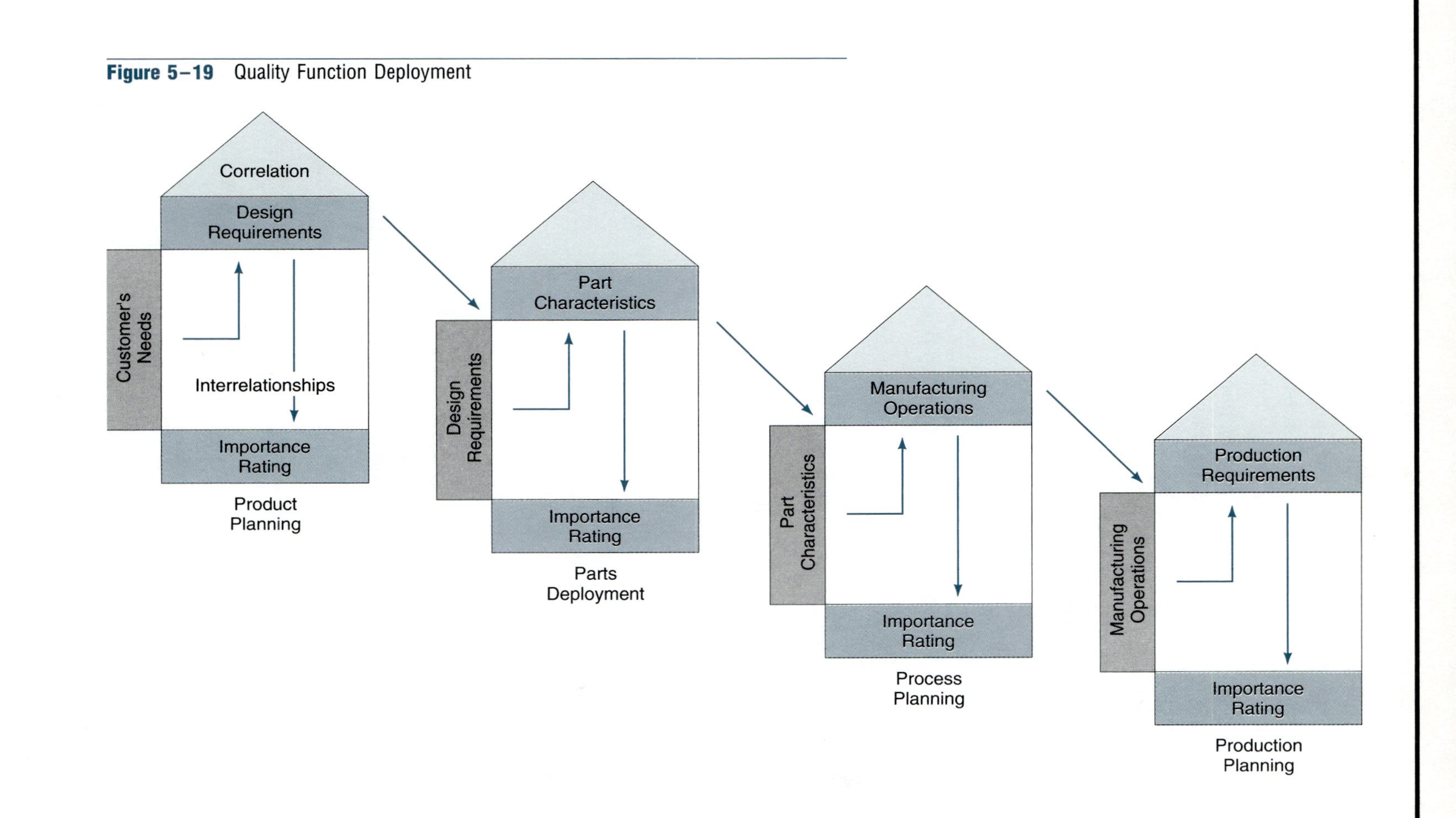

not discussed in this book are like tools in a mechanic's tool box. One of the biggest challenges for the quality professional is to match the task to be accomplished with the proper tool. Trying to complete a job with the wrong tool is a frustrating endeavor. The analogy of trying to unscrew a tight bolt with a pair of pliers provides some idea of what it is like to employ the wrong method on a quality improvement assignment.

Some basic steps that you can follow to help in an improvement process or problem-solving activity are outlined here.

1. Decide which problem will be addressed first (next).

   *flow charts *check sheets
   *Pareto charts *brainstorming
   *ogive charts *quality circle

2. Analyze the effects of the problem to determine possible causes.

   *run charts *histograms
   *control charts *probabilities
   *scatter diagrams *tally sheets
   *Pareto charts

3. Determine probable causes.

   *fishbone diagram *brainstorming

4. Develop and implement corrective action.
5. Analyze the effects of corrective action to quantify improvement.

   *run charts *histograms
   *control charts *probabilities
   *scatter diagrams *tally sheets
   *Pareto charts

6. Repeat steps 3, 4, and 5.

It is important not to implement too many corrective actions at one time. It is imperative that improvement be assignable to an action. If too many "fixes" are made at one time, the one that actually caused the improvement may be hard to determine.

KEY TERMS

Appraisal costs Those costs associated with determining the condition of the product the first time through.
External failure costs Those costs associated with shipping a defective product to the customer.
Internal failure costs Those costs associated with locating the defective product before it is shipped to the customer.
Philosophy The basic belief in a certain management style.
Prevention costs Those costs associated with keeping internal and external failure costs to a minimum.

METHODS TEST

1. Discuss in a short paragraph the concept of quality cost.
2. Discuss in a short paragraph the use of a Pareto analysis.
3. Discuss in a short paragraph the use of a fishbone analysis.
4. Discuss in a short paragraph the concept behind and use of statistical quality control.
5. Discuss in a short paragraph the concept of design of experiments.
6. Discuss in a short paragraph the use of quality circles.
7. Discuss in a short paragraph the concept of self–quality control.
8. Discuss in a short paragraph the use of quality audits.
9. Discuss in a short paragraph the concept of just-in-time delivery.
10. Define CIM.

NOTES

1. Joseph Juran and Frank Gryna, Jr., *Quality Planning and Analysis,* 2d ed. (New York: McGraw-Hill, 1980), 13.
2. Juran and Gryna, 14–16.
3. Juran and Gryna, 14–16.
4. Juran and Gryna, 371.
5. Juran and Gryna, 14–16.
6. Juran and Gryna, 14–16.
7. Juran and Gryna, 14–16.
8. Frank Kaplan, *The Quality System* (Radnor, PA: Clinton Book Co., 1980), 18–20.

9. *DataMyte Handbook,* 2d ed. (Minneapolis, MN: DataMyte Corp., 1986), 1–3.
10. Juran and Gryna, 20.
11. Harvey Charbonneau and Gordon Webster, *Industrial Quality Control* (Englewood Cliffs, NJ: Prentice-Hall, 1978), 207.
12. Kaoru Ishikawa, *What Is Total Quality Control?* (Englewood Cliffs, NJ: Prentice-Hall, 1985), 84.
13. Dorsey Talley, *Management Audits for Excellence* (Milwaukee, WI: ASQC Press, 1988), 42.
14. *DataMyte Handbook,* 1–23.
15. *DataMyte Handbook,* 1–23.
16. W. Edwards Deming, *Out of the Crisis* (Cambridge, MA: MIT Press, 1985), 238–41.
17. *DataMyte Handbook,* 1–23.
18. Masaaki Imai, *Kaizen, the Key to Japan's Competitive Success* (New York: Random House, 1986), 69.
19. Richard Shores, *Survival of the Fittest* (Milwaukee, WI: ASQC Quality Press, 1988), 249–50.
20. Bonnie Small, *Statistical Quality Control Handbook,* 2d ed. (Indianapolis, IN: AT&T Technologies, 1985), 121–41.
21. Eugene Grant and Richard Leavenworth, *Statistical Quality Control,* 5th ed. (New York: McGraw-Hill, 1980), 51–54.
22. Small, *Handbook,* 6.
23. Irving Burr, *Elementary Statistical Control* (New York: Marcel-Dekker, 1979), 3.
24. Jerome Braverman, *Fundamentals of Statistical Quality Control* (Reston, VA: Reston Publishing Co., 1981), 131.
25. Small, *Handbook,* 13–16.
26. Small, *Handbook,* 15.
27. Small, *Handbook,* 101–12.
28. Juran and Gryna, 122.
29. Juran and Gryna, 99–116.
30. Juran and Gryna, 107.
31. General Atomic Project 2117, *Quality Assurance System Audit* (General Atomic Company, 1982), 23–28.
32. Talley, 42.
33. David Lu, *Kanban, Just-in-Time at Toyota* (Cambridge, MA: Productivity Press, 1986), 65–69.
34. Lu, 64.
35. Imai, 69.
36. Ishikawa, *What Is Total Quality Control?,* 84.

ISO-9000—WHAT IS IT AND FROM WHERE DID IT COME?

6

ISO-9000 is a widely misunderstood standard that many companies are trying to implement today. After having gone through the effort to lead a manufacturing facility into compliance with and certification to ISO-9001 and providing direction and resources to bring a chemical manufacturing facility into compliance with ISO-9002, we can say without hesitation that the ISO-9000 series of standards is *not* a quality control standard.

What, then, is ISO-9000 if it is not a quality control standard? The standard is a guideline for the design, manufacture, sale, and servicing of a product; to successfully implement this standard, a group must view the ISO-9000 series as such. This program puts in place a system that develops consistency of purpose and involves everyone in the company.

OBJECTIVES

After completing Chapter 6, the reader should be able to:

- Identify the differences between ISO-9000, -9001, -9002, -9003, and -9004.
- Identify the applications for -9001 as opposed to those for -9002.
- Discuss the developmental history of the ISO-9000 series of standards.
- Discuss the differences between the ISO-9000 series standard and a traditional quality standard.

6–1 FROM WHERE DID IT COME?

During the early 1940s, World War II (WWII) raged throughout most of the world. To stop the march of Germany, Italy, and Japan, many countries (U.S., Great Britain, France, and others) united against the Axis countries. As the countries fought side by side, they encountered major problems. Not only did the Allied countries have different languages, customs, and religious beliefs with which to contend, but each country used different munitions, vehicles, and units of measure. The problems centered around the inability to share items such as bullets by forces from different countries fighting on the same side.

The incompatibility of components assembled by the Allied forces in WWII showed the military the need to develop a means of standardization to draw on resources from many nations. To keep this from becoming a future issue, the first *military standards* were developed.

Industry soon realized the merit of standardization and followed the lead of the military. Unfortunately, every industry in every country developed its own standards. Many of the standards that are used today are direct descendants of these early standards. The following is a list of a few of the standards that were developed over the years.

Year	Standard	Developer
1963	MIL-Q-9858A	U.S. military
1969	AQAP	NATO
1971	ASME BOILER CODE	Mechanical engineers
1973	DEFSTAN 05	UK modification
1975	CSA-Z299	Canadian standard
1975	AS1821/22/23	Australian standard
1979	BS5750	British standard
1985	CSA-Z299 (revision)	
1985	AS1821/22/23 (revision)	

The progression of standards has been along a normal course of development. One common element among many of the military, petroleum, and governmental standards was that the standards were heavily inspection oriented and focused on the end product. The standards tried to fit everyone into somewhat the same mold by employing armies of inspectors to look over other people's work. In other words, the standards focused on what the end product should be while ignoring the processes used to achieve the end result.

During the 1970s and 1980s, the world of quality control made normal changes and evolved from reactive (inspection-dominant) organizations to proactive (system-oriented) quality assurance organizations. The focus changed from the end result (the product) to the process by which the product was produced. The theory behind this change is that if the process used to produce the product is properly maintained and developed, the product being produced by the process will be consistent and the quality can be improved. This theory was set forth by the leaders in quality, such as Juran, Deming, Crosby, and Feigenbaum. This was the manufacturing approach taken by Japan in that country's climb to become an industrial power.

The problem was that many of the standards did not make the transition from product orientation to process orientation. Some countries saw the need and tried to make the change in their own standards. As markets became global, companies found themselves having to meet many standards for many countries. In some cases, the standards were conflicting, and in most cases they were confusing. In an effort to eliminate some of the confusion, the International Organization for

Standardization convened to develop an international quality system standard and issued the ISO-9000 series in March 1987.

The ISO-9000 series standards have become the most widely recognized and accepted standards in the world. Currently it is mandatory in some cases to possess *certification* to the ISO-9000 series to participate in the European Community (EC), the largest continuous market in the world. EC countries control oil production in all sectors of the North Sea. Other countries, including Canada, are adopting these standards. Many of the countries have adopted the ISO-9000 standards and have written compatible national standards such as the following:

ISO-9000 clones

1987 ISO-9000 series
1987 BS 5750 (revision)
1987 AS 3900
1987 NZS 5600 series
approximately 20 others

This chapter applies to the 1987 version of the ISO-9000 series standard. In March 1993, the first recommended revisions to the ISO-9000 series standard were presented to the member countries for ratification. The draft standard was referred to as ISO/DIS-9000 through ISO/DIS-9004. The member countries evaluated the standard and, in September 1993, voted to accept the modifications to the ISO-9000 standard. The modified standard was scheduled to be issued in the spring of 1994, with the U.S. version being issued through the American National Standards Institute. Although some minor editorial changes were expected to occur between the ISO draft and the formal issue of the ISO-9000 1994 series standard, no major content changes or additional requirements were expected.

The wording of ISO-9001 and 9004 as cited in this chapter is that of the 1987 versions, since the revisions incorporated by the 1994 version have not altered the intent or basic requirements of the standard. The major revisions have provided further clarification of the 1987 standard or added requirements exceeding those originally issued. In theory, the standard has been modified to stress even further the purpose of customer satisfaction. The introduction in ISO-9001 has been modified to state, "For the purpose of a supplier demonstrating its capability, and for the assessment of such supplier capability by external parties." Whereas the 1987 version stated that the scope was "aimed primarily at preventing nonconformances at all stages . . . ," the 1994 version states that it is "aimed primarily at achieving customer satisfaction by preventing nonconformity at all stages. . . ."

6–2 WHAT IS ISO-9000?

The ISO-9000 standard is a five-part standard that was written to address good business practices for every part of a business. The standard, as written, is not industry specific, but a general guideline for a good, efficient business operation. What is unique about the standard is that it allows a company to evaluate each *element* of the standard and decide how to meet effectively the intent of that element. This means that two companies may look at an element such as design control and decide to implement programs that are radically different in content and documentation, with both meeting the intent of the standard concerning design control. The standard is a guide to good business practices. It is up to the individual company to decide what makes sense for that organization in the way the elements are implemented. The ISO-9000 series standard provides companies in search of suppliers some assurance of consistency, quality of goods/services, and sound business practices on the part of potential suppliers having ISO-9001 or ISO-9002 certification. Companies that gain certification are awarded a pedigree attesting to their commitment to quality goods/services and sound business practices.

6–3 HOW IS THE STANDARD STRUCTURED?

The ISO-9000 series is produced in five parts (ISO-9000, -9001, -9002, -9003, and -9004). The entire collection of standards is identified as the ISO-9000 series standards. ISO-9000 and -9004 are basic guidelines. ISO-9001, -9002, and -9003 are the standards to which companies may apply for certification. Based on their similarities of purpose, ISO-9000 and -9004 are discussed first, followed by ISO-9003, -9002, and -9001.

ISO-9000 is titled "Quality Management and Quality Assurance Standards—Guidelines for Selection and Use." This section is an advisory document that explains how the standard is divided and gives guidelines for companies to determine which classification applies to their line of business and how to implement the systems. The purpose of ISO-9000 is to:

- Explain how the standard is divided
- Give guidelines to use in determining which of the three classifications (-9001, -9002, or -9003) is applicable to a given business
- Give guidelines on how the systems may be implemented

ISO-9004 is called "Quality Management and Quality Systems Elements—Guidelines." This section is the second *advisory document* in the series. It provides detailed advice to businesses on overall quality management and the quality system elements within the ISO-9000 series. In other words, this section helps determine the intent of the elements of -9001, -9002, and -9003. In addition to information relevant to the requirements of the three standards, ISO-9004 provides guidance in other areas such as marketing, product safety and liability, and quality costs. This section has at the present time two subsections—9004.1, which explains the ISO-9000 series for manufacturing, and 9004.2, which explains the same standards for service companies. The following is a summary of the elements of ISO-9004:

- The second advisory document in the series
- Provides detailed advice on overall quality management and quality system elements within the ISO-9000 series
- Provides guidance in other areas of business

ISO-9003 is titled "Quality Systems—Model for Quality Assurance in Final Inspection and Test." This standard is used when conformance to specified requirements is to be ensured by the supplier solely by final inspection and testing. In general, an ISO-9003 quality system is relevant only to a fairly simple product or service; certification to this standard is not widely accepted. The following is a summary of the elements of ISO-9003:

- Used when conformance to specified requirements is to be assured by the supplier solely by final inspection/test
- Stipulates requirements for the following system elements:
  - Management responsibility
  - Quality system
  - Document control
  - Product identification
  - Inspection and testing
  - Inspection, measuring, and test equipment
  - Inspection and test status
  - Control of nonconforming product
  - Handling, storage, packaging, and delivery
  - Quality records
  - Training
  - Statistical techniques

ISO-9002 is called "Quality Systems—Model for Quality Assurance in Production and Installation." The ISO-9002 standard is used when the supplier is responsible for ensuring conformance to specified requirements during production and installation. It incorporates the final inspection and test requirements of ISO-9003, but significantly expands the detail of the ISO-9003 clauses. In addition, ISO-9002 adds elements that are not included in ISO-9003. The following is a summary of ISO-9002:

- Used when conformance to specified requirements is to be ensured by the supplier during production and installation
- Incorporates the final inspection and test requirements of ISO-9003, but significantly expands on the ISO-9003 clauses
- Adds the following requirements to those of ISO-9003:
  - Internal auditing
  - Contract review
  - Purchasing
  - Process control
  - Corrective action
  - Purchaser-supplied product

ISO-9001 is titled "Quality Systems—Model for Quality Assurance in Design/Development, Production, Installation, and Servicing." This is the most complete model available for quality assurance systems. The wording of the clauses in this standard is identical to that in ISO-9002, but adds two more quality system elements to the mandatory requirements—design control and servicing. The following is a summary of ISO-9001:

- The most complete model for quality assurance systems
- Wording is identical to that in ISO-9002, but adds two more elements to the mandatory requirements—design control and servicing

What About the 1994 Version? The 1994 version of ISO-9002 incorporates the element of servicing. In the 1987 version of the standard, any company performing *either* the design of the product or the service after the sale was required to certify to ISO-9001. In the 1994 version, a company performing service after the sale but having no design responsibility must certify to ISO-9002. This is one of the major changes in the 1994 version of the standard. Companies already certified to ISO-9001 who perform servicing without design responsibility should ask their *registrar* (a third party responsible for reviewing programs for compliance with ISO standards) how the new standard affects their certification. The 1994 version also expands the introduction to define further the purpose of this standard. This definition states that the ISO-9000 series standards are to "specify requirements which determine what elements quality systems have to encompass, but it is not the purpose to enforce uniformity of quality systems." The standards are designed as "generic, independent of any specific industry or economic sector." The introduction stresses that each organization must develop its quality system with regard to its specific needs, objectives, products, and services.

As noted earlier, the entire collection of standards is identified as the ISO-9000 series standard. Basic guidelines and advisory information are presented in ISO-9000 and -9004. ISO-9001, -9002, and -9003 are the categories for which companies may apply for certification. Which one applies to a company is totally dependent on the nature of its business. Certification to ISO-9003 is not widely accepted, and many registrars will not consider a request for ISO-9003 certification. Certification to ISO-9002 is for companies that do not design the product that they produce and do not service the product on an ongoing basis (for example, service on a copier or mainframe computer) after the sale. Certification to ISO-9001 is for companies that have the design responsibility for and/or service the product after the sale. One major misconception about the three categories is that a company starts at ISO-9003, then moves to ISO-9002; as the quality system improves after further enhancement, the next step is ISO-9001. ISO-9001 is not better than ISO-9002 and ISO-9003; it is simply written for companies that have control of the design and/or service.

In summary, the sections of the ISO-9000 series are as follows:

ISO-9000 Guideline for the selection and use of ISO-9001, -9002, or -9003

ISO-9001, -9002, -9003 Models for programs; which standard applies depends on the type of activity performed by the organization.

ISO-9004 Handbook for implementation of quality management and quality system elements. This standard is a guideline only; although it is helpful, implementation of the guidelines is not mandatory.

6–4 WHAT ABOUT THE 20 ELEMENTS?

ISO-9001 consists of 20 elements that make up the quality system. ISO-9002 consists of 18 of the 20, whereas ISO-9003 consists of 12 of the elements. To further define how the specification is structured, the 20 elements of ISO-9001 (which include all elements of -9002 and -9003) can be divided into the following three categories, based on the implementation activities:

- Management activities
- Companywide activities
- Specific requirements

Each of the 20 elements can be put under one of these categories, depending on the responsibility for *implementation.*

MANAGEMENT ACTIVITIES

4.1 Management Responsibilities
4.2 Quality System

COMPANYWIDE ACTIVITIES

4.5 Document Control
4.8 Product Identification and Traceability
4.12 Inspection and Test Status
4.13 Control of Nonconforming Product
4.14 Corrective Action
4.16 Quality Records
4.17 Internal Quality Audits
4.18 Training

SPECIFIC REQUIREMENTS

4.3 Contract Reviews
4.4 Design Control
4.6 Purchasing
4.7 Purchaser-Supplied Product
4.9 Process Control
4.10 Inspection and Testing
4.11 Inspection, Measuring and Test Equipment
4.15 Handling, Storage, Packaging and Delivery
4.19 Servicing
4.20 Statistical Techniques

With the elements properly divided, responsibility for the activities can be easily defined. The implementation process from this point becomes an individually structured exercise that companies must adjust to fit the organizational environment. Each of the elements has specific requirements for a manufacturer to fulfill to meet the intent of the

Figure 6–1 Elements of 1994 ISO-9001 Specifications

MANAGEMENT ACTIVITIES

- 4.1 Management Responsibilities
- 4.2 Quality System

COMPANYWIDE ACTIVITIES

- 4.5 Document Control
- 4.8 Product Identification and Traceability
- 4.12 Inspection and Test Status
- 4.13 Control of Nonconforming Product
- 4.14 Corrective Action
- 4.16 Quality Records
- 4.17 Internal Quality Audits
- 4.18 Training

SPECIFIC REQUIREMENTS

- 4.3 Contract Reviews
- 4.4 Design Control
- 4.6 Purchasing
- 4.7 Purchaser-Supplied Product
- 4.9 Process Control
- 4.10 Inspection and Testing
- 4.11 Inspection, Measuring, and Test Equipment
- 4.15 Handling, Storage, Packaging, and Delivery
- 4.19 Servicing
- 4.20 Statistical Techniques

standard. However, the requirements of the standard consist of good business practices and common sense. The following section is a breakdown of the requirements of the ISO-9001 standard.

ISO-9001 SERIES REQUIREMENTS: CHANGES IN THE 1994 VERSION

4.1 Management Responsibilities

- The company quality policy is defined and understood.
- All organizational responsibility, authority, and interrelations are defined.
- Management has a continual review of the system.

Quality Policy (NOTE: At this writing, the ISO-9000 Series Standard has not cleared the ratification process. Therefore, we refer to "ISO/DIS-9000," for "draft international standard.") ISO/DIS-9001 has expanded the policy requirement by stating that "policy shall be relevant to the supplier's organizational goals and the expectations and needs of its customers." This addition further stresses the importance of customer satisfaction in the goals and policies of organizations complying with the ISO-9000 series standard.

The 1994 version has also included the terminology "supplier's management with executive responsibility for quality. . . ." This statement appears to further elevate the importance of executive-level involvement in the quality system. Companies attempting certification should ask their registrar for their definition of "executive responsibility."

Responsibility and Authority The 1987 version refers only to preventing the occurrence of product nonconformity. ISO/DIS-9001 has expanded this requirement to include both process and quality system nonconformances in addition to those occurring with the product. The same intent has been incorporated further in this element with "identify and record any product, process and quality system problems," whereas the 1987 version referenced only the product.

Verification Resources and Personnel This section has been modified from the original to "Resources" only. The standard now includes a requirement to assign "trained personnel for management, performance of work . . . including internal audits." The requirement of trained personnel for management functions was not specifically addressed in the 1987 version of 9001.

The 1987 version stated that design reviews and audits of the quality system, process, and/or product were required to be carried out by personnel independent of those having direct responsibility for the work being performed. ISO/DIS-9001 has moved the independence requirements from the resources element and into the applicable elements themselves.

Management Representative This entry has been modified to state "supplier's management with executive responsibility for quality shall appoint a member of its own management . . . to have defined authority for the quality system." This is the second instance in which the phrase *executive responsibility for quality* has been added. The exact definition of "executive" should be determined by each organization's registrar. This representative now has the responsibility of "reporting on the performance of the quality system to the supplier's management for review and as a basis for improvement of the quality system." This further encourages a proactive approach to the quality system in review and improvement. By placing increased responsibility for reporting with the management representative, the standard again stresses the importance of analyzing the quality system and its conscious control. A note has also been added to this section regarding the representative's responsibility for liaison with external bodies regarding the quality system.

Management Review Wording has been added to the "Management Review" section, again referencing "management with executive responsibility . . ." to review the quality system at "defined intervals sufficient to ensure" continuing suitability. The inclusion of the word *defined* indicates that this activity may need to be documented by a schedule.

4.2 Quality System

- The quality program is defined, documented, and implemented.

The standard calls for the "preparation of documented quality system procedures and instructions." Although suggested by ISO-9004, the 1987 version of ISO-9001 did not require a *quality manual.* ISO/DIS-9001, however, does require that a quality manual be developed and maintained to include or reference the documented procedures that form part of the quality system. The amount of documented procedures is up to the discretion of the organization, based on the skills and training of the affected work force. With both versions, the procedures must be developed in accordance with the requirements of the standard.

Quality planning has been addressed as a separate section in ISO/DIS-9001 and is required to be consistent with all other requirements of the quality system. Documented procedures are required to define how the requirements for quality will be met.

4.3 Contract Review

- The requirements of the contract are defined.
- Differences between the contract and the tender are resolved.
- The company is able to meet the requirements.
- Records of the review are maintained.

The "Contract Review" element has been revised to address verbal orders. The same requirements as for written contracts are still applicable in ensuring a common understanding and ability to deliver.

The standard now requires that the mechanism addressing amendments to contracts and the manner in which such information is transferred to the functions concerned be identified.

4.4 Design Control

- Identify and assign responsibility for each design and development activity.
- Identify the organization and its interfaces.
- Identify the design inputs.
- Identify the design outputs.
- Identify the design verification procedures.
- Address the control of design changes.

Design Input The 1994 version of the standard adds that "Design input shall take into consideration the results of any contract review activities."

Design Verification The 1987 version of ISO-9001 requires design verification by means such as holding and recording design reviews to establish that design output meets design input. In ISO/DIS-9001, this requirement has been expanded to require that "At appropriate stages of design, formal documented review of the design results shall be planned and conducted. Participants at each design review shall include representatives of all functions concerned with the design stage being reviewed as well as other specialist personnel as required. Records of such reviews shall be maintained."

This addition expands the responsibility for design reviews to departments and other functions that may be affected by decisions made during the design review stage. In typical operations, design reviews are solely the responsibility of the design department. This addition to the standard requires an organization to assess the potential impact on additional departments, such as manufacturing, purchasing, and quality, for inclusion in such reviews.

Design Output Under design output, the identification of characteristics crucial to the safe and proper functioning of the product have been detailed to include "operating, storage, handling, maintenance and disposal requirements." The design output is also required to "include a review of design output documents before release."

Design Validation ISO/DIS-9001 incorporates the requirement of design validation by stating that "Design validation shall be performed to ensure that product conforms to defined user needs and/or requirements. A note has also been included stating "Design validation follows successful design verification."

Of all revisions incorporated in ISO/DIS-9001, the addition of design validation within the "Design Control" element may be the most dramatic. The purpose of validation is to take the requirements of design verification one step further to ensure that the final configuration conforms to the needs and requirements specified by the customer.

For any organization that utilizes *prototype* testing (an experimental model or design created prior to release for production) as a method of design verification, this addition can be met fairly easily if such testing is conducted under simulated conditions. In cases, however, where calculations or comparisons are used as methods of design verification, additional activities may be required to meet the intent of design validation.

4.5 Document Control

- Address pertinent issues where necessary.
- Make sure obsolete documents are removed from use.
- Make sure document changes are reviewed and approved.
- Current document revision levels are identified.

This element has been updated to address data in different forms such as "hard copy media, or . . . electronic or other media." The standard has further defined the need to retain obsolete documentation, with requirements for proper identification to preclude unintended use.

4.6 Purchasing

- Perform assessments of subcontractors.
- Purchase orders are documented.
- Purchased product quality is verified.

The first paragraph now includes the word *evaluate* prior to selecting subcontractors. Although this was implied in the 1987 version of the standard, evaluation of subcontractors is now a specific requirement.

ISO/DIS-9001 has added a paragraph to the "Purchasing" element under verification of purchased product that states, "Where the supplier verifies purchased product at the subcontractor's premises, the supplier shall specify verification arrangements and the method of release in the purchasing documents." In many industry-regulated contracts, it is common practice to incorporate third-party or source inspection of product at subcontractors' locations. The standard now requires that when such activity is to be performed, purchasing documents must indicate the specifics of the verification arrangements and the method of release on product approval.

4.7 Purchaser-Supplied Product

- Procedures exist for the receipt, inspection, storage, or maintenance of purchaser-supplied product.

This element has been retitled "Control of Customer-Supplied Product." The only change was to require documented procedures to address this element if applicable.

4.8 Product Identification and Traceability

- Products are to be identified during all stages of production.
- Traceability is maintained and documented where required.

The only change to this element in the 1994 revision is to require documented procedures to identify and trace product, where applicable. The wording has also changed from specifying typical methods of identification to the broader "suitable means."

4.9 Process Control

- Documented work instructions are maintained where necessary.
- Special processes are qualified and documented as appropriate.

Although there have been some minor wording changes in this element, the most important change is the requirement for "suitable maintenance of equipment to ensure continuing process capability." This includes maintenance of equipment such as welding machines, furnaces, and automated processing equipment.

Special Processes In the 1994 version of the standard, the term *special process* has been further defined by a note, thus: "Such processes requiring pre-qualification of their process capability are frequently referred to as special processes."

4.10 Inspection and Testing

- Receiving inspection/conformance verification is performed.
- Positive recall for urgent release of product is maintained.
- In-process inspection/verification is performed.
- Final inspection and testing are performed.
- Documentation of tests/results is maintained.

ISO/DIS-9001 requires that the quality plan or documented procedures define "The required inspection and testing and the records to be established. . . ." The note regarding determination of the nature and amount of receiving inspection has been removed, but its content is added as a full-fledged section in the element.

The inspection authority responsible for the release of product must now be identified in the records.

4.11 Inspection, Measuring, and Test Equipment

- Control and calibration of inspection, measuring, and test equipment are maintained.
- Procedures for calibration are maintained.
- Documentation of calibration results is maintained.
- Positive recall is maintained.

With the exception of the added requirement for documented procedures to support this element, the only change was to specify "test software" in the body of this element.

4.12 Inspection and Test Status

- The status is identified at all stages of processing.

The only change to this element in the 1994 version is to add maintenance of inspection and test status during servicing.

4.13 Control of Nonconforming Product

- Process is documented.
- Responsibility for review and authority for disposition are defined.
- Procedure for reinspection of repaired/reworked product is maintained.

Only minor editorial changes were made to this element. Again, the 1994 revision requires documented procedures for this element. References to "purchaser" have been changed to "customer," and the element now references repair or rework in accordance with the "quality plan" or documented procedures.

4.14 Corrective Action

- Identification of the cause of nonconforming product and the corrective action taken is documented.
- The potential cause of nonconforming product is identified and eliminated.
- Controls are applied to ensure that corrective action is taken and is effective.

Corrective and Preventive Action The 1987 version of ISO-9001 addresses this element only as "Corrective Action." ISO/DIS-9001 has expanded the element title to "Corrective and Preventive Action." Although preventive action was addressed in the original issue of the standard, the 1994 revision has pulled all references in the "Corrective Action" section and defined *preventive action* in a section of its own. The section on preventive action states that procedures shall include:

> a) the use of appropriate sources of information such as processes and work operations which affect product quality, concessions, audit results, quality records, service reports and customer complaints to detect, analyze and eliminate potential causes of nonconformities; . . . b) determining the steps needed to deal with any problems requiring preventive action; c) initiating preventive action and applying controls to ensure that it is effective; d) ensuring that relevant information on actions taken including changes to procedures is submitted for management review.

Here again the standard has been revised to force a more proactive approach to preventing nonconformities in products, processes, and the quality system. The last statement requires that actions taken be submitted to management for review.

4.15 Handling, Storage, Packaging, and Delivery

- Deterioration and damage is avoided during handling.
- Product is stored in a suitable location.

- The condition of the stored goods is checked at regular intervals.
- The responsibility for the receipt and release of product is documented.
- Packaging is sufficient to ensure conformance to requirements.
- The quality is protected following final inspection/testing.

Handling, Storage, Packaging, Preservation, and Delivery Originally issued as "Handling, Storage, Packaging and Delivery" in the 1987 version, ISO/DIS-9001 has expanded this element to include requirements for preservation of the product. This is stated as "Appropriate methods for preservation and segregation of the product shall be applied when such product is under the supplier's control." In support of the original intent of this element, it would be of no benefit to have the most strictly designed and manufactured product if the quality were destroyed by lack of proper preservation.

4.16 Quality Records

- Procedures are established for the identification, collection, indexing, filing, storage, maintenance, and disposition of quality records.
- Records of quality control data are to be kept to prove the quality system is effective.
- Records are readily retrievable.

The title of this element has been revised to "Control of Quality Records." Documented procedures are now required, along with retention "to demonstrate conformance to specified requirements. . . ." The standard also notes that records can be hard copy, electronic, or other media.

4.17 Internal Quality Audits

- Internal quality audits are planned and documented.
- Results of the audits are documented.
- Responsibility for follow-up is shown.
- Timely corrective action is taken by management personnel responsible for the area audited.

Along with some minor wording revisions, the standard has added the requirement for auditing activities to "be carried out by personnel independent of those having direct responsibility for the activity being audited." This requirement was previously addressed in the management responsibility element.

4.18 Training

- Training requirements have been identified and documented.
- Personnel should be qualified by experience/training.
- Training records are maintained.

No changes were made to this element in the 1994 version, with the exception of the requirement for documented procedures.

4.19 Servicing

- The performance and verification of servicing the product meet the specified requirements.

The 1994 version requires documented procedures to address this element when applicable. Such procedures must address the reporting aspect with regard to servicing activities.

In the 1994 version, however, a company that performs servicing of the product after the sale will certify to ISO-9002, provided that it has no responsibility for the design of the product.

4.20 Statistical Techniques

- A program for application of statistical techniques is in place to ensure consistent quality of the product.

The only change to this element is the requirement for documented procedures when statistical techniques are determined to be applicable.

With a proper understanding of ISO-9000, -9004, and -9001, the implementation process becomes much easier. When the intent of the standard is clear, the process of preparing for certification becomes an exercise in determining how to "do business well."

SUMMARY

The ISO-9000 series standard is a guideline for the design, manufacture, sale, and servicing of a product. It is a generic document, independent of any industry or economic sector, that allows each organization to determine the best method of implementation. The standard addresses 20 elements that make up the quality system. These elements are divided into management responsibilities, companywide activities, and specific requirements that may be addressed by a single department or activity.

The ISO-9000 series standard is produced in five parts as follows:

ISO-9000: "Quality Management and Quality Assurance Standards—Guidelines for Selection and Use." This section explains how the remainder of the standard is divided, with guidelines on which classification is applicable to a business.

ISO-9004: "Quality Management and Quality Systems Elements—Guidelines." The second advisory document in the series provides detailed advice to businesses on overall quality management and the quality system elements within the ISO-9000 series. In addition, ISO-9004 provides guidance in areas such as marketing, product safety and liability, and quality costs.

ISO-9003: "Quality Systems—Model for Quality Assurance in Final Inspection and Test." Used for organizations that provide fairly simple products, this standard covers conformance to requirements ensured solely by final inspection and test. ISO-9003 addresses:

* Management responsibility
* Quality system
* Document control
* Product identification
* Inspection and testing
* Inspection, measuring, and test equipment
* Inspection and test status
* Control of nonconforming product
* Handling, storage, packaging, and delivery
* Quality records
* Training
* Statistical techniques

ISO-9002: "Quality Systems—Model for Quality Assurance in Production and Installation." Incorporating the final inspection and test requirements of ISO-9003, ISO-9002 adds requirements for:

* Internal auditing
* Contract review
* Purchasing
* Process control
* Corrective action
* Purchaser-supplied product

ISO-9001: "Quality Systems—Model for Quality Assurance in Design/Development, Production, Installation and Servicing." This is the most complete model for quality assurance systems. The wording of the clauses is identical to that in ISO-9002 but adds the following elements to the mandatory requirements:

* Design control
* Servicing

ISO-9000 and ISO-9004 are advisory documents; ISO-9001, -9002, and -9003 are the categories for which companies may apply for certification.

The 1994 version is the latest standard. Although this new version remains substantially the same in theory, there have been modifications, clarifications, and enhancements to many sections of the standard. The major changes are in the areas of design control; quality system; purchasing; process control; corrective action; and handling, storage, packaging, and delivery. Companies that service the product after the sale but have no responsibility for the design will certify to ISO-9002 instead of ISO-9001 under the 1994 version.

KEY TERMS

Advisory document A publication providing supplementary or informational material that may be helpful but is not mandatory.

Certification A written statement by an authorized body that attests that a specified piece of equipment, process, or person complies with a specification, contract, regulation, or company policy.

Corrective action Action taken to prevent recurrence of a discrepancy.

Element A section, component, or category describing the scope, intent, and requirements of an activity.

Implementation The execution of activities devised to meet requirements.

Military standards A group of documents issued and controlled by the government and used by both military and civilian industry to describe requirements for the manufacture of product.

Preventive action Action taken to prevent occurrence of discrepancy.
Prototype An initial unit or design produced prior to release for production.
Quality manual A collection of procedures and activities that set standards for the quality system.
Registrar The person or organization (third party) responsible for the independent review and analysis of a program or activity in regard to compliance with stated ISO-9000 requirements.

ISO-9000 TEST

1. Discuss in a short paragraph the differences among ISO-9000, -9001, -9002, -9003, and -9004.
2. Identify the applications for ISO-9001 as opposed to ISO-9002.
3. Discuss in a short paragraph the developmental history of the ISO-9000 series of standards.
4. Discuss in a short paragraph the differences between the ISO standard and a traditional quality standard.

PART II: NONDESTRUCTIVE TESTING

Inspection was the first tool employed in quality control. Genesis 1:31 states, "And God saw all that He had made, and behold, it was very good. . . ." Inspection is now and will continue to be an important tool for the quality technician, but it is not the only tool. Used properly, it is a powerful tool; used improperly, it is costly and of little value.

As industry searches for new plateaus of excellence, the use of inspection varies. Inspection is a means of gathering information about the product's conformance to requirements, and the process that created that product. However, it must be a part of a larger system. Where in the process is the inspection to be made? What are the critical parameters to be measured? What is done with the data after the inspection? The system must coordinate these and other data to achieve the most consistent finished product at the lowest possible cost.

This section gives the fundamental requirements of the inspection. It will expose you to the physics behind some of the most widely used nondestructive testing (NDT) methods. This is done not to make you a technician, but to give you a background in which types of NDT to use

PART OBJECTIVES

In this section you will be introduced to the basics of the following:

- The history of inspection
- The evolution of inspection
- The need for inspection
- The uses and misuses of inspection
- Basic requirements for inspections
- The physics behind some of the more popular methods of nondestructive inspection (NDT)
- Where each NDT method would be used and why
- Some of the codes and standards behind inspection

in different circumstances. Using the wrong type of NDT in an application can be costly and dangerous.

As we explore the world of inspection, you will find there are rules that must be followed. These rules come in the form of codes and standards. They are in place to ensure the proper training of personnel, proper performance of the inspection, and proper documentation.

CHAPTERS IN THIS PART

INSPECTION PROCEDURES, CODES, AND STANDARDS

7

OBJECTIVES

After completing Chapter 7, the reader should be able to:

- Display knowledge of the evolution of inspection.
- Distinguish among inspection, audit, and surveillance.
- Demonstrate knowledge of the basic elements of an inspection procedure.
- Illustrate the pros and cons of 100 percent inspection.
- List the reasons for inspection codes.

During the evolution of quality from the time of the craftsperson to mass production, inspection has undergone many changes. The artisan inspected his or her craft at all times. The work was dictated by these inspections. For example, if a gunsmith were making a gun stock, he or she would look at the wood and shave a little off from one side, then look at it again and perhaps remove some from the other side. Every operation was dictated by the inspection of the last operation.

As mass production gained in popularity, inspection played an ever-increasing role in production. However, now the sequence of events was set. No longer was the sequence of events dependent on the inspection. The assembly line moved in a precise manner. The purpose of inspection changed to finding parts that did not conform to standards. As more and more parts were manufactured, more and more inspectors were needed to find the nonconforming parts. If quality was to be improved, inspection had to be intensified to find more of the bad parts.

In some companies, quality control departments were called inspection departments. The inspectors acted as police officers. This set up an adversarial relationship between quality control and production because rejection of bad parts made it more difficult for production workers to meet the quotas set by management. Management was attempting to inspect quality into parts. This *cannot* be done.

Inspection is an important tool for quality control, but it is not the whole toolbox. It will help to detect nonconformance in parts. However, the quality technician must remember that a nonconformance is only a symptom of the real problem. If the problem is fixed, the symptom will go away. If only the symptom is treated and not the problem, another symptom will soon follow. Monitoring the parts simply makes inspection a more skillful tool to modify the process.

For many years management has recognized the need to examine or test a product before shipping. Until recently, however, early detection of defects, and the money spent to do it, have not always been considered a wise investment. However, a study of quality costs shows that an investment in defect identification and prevention usually pays off handsomely to the manufacturer.

An example of this would be if a drill bit were worn to the extent that it drilled an undersized hole. When inspection detects the undersized hole, at least two things should happen: (1) correct the undersized hole, and (2) more important, replace the drill bit that caused the undersized hole. In many cases, much time and effort are spent to find the undersized hole and very little to find and change the drill bit.

There is a current trend to minimize inspection. In most cases it is necessary and long overdue. It is important for the quality professional to know what inspection will and will not do, and what types of inspection are available. This section attempts to: 1) give a background in different types of inspection, 2) identify where each type would be used, 3) provide basic requirements for inspectors and inspector training, and 4) list the requirements for inspection procedures.

7–1 INSPECTION, AUDIT, AND SURVEILLANCE

Inspection is a highly defined, close examination of a product or process. The term *inspection* usually refers to the examination of an object or product.

Audit is an inspection of an organization's adherence to the established quality standards. As an example, a company would audit an inspection process to make sure that the procedure is current and is being followed.

Surveillance is a loose inspection process that uses some of the techniques of the audit function and some of the inspection. The procedures used in the surveillance are much less concise than for the inspection or audit.

7–2 TYPES OF INSPECTION

Incoming Inspection This type of inspection is also referred to as *vendor inspection.* It is the inspection of parts or materials received from another source. The extent of this inspection and the sample size depend on many things[1]:

- Importance to the serviceability of the finished part
- Past performance rating of the vendor
- Quality system used by the vendor
- Number of parts being received
- Applicable standards
- Customer requirements

Inspectors employed by the purchasing company sometimes maintain an office at the vendor's facility. This *source inspection* allows the company purchasing the parts to inspect the product before it leaves the vendor's plant. This allows for defects to be quickly rectified without the delay of returning the product to the vendor's plant. It also allows for minimal inspection on receipt of the product. It is often useful where the warehouse space is limited or in conjunction with a "just in time" program. This type of inspection is done when cost justifies the return.

The degree of detail of the inspection depends on the critical nature of the part, complexity of the inspection, and the economics of performing the inspection. If the part is essential to the safe operation of the finished product, the inspection will be extensive. On the other hand, if the operation of the finished product will not be greatly affected by this part, the inspection will be less critical.

Process Inspection The procedures for *process inspection* vary widely, depending on the type of manufacturing being accomplished. However, this type of inspection has two purposes[2]:

- To ensure the integrity of the part
- To give data for manufacturing process modification

In the case of a plant where widgets are made, the inspector checks samples of widgets at various stages of manufacture. The resulting data from these inspections would then be analyzed and the manufacturing process changed to bring the data back in line with requirements. The earlier in the process this happens, the less costly it is to the company.

In the case of state-of-the-art continuous manufacturing processes such as extrusion, the process itself is inspected. These highly automated and computer-controlled processes produce a consistent product if the process is set up right. Thus, the inspection is directed toward the process itself. If the process parameters are right, the product will be right. Product inspection in this type of manufacturing process becomes a quality assurance function.

In the case of a job shop (see Chapter 1), the product is inspected to the standards required. If safety is a consideration, the inspection is more critical. The process used to build the product to that point is also examined to ensure that proper steps were taken to guarantee part integrity and traceability.

Final Inspection The *final inspection* is performed to give a confidence level about acceptability standards for product leaving the plant. This is often done by some automated process at 100 percent. In the case where public safety is involved, 500 percent or 600 percent inspection may be required by code.

7–3 INSPECTION PROCEDURES

An inspection process should have a written set of steps that are to be followed, called a *procedure.* The procedure is a detailed progression outlining steps and resources used in the inspection. Generally, the sequence starts by locating gross (large or easily identified) nonconformances and works toward finding the less obvious defects.

For inspection to be successful, it should follow a written procedure that possesses the following criteria:

- It should have a clear definition of the test progression.
- It should define the reference standards.
- It should closely define the acceptance/rejection criteria.
- It should detail the required test records (documentation).[3]
- It should detail the path of nonconforming product.
- It should outline and provide for a timely critique of the test results report.

Clear Definition of the Test Progression This portion of the procedure calls out the frequency and sample size for the inspection. It also specifies the section of the part to be inspected and the steps required. It is often as important that the steps be followed in the proper order as it is that the part conform to standards. In addition, this part of the procedure names the instruments used in the inspection.

Reference Standards This portion of the procedure names standards to be used in the inspection. There may be many types of written standards used. There may be one for the calibration of the inspection equipment. Another standard may set the acceptance criteria for the part. Sample size may be determined by a third standard. Nondestructive inspections often use what is called a reference standard to calibrate the test. This is a physical piece, a product sample, with known defects of known size. These standards are used to set up the inspection equipment. All these standards should be called out in the procedure.

Clear Accept/Reject Criteria The procedure must specify in detail what is acceptable and what is cause for rejection. This is usually done by using a defect classification system. There are often three broad classifications, or a variation of them.[4]

- A *critical defect* is one that will cause a hazardous or unsafe condition and adversely affect the operation of the product.
- A *major defect* is one that may result in failure of the product.
- A *minor defect* is one that will not reduce the usability of the product.

There may be variations on or subdivisions of these categories. However the company chooses to classify defects, one fact remains—all defects must be classified. This classification structure is normally used on all inspection procedures. That puts a critical defect in one inspection in the same category as a critical defect from another inspection.

The inspection procedure should call out the *anomalies* that are classified as *defects.* It should then classify them as outlined in the previous paragraphs. Then the procedure should list the number and classification of allowable defects.

Required Test Records This portion of the procedure outlines what and how paperwork is to be completed. It specifies the number of copies and to whom they are sent. If the product is manufactured to code or regulatory requirements (API, ASME, ISO, and the like), the procedure identifies what stamps, monograms, or other types of marking are to be applied to the product.

Nonconforming Product Path Product that does not conform to standards must be rerouted through the system for rework or scrapped. The procedure should outline this rerouting so that there is no confusion on the inspector's part concerning the disposition of the product.

Timely Critique of Test Results The procedure should outline the sequence of events that follow a failure report. Management personnel must be made aware of trends and unusually serious failures. Failure in this section of the procedure will cause a decrease in the improvement rate. It stops the flow of information back to the production department so that process improvements can be made. Failure of this portion of the procedure can also cause morale problems among inspectors as they see nothing happening as a result of their efforts.

7–4 INSPECTOR TRAINING AND SUPPORT

The key service performed by inspectors is measuring production/operation performance against product design characteristics. The ability to do this effectively does not just happen. Inspector training is critical to successful inspection. In many cases, the inspection procedure is complex, and the equipment needed for testing may be sophisticated.

Management Support Many industries have made giant strides in improving the inspector's situation. Good management realizes that reading instruments with accuracy and consistency depends on many criteria. The following are some considerations for accurate and consistent inspection results[5]:

- Remove the workers from piecework quotas. Time/output requirements reduce an inspector's ability to be thorough and consistent.
- Maintain calibrated tools. An inspection will be only as accurate as the gauges being used.
- Provide a reasonable area in which to work. Proper lighting, work space, and environment are directly related to the accuracy of the results.
- Provide a place and sufficient time for record keeping. Accurate results are unusable unless properly documented. Test results should be used to track the performance of the process used in producing the product.
- Develop a reporting channel to upper management other than the production supervisor. Inspection results and process performance should be reported directly to upper management without being filtered through several levels of management.
- Ensure that inspectors are exempt from personal reprisals. Note that an inspection is a vehicle used to determine the level of adherence to the specifications. The inspector is not usually involved in setting the specifications or in the production of the product. Do not kill the messenger for delivering an unfavorable message!

- Develop a good follow-up system. One of the more important uses of the inspection function is as an early warning system of problems that are developing in a process. For the quality department to be effective, this warning should be followed up with a corrective action system to correct the deficiencies discovered by inspection.

To minimize errors caused by long, tedious procedures, the inspector must have management's support. Inspection stations must be well lighted and comfortable. This is often a challenge, because inspection stations must be kept in the product flow. Quotas may force inspectors to compromise their inspections for the sake of time. The money spent by a company in this area is well spent, but it is hard to determine the return on this investment because it is impossible to count errors not made.

Inspector Training Companies expend great effort to train inspectors in specific areas of inspection. In many cases, extensive training and experience are required by code. An example of this would be in the area of nondestructive inspections, such as ultrasonic inspection, radiographic inspection, and eddy current inspection. Dimensional inspectors must be trained in the use of measuring equipment. Sophisticated equipment, such as coordinate measurement machines (CMM), require inspectors to be not only well trained in measurement techniques, but computer literate as well.

To minimize the expense of inspection, companies must cross-train inspectors to do many inspections using various types of equipment. This demands that inspectors be self-motivated and intelligent individuals. Often, inspection procedures are long, complicated, and tedious. This demands a person who is focused and who can remain at the same task for long periods of time.

Tools used by inspectors are often designed to measure a certain property of the product or subassembly so the performance of the process used to make the product can be evaluated and tracked. The quality professional then applies statistical tools in these areas to make a judgment as to the process performance. Statistical methods, such as control charts and graphs, can be applied to the inspection results to detect problems early in a production process. Careful attention must be paid to calibrated tooling, specified tolerances, and sample selection.

After interchangeability, product costs, process capability, and customer acceptance are considered, the tolerance determined must not be changed by a generous inspector. To do so is, in effect, redesigning the product without authority. Likewise, inspection departments often receive verbal changes from engineering. The drawings are then marked up, or the inspector makes notes in his or her "little black book" to keep track of changes. This practice results in two sets of specifications—one official, one unofficial—leaving the customer ignorant of what to expect. This will eventually lead to serious problems for the company.

Of course, the solution is for inspectors to accept official tolerances as correct, to be changed only by proper authority and procedures (e.g., a written change notice from engineering). An inspector has more than enough responsibility in his or her own job without taking on the responsibilities of design, production, and quality engineering.

7–5 INSPECTOR'S RESPONSIBILITY

What is the responsibility of an inspector? This varies widely from company to company, depending on the management. However, in every case it is the responsibility of the inspector to evaluate the efforts of others by visual, mechanical, electrical, or chemical measurements.

An inspector must make every effort to ensure that the measurements taken are accurate. The first step in doing so is to use the proper instrument for the job. Too often, the handiest instrument is the instrument used, which may not have the discrimination needed to evaluate the product properly. This could destroy the user's confidence in the result. Frequently, an inspector does not realize the long-term effects of inaccurate measurements. In some cases, incorrect measurements could result in a decision to stop production because product improvement would seem unfeasible. This could result in the loss of many jobs, perhaps even the inspector's own.

A professional inspector is key to the successful manufacturer's operation, but the inspector's usefulness is totally dependent on his or her ability to take accurate measurements, within assigned tolerances, with timely reporting of the results.

7–6 100 PERCENT INSPECTION

No inspector ever inspected quality into a product. The idea of 100 percent sorting being 100 percent effective for screening out bad units is false. The fact that 100 percent sorting is from 80 percent to 90 percent effective has been proved time and time again by experimentation with qualified inspection personnel. The number of defective units that get past an inspector is a percentage of those submitted to the inspector. The more defectives that make it to a final inspector, the more defectives will be shipped.

One alternative to sorting good product from bad is to statistically sample the product and make accept or reject decisions about the lot based on the sample. This method has gained wide acceptance in manufacturing because of the accuracy and speed of decision making. However, this method has some problems that must be overcome. All the product in a lot is accepted or rejected based on a sample; this is not a method of screening good product from bad. As in any sampling method, compensation must be made for statistical variation and sampling error.

A common question is, if sampling has an error rate, why not go to 100 percent inspection? This is a valid question that needs to be addressed. The 100 percent inspection has two major problems; the first is the cost in time and money. If a small number of items (10 or fewer) were to be inspected, such a program might be cost effective. But if every BB made by Daisy Air Guns were to be inspected, the cost of inspection would outweigh the cost of production. The time needed to inspect every BB would slow production to the point that Daisy would not make money. The second reason that 100 percent inspection is not used much is that it is only about 80 percent effective. That is, if a large amount of product contained 100 defective pieces, a 100 percent inspection of the product would eliminate only 80 of the 100 bad pieces. The reason: inspector fatigue and boredom.

Try a personalized 100 percent inspection. Allow yourself one minute to go through the last paragraph only once. Count the number of times that the letter *e* is present. Start with "A common question . . . " and end with " . . . fatigue and boredom."

How many times was the letter *e* used? Try recounting the *e*'s in the same paragraph within the same time limit. Did the number change? In most cases, your count will change. If it did not, you have what is called "high repeatability."

The letter *e* occurs 90 times. If this was the number you found, you are very good! If you found more or less than that number, do not be upset. Finding wrong numbers is common. For this reason, 100 percent inspection is not as accurate as sampling. In some industries where 100 percent testing is used, the inspector is checking circuit boards 8 to 10 hours a day, 5 or 6 days a week. These industries are the types that are likely to incorporate sampling inspection. Inspection cannot be the total quality effort in a company; it should be used as a tool to evaluate the performance of the process producing a product and as an early warning system that trouble is brewing.

A professional inspector must be knowledgeable and confident in his or her knowledge. Since most people tend to identify with their work and to resent being told that it is substandard, the inspector must be able to perform testing and measuring with finesse and courtesy. Most processes produce more good products than bad. Inspecting samples rather than 100 percent of the product produced by a process reduces possible inspection mistakes caused by such factors as fatigue and boredom. In the case of destructive testing, sampling becomes a necessity, since all product tested is destroyed.

7–7 INSPECTION CODES AND STANDARDS

The area of inspection finds many written codes and standards. These codes range in scope from determining sample size to evaluating inspection systems. These codes attempt to set guidelines in which to operate.

There are very few coding organizations that carry any authority. In most cases, the authority is the customer's ability to withdraw business if the codes are not followed. This can be a great influence on a contractor. For example, a defense contractor found to be inspecting outside the constraints of MIL-STD-410D could be put out of business if the contract were withdrawn.

The more critical the inspection, the more stringent the code. Nondestructive inspections such as those in the following chapters are heavily coded. The SNT-TC-1A is the code that governs the training of nondestructive testing technicians. MIL-STD-410D was derived from SNT-TC-1A. Although the training and experience requirements of these codes are stringent, compliance with the code is left to each company to decide. The coding body is the American Society for Nondestructive Testing (ASNT). The ASNT has no authority to police anyone under SNT-TC-1A.

There are many standards outlining the requirements of nondestructive examinations. All call out basic requirements for the test, as well as training and experience for the technician. In the chapters that follow, you will begin to understand why these codes exist. The inspection field is vast and necessary to the manufacturing industry. Without these codes, unscrupulous individuals could exploit the public. The challenge, then, lies in the ability to determine which codes apply and how to interpret them.

SUMMARY

Inspection is an important tool in the quality professional's toolbox, but it is not the whole toolbox. It is important to know that vendor inspections are done on incoming product. But it is more important to make vendors accountable for the result of those inspections and for altering their processes to fix any problems that the inspections reveal.

Formal inspections must have a written procedure. This ensures that the customer's critical parameters are being checked and that inspectors have actual standards for how each piece is to be checked.

There is a long-overdue move away from inspection by inspectors to one of continuous inspection by the persons manufacturing the product. This approach allows changes to the process at the time anomalies are noted, before they become defects. Still, there will always be a need for formal inspection. As inspection equipment becomes more and more sophisticated, trained individuals will be needed to operate it. But the extent of the inspection must be reduced to streamline the manufacturing operation.

Always remember, as an inspector your job is to find symptoms to fix the process, not fix blame.

KEY TERMS

Anomalies Deviations from the requirements or standards that may or may not cause rejection.

Audit An inspection of an organization's adherence to the established quality standards.

Critical defect One that will cause a hazardous or unsafe condition and adversely affect the operation of the product.

Defect Deviation from the requirements or standards that cause rejection.

Discrimination The smallest increment a measuring instrument can accurately measure.

Final inspection Inspection of the final assembly.

Inspection A highly defined, close examination of a process or product.

Inspection procedure A detailed progression outlining steps and resources used in the inspection.

Major defect One that may result in failure of the product.

Minor defect One that will not reduce the usability of the product.

Process inspection Inspections performed at specific intervals during the manufacturing process.

Source inspection Inspection performed at the vendor's facilities by the customer.

Surveillance A loose inspection process.
Vendor inspection Inspection of incoming products.

INSPECTION PROCEDURES, CODES, AND STANDARDS TEST

1. How did inspection change with the advent of assembly lines and mass production?
2. What is the purpose of inspection?
3. Define *inspection*.
4. Differentiate among inspection, audit, and surveillance.
5. What factors influence the extent of incoming inspection?
6. What are the purposes of in-process inspections?
7. What type of inspection most often uses 100 percent inspection?
8. What is a written, detailed progression that outlines the steps and resources used in the inspection?
9. What is a defect classification system?
10. Who is responsible for maintaining calibrated inspection tools?
11. Why is 100 percent inspection generally not done as a process inspection?
12. What are the reasons for codes and standards?

NOTES

1. J. Juran, *Quality Control Handbook,* 3d ed. (New York: McGraw-Hill, 1974).
2. Juran.
3. Juran.
4. U.S. Department of Defense, MIL-STD 105-D, 1974, 3–29.
5. Juran.

MAGNETIC PARTICLE TESTING

8

Magnetic particle testing (MT) is one of the most popular types of nondestructive testing (NDT) used on ferromagnetic materials. It is relatively inexpensive and does not demand much training for technicians to qualify.

MT is a method of NDT in which a magnetic field is induced into a ferromagnetic specimen and iron oxide particles are applied to detect surface and near-surface discontinuities. It ranges in sensitivity to accommodate the needs of tests. The two major limiting factors are material types and flaw depth.

The specimen to be inspected must be made of a *ferromagnetic* material. That is, it must be able to hold a magnetic field. Metals such as iron, steel, nickel, and cobalt can be easily inspected by the use of MT. Metals such as aluminum, brass, copper, magnesium, bronze, lead, titanium, and stainless steel are not able to retain a magnetic field well enough to be suitable for MT. If a material can be attracted by a magnet, it can be inspected using MT.

MT is used to find surface flaws and flaws that are just below the surface. A flaw more than 1/8 in. below the surface will not easily be detected by MT. If deeper flaws are suspected, a different method of NDT should be used.[1]

To understand MT more fully, one must have an understanding of magnetism. Therefore, let us briefly discuss magnetic theory.

OBJECTIVES

After completing Chapter 8, the reader should be able to:

- List the advantages and limitations of magnetic particle inspection.
- Explain the principle of operation in magnetic particle inspection.
- Answer questions concerning characteristics of flux lines.
- Discuss the magnetic domain theory.
- Know the definitions of basic magnetic vocabulary.
- Describe how to orient flux lines to locate suspected defects.
- List the precautions needed for certain MT techniques.

8–1 MAGNETIC THEORY

Magnetism is the ability of some metals, mainly iron and steel, to attract other pieces of iron and steel. It is thought that ferromagnetic materials are made up of small molecular magnets called "domains."

Figure 8–1 (a) Random Orientation of Molecules in an Unmagnetized Specimen (b) Magnetically Aligned Molecules in a Magnetized Specimen

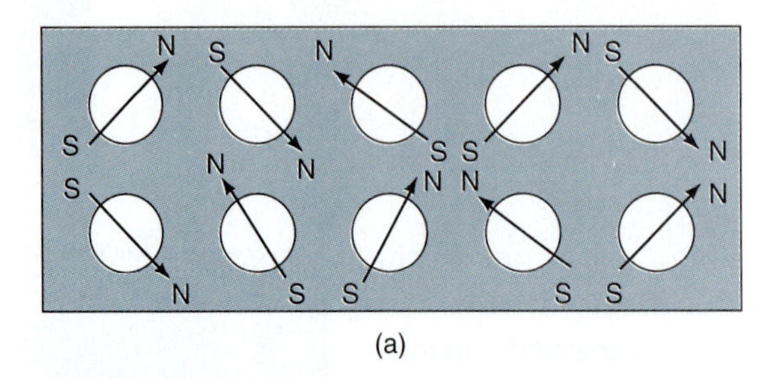

(a)

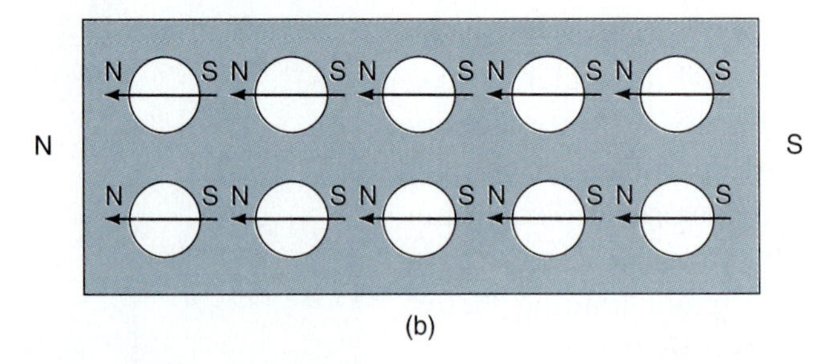

(b)

In a demagnetized ferromagnetic material, domains are in a random alignment. Each domain has a north and a south pole. However, the domains point in all different directions, and that causes the magnetic polarity of the domains to cancel each other out. When the material is "magnetized," the domains are aligned so that all the north poles point in the same direction and all the south poles point in the other direction. This causes the entire specimen to take on a polarity. As more and more of the domains are aligned, the magnet becomes stronger. Thus, it appears that the force of these domains is additive. Figure 8-1 shows random orientation of domains and aligned domains.

When all the domains are aligned, if the magnet is cut in half there will be two magnets of the same polarity. Every time the magnet is cut, the pieces will exhibit the same polarity (Figure 8-2).

The domains exert a force on one another. This force is called *flux*. These are lines of force known as *flux lines*. The more lines of flux per unit area, the stronger the attraction. The number of flux lines per unit area is called *flux density* and is measured in *gauss*. There are some characteristics of flux lines that always hold true:

- They appear to leave the magnet's north pole and enter the south pole.
- They always form a closed loop.
- They do not intersect or merge.[2]

Figure 8–2

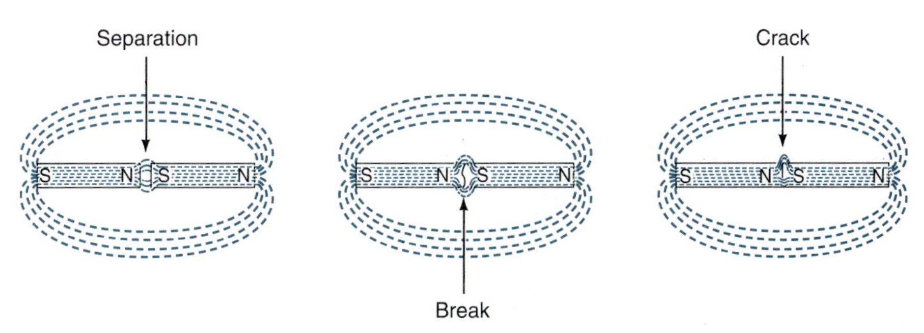

The magnetized part has these flux lines leaving the north pole, going through the air, and reentering the magnet at the south pole. This is called *flux leakage.* The space around a magnet in which the lines of force act is the *magnetic field.* The leaving and entering of the flux lines causes poles to be formed. If the flux lines do not leave the magnet, as in a round magnet, no poles are formed. Figure 8-3 shows a horseshoe magnet and a round specimen with no ends, and therefore no poles and no magnetism.

Iron is attracted to a magnet only at the poles. Therefore, if there are no poles, there is no attraction. This is the key to MT. For an indication to appear, there must be flux leakage to attract the iron oxide particles. Because flux lines never intersect or merge, the flux lines are forced to leave the specimen in Figure 8-4. That causes a polarity to be established on each side of the discontinuity and an attraction to the indication medium. This can also be understood as a new set of north and south poles.

Figure 8–3

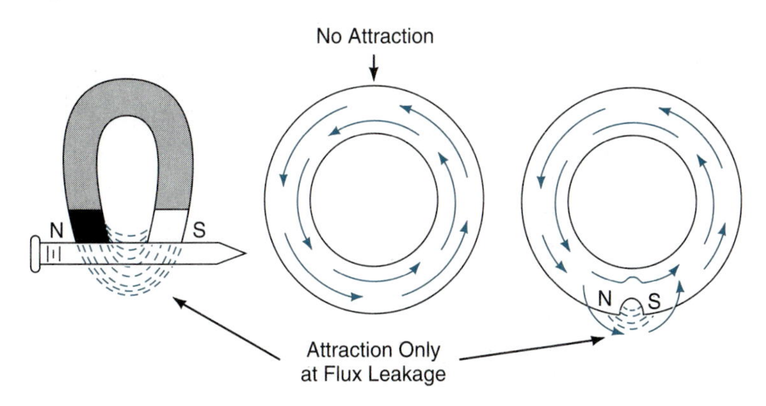

Figure 8–4 Flux Leakage Causes Particle Build-Up

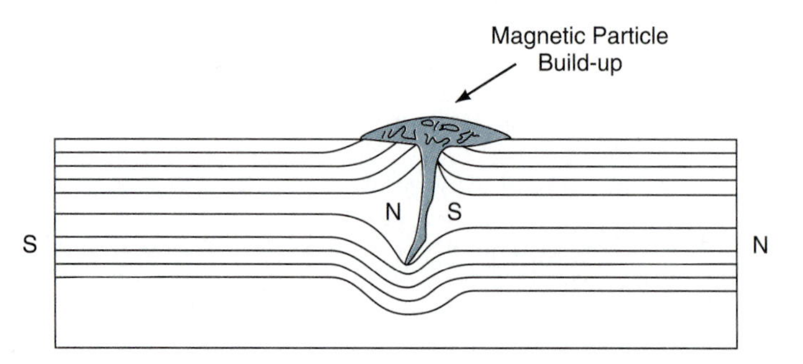

The larger the flaw, the more lines of flux will be forced out of the specimen, causing a greater attraction of particles, which in turn gives a more vivid indication. If a flaw is located deep in the specimen, however, flux leakage will not take place and there will be no indication (Figure 8-5).

It should be evident by this time that the characteristics of the metal play an important part in the quality of indication yielded. Some of these characteristics are discussed in the sections that follow.

Permeability *Permeability* is the ease with which a material can be magnetized.[3] It reflects the ease of flux spread through a material while it is being magnetized and how easy it is to align the domains. A high permeability indicates a material that is easy to magnetize; a low permeability indicates a material that is hard to magnetize.

It is simple to align the domains of a specimen with high permeability, but it is also easy to misalign the domains. This means that the specimen exhibits magnetic lines of flux while the magnetizing force is

Figure 8–5 Without Flux Leakage, There Is No Attraction

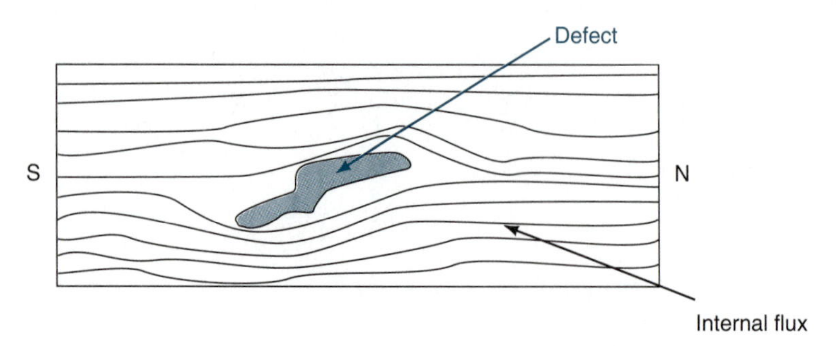

being applied but will not hold the alignment well. A specimen with high permeability has low retentivity.

Retentivity *Retentivity* is the ability of a material to retain a magnetic field after the magnetizing force has been removed.[4] A highly retentive material makes a good permanent magnet. After it is magnetized, it exhibits residual magnetism.

Residual Magnetism *Residual magnetism* is the magnetic field that remains after the magnetizing force has been removed.[5] Note that in all cases, the residual magnetism of a material is always lower than the magnetism exhibited during magnetization. This occurs because of the magnetic flux exhibited by the magnetizing force itself.

Metals that are well suited to MT are those with a high retentivity. When the magnetizing force is removed, the specimens tend to exhibit a magnetic field. Because these specimens have a high retentivity, they have a low permeability. As a result, it is often hard to magnetize such specimens. Metals that are hard to magnetize are also hard to demagnetize. *Demagnetization* is the act of misaligning the domains. It is hard to misalign the domains in a specimen with low permeability.

8–2 PRODUCING A MAGNETIC FIELD

There are three ways to produce a magnetic field in a ferromagnetic specimen.

1. *Centrifugal force tends to align the domains.* Magnetic fields can appear in a spinning shaft in a jet engine. Even the shafts of large ships have been magnetized by the spinning motion of normal operation.
2. *Exposure to another magnet.* A weak magnetic field may be installed by stroking a ferrous specimen with a permanent magnet. The force of the aligned domains exhibits an aligning force on the ferrous part, and the part begins to exhibit a magnetic field of its own.
3. *Pass electrical current through a conductor.* Any time current is passed through a conductor, magnetic lines of flux are established. If the conductor is a ferromagnetic material, it will retain some of the force after the current is removed.

Electrical current is the most widely used method of creating magnetic fields in industry. Therefore, we need to review this phenomenon briefly.

Figure 8–6 Flux

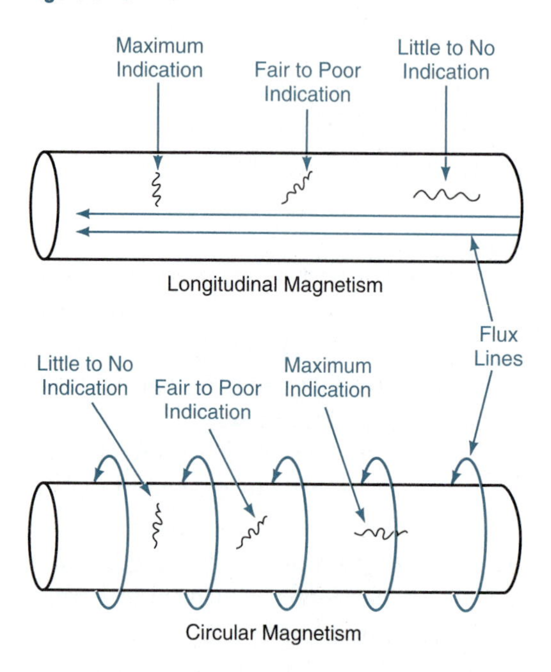

Any time current passes through a conductor, a magnetic field is established in and around that conductor.[6] This is *Faraday's law,* and a vital rule to remember. It makes it possible to inspect using MT, eddy current, and various other types of flux leakage inspection methods.

Lines of flux are exhibited at 90° to the current flow. To determine the direction of flux lines, the right-hand rule may be used. Simply stated, grasp the conductor with the right hand so that the thumb points in the direction of current flow. The fingers will point in the direction of the magnetic lines of flux. Using this rule, it is possible to determine the direction of the lines of force.

Another useful rule to remember is that maximum sensitivity is achieved when the defect is oriented 90° to the lines of flux.[7] This gives maximum flux leakage. Remember that attraction takes place only at the poles, and poles are formed only where there is flux leakage. Therefore, the more flux leakage, the more attraction of particles. Note in Figure 8-6 the defect orientation in relation to the flux lines.

An inspector should look for defects that are oriented parallel to the current flow or 90° to the flux lines. Therefore, the inspector must install flux lines to give the best indication of the suspected defects. This demands that the inspector be familiar with the circumstances surrounding the specimen to be inspected. For example, if a crankshaft

Figure 8–7

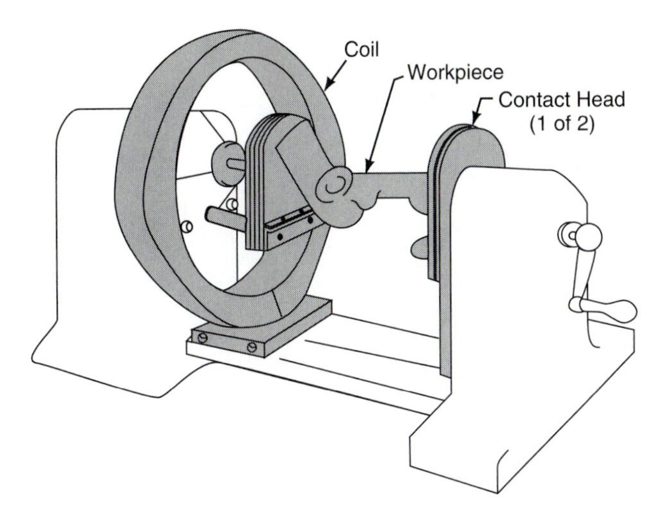

from an engine is being inspected during an overhaul, the inspector must magnetize the part so that cracks in the fillet of the crank-throw journals will be detected. On the other hand, if a machined shaft is being inspected for heat checks after grinding, the inspector orients the flux lines in a different direction.

8–3 TEST TECHNIQUES

There are basically two directions in which to magnetize a specimen—longitudinally or circumferentially. The direction is determined by the suspected flaw, as previously stated. There are several ways to achieve the desired flux orientation.

Circumferential Magnetism There are basically three ways to achieve circular magnetism. Remember that circular magnetism is used to locate lengthwise cracks.

1. *Direct contact (head shot) on a wet, horizontal, magnetic particle machine.* The specimen is placed between the electrodes (heads) of the machine, and current is passed through the part (Figure 8-7). An inspector must take great care with this technique. If not enough current passes through the part, there will be insufficient magnetism. If too much current is passed through the part, arcing at the point of contact will cause localized overheating, which may result in rejection of the

Figure 8–8

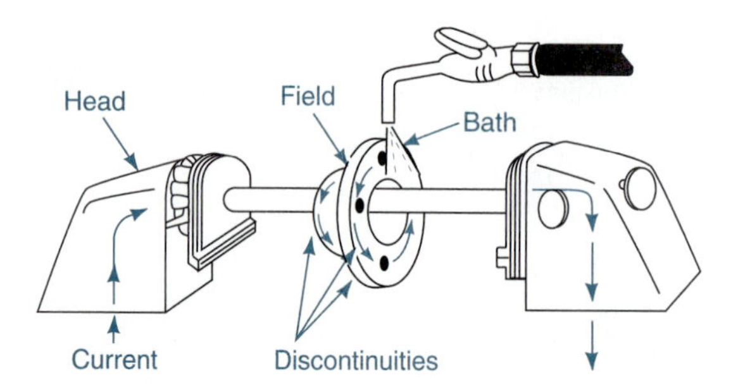

part. For this reason, current should not exceed 1000 amps per square inch of contact area.

When using a head shot, the flux density is greatest at the outside surface. This makes it useful for checking the outer surface of a machined shaft. However, it is not well suited for inspecting the inside diameter of pipe or tubing. This is especially true of thick-wall tubing and pipe.

Head shots can be made using the *wet residual technique* or the *wet continuous technique*. In the wet residual technique, current is applied to the part and then to the oxide particle bath. Oxide particles are attracted to the discontinuity by the leakage of the residual flux lines. In the continuous technique, the bath is flushed over the part. As the bath runs off the part, current is applied. This allows an increased number of oxide particles to be attracted to the part.

2. *Central conductor on a wet horizontal machine.* This technique again employs the wet horizontal machine. This method requires that the part being inspected be suspended from a nonferrous conductive bar and that the bar be placed between the heads of the machine. Current is then passed through the bar, and the resulting magnetic field is induced into the suspended part. Remember that a magnet can be created by exposure to a strong magnet (see previous discussion). The strongest flux will be on the bar's surface, which is located on the specimen's inside diameter. This allows an inspection to be made on the inside diameter of the specimen. (Figure 8-8)

   This technique is useful when inspecting specimens that are hard to orient between the heads of the machine (such as gears). Many parts can be magnetized at once using this technique. It is also acceptable to inspect the outside diameter

Figure 8–9

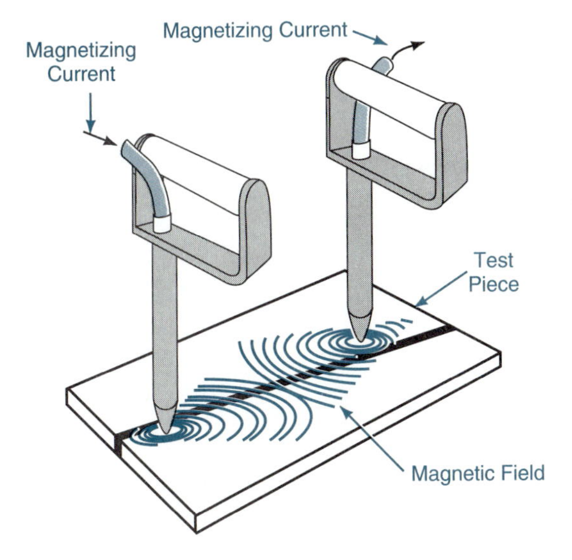

Single Prod Contacts

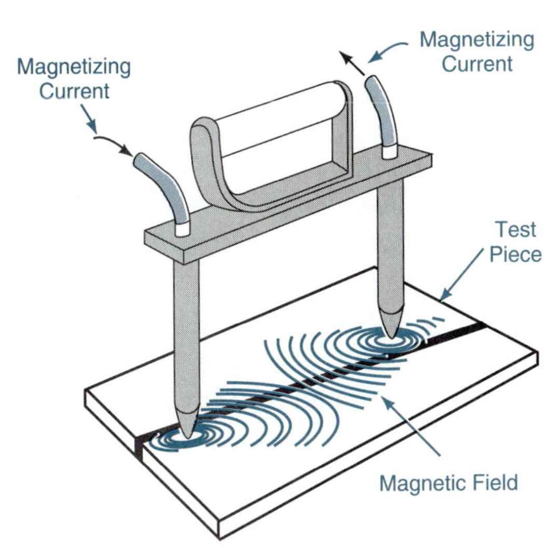

Double Prod Contacts

of the part if the wall thickness is not too great. Since the current flows through the central conductor, there is no danger of damaging the part by excessive current.

3. *Prod technique.* The prod technique is widely used on plate steel. Portable prods (Figure 8-9) offer a magnetizing force for large specimens. One prod acts as the positive and the other the

Figure 8–10 Longitudinal Magnetization Locates Traverse Discontinuities. Inspect for Particle Indications Showing Traverse Discontinuities.

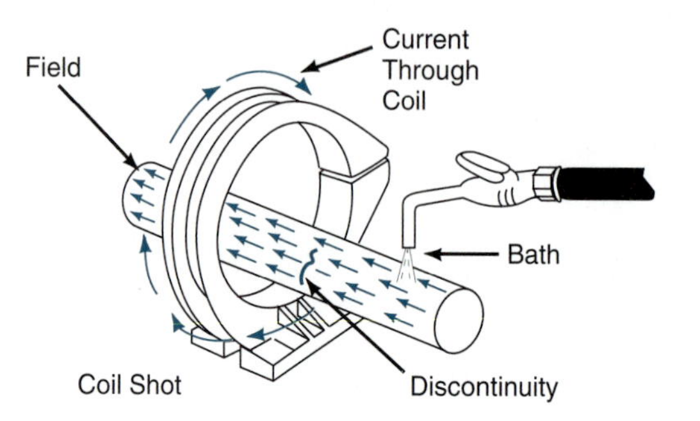

negative. Current passes from the negative prod through the specimen and into the positive prod. This causes a magnetic flux to be installed in the area around the prods.

Prods should be placed 6 to 8 in. apart. Defects are most easily located when they are in line with the prods. The inspector must take care with this technique to prevent arcing at the prods. This will cause localized overheating, which can damage the part.

Longitudinal Magnetism To get magnetism through the length of a part, current must flow around it. There are two basic ways to induce this type of magnetism.

1. *Solenoid or coil wrap.* If a long part is to be magnetized throughout its length, it is most expedient to wrap a coil of wire around the part and pass a current through the wire. The wire can be stretched out, as in Figure 8-10, or drawn into a solenoid.

   The field strength is proportional to the amount of current flowing through the wire times the number of turns of wire around the specimen. These are called "ampturns." The greater the ampturns, the greater the field strength.
2. *Yoke technique.* The coil theory can be extended to include the magnetic field induced from a temporary magnet called a yoke. In Figure 8-11, the coil induces a magnetic field into the "extenders." The field is transmitted through the extenders and into the specimen. This gives a longitudinal magnetism.

Figure 8–11

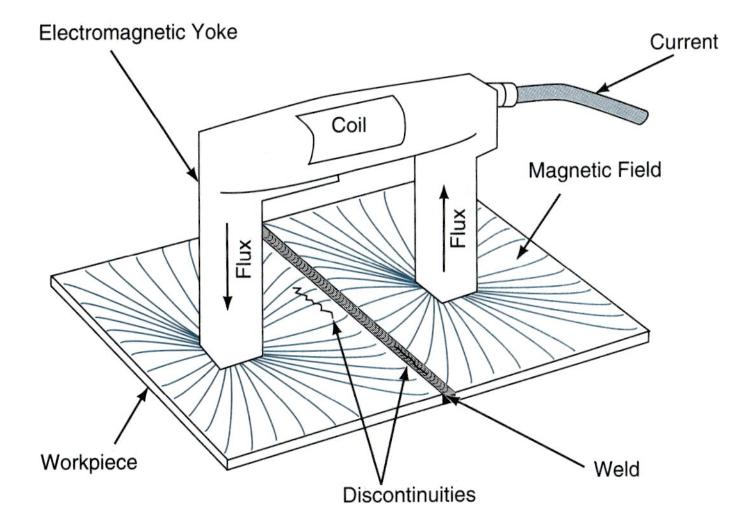

Notice that in both the longitudinal magnetism techniques, no current is passed through the specimen. This makes it much safer to use this type of magnetic field whenever possible.

SUMMARY

Magnetic particle testing (MT) is a method of nondestructive testing of ferrous parts for the detection of surface or near-surface discontinuities. Its effectiveness is limited to materials that can be magnetized. Products made of brass, bronze, stainless steel, and the like, must be inspected using another method.

The part is magnetized so that the flux lines are at a 90° angle to the suspected defect. Iron oxide particles are applied to the part, either as a fine powder or in a solution of light oil or solvent. If a defect is present, flux leakage will result, causing a north pole and a south pole to be established. This polarity attracts the iron oxide particles, which outline the defect. Indications of a defect may be more readily seen by the use of fluorescent particles or particles that contrast with the background of the part.

Training for this method is fairly simple, and the cost of the equipment is moderate. It is not considered a fast test and requires extensive demagnetization and cleaning after the test. It may not be a wise choice for assembly line application. Explore flux leakage detection or eddy current testing for such applications instead. MT is a possible choice for overhaul and life-cycle testing or welded vessels.

KEY TERMS

Demagnetization The act of misaligning the domains in a ferromagnetic material.

Faraday's law Any time current passes through a conductor, a magnetic field is established in and around the conductor.

Ferromagnetic Of or relating to substances with an abnormally high magnetic permeability, a definite saturation point, and appreciable residual magnetism.

Flux Lines of force surrounding a magnetic field.

Flux density The number of flux lines per unit area.

Flux leakage Characterized by the exit and entrance of flux lines in a ferromagnetic material.

Gauss The unit of measure of flux density.

Magnetic field The space around a magnet in which the lines of force act.

Permeability The ease with which a material can be magnetized.

Residual magnetism The magnetic field that remains after the magnetizing force has been removed.

Retentivity The ability of a material to retain a magnetic field after the magnetizing force has been removed.

MAGNETIC PARTICLE TEST

1. Define a domain as used in this chapter.
2. Define flux density.
3. List the characteristics of flux lines.
4. Where is iron attracted to a magnet?
5. Define permeability.
6. Define retentivity.
7. What is the relationship of permeability to retentivity?
8. What are the three ways to produce a magnetic field?
9. What is the flux/defect relationship that offers the best sensitivity?
10. How is circumferential magnetism produced?
11. What cautions must be taken with a "head shot"?
12. How is longitudinal magnetism produced?

NOTES

1. *Nondestructive Testing: Magnetic Particle* (CT-6-3), 2d ed. General Dynamics Convair Division, 1977.
2. *NDT: Magnetic Particle.*
3. *NDT: Magnetic Particle.*
4. *NDT: Magnetic Particle.*
5. *NDT: Magnetic Particle.*
6. *NDT: Magnetic Particle.*
7. *NDT: Magnetic Particle.*

DYE PENETRANT TESTING

9

Dye penetrant testing (PT) is a method of nondestructive testing (NDT) used to locate defects open to the surface. It can be used on almost any nonporous surface.[1] Metals, glass, or plastics can all be tested using dye penetrants. The major limiting factor is that the defect must be open to the surface. PT is an easy test to perform and requires little training for technicians. It is inexpensive and gives results in a short amount of time.

PT uses the principle of *capillary action*. This is the action by which the surface of a liquid, where it is in contact with a solid, is elevated or depressed. After cleaning all foreign material from the part, the penetrant is applied. As the penetrant contacts the surface of the part, capillary action draws it into any defects (Figure 9-1). The excess penetrant is removed, leaving only the penetrant drawn into the defect (Figure 9-2). Then, a *blotting agent* is applied to the surface of the part, which draws the penetrant to the surface and outlines the defect in a contrasting color[2] (Figure 9-3).

OBJECTIVES

After completing Chapter 9, the reader should be able to:

- List advantages and limitations of dye penetrant inspection.
- Demonstrate a knowledge of the principles of operation by answering questions concerning the characteristics of penetrants.
- Discuss the major categories of penetrants.
- Answer questions concerning factors influencing sensitivity.
- Demonstrate knowledge of cleaning requirements, cautions, and variations of techniques.

9–1 CATEGORIES OF PENETRANTS

Penetrants are categorized in two ways: (1) by the type of dye contained in the penetrant, which may be visible or fluorescent, and (2) by the process.[3] By "process," we mean the method used to remove excess penetrant. This depends on what type of base is used for the penetrant. It may be water washable, where the excess penetrant is removed with a gentle water spray. It may be solvent removable. In this case the excess penetrant is removed by the use of a highly volatile solvent cleaner. This type cannot be removed with water. Last is the

Figure 9–1

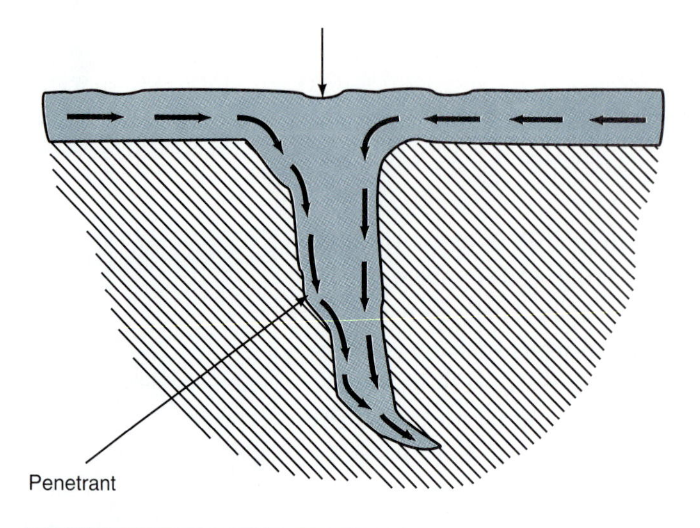

Figure 9–2

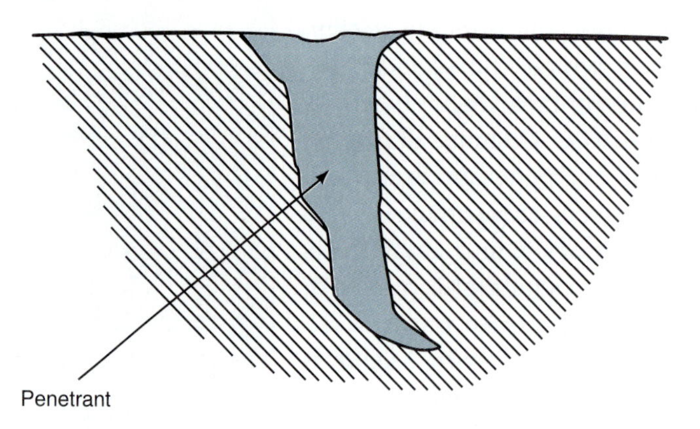

postemulsifiable, water-washable penetrant. This type is not water-washable until it is treated with an emulsifier.

A water-washable, fluorescent dye penetrant is a common classification of dye penetrant. One may also find a postemulsifiable, visible dye penetrant. Review the previous paragraph and make sure you understand the classifications of each type.

9–2 SELECTION OF PENETRANT TYPE AND PROCESS

There are several areas that should be considered when choosing the type and process of PT to be used. The quality professional must

Figure 9–3

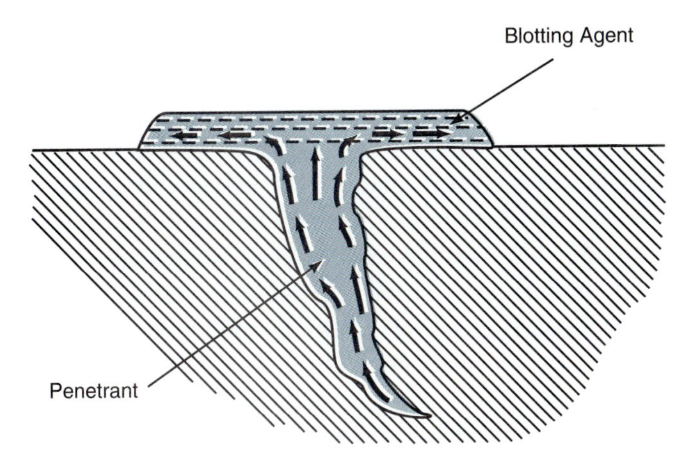

consider training of personnel, time required for testing, and cost of each test. Some of the areas that influence these are discussed in the sections that follow.

Sensitivity Required *Sensitivity* is the ability of the test to indicate flaws of different sizes. Is the type of defect large or small? There are two areas that influence sensitivity—the viscosity of the penetrant and the seeability of the results. The viscosity indicates the ease with which the penetrant is drawn into the defect. The lower the viscosity, the more easily it enters. This also means that it is easily removed from the defect and may be washed away when removing the excess penetrant. This can be offset by the use of postemulsifiers, which will be discussed later in the chapter. Seeability is determined by whether a visible or fluorescent penetrant is used. The visible penetrant often uses a red indication on a white background. The most seeable is the fluorescent indication. A near-ultraviolet light in a darkened booth gives an extremely brilliant indication. The more seeable, the higher the cost of the equipment.

Number of Articles to Be Tested If only one or two specimens are to be tested every now and then, it is not economically feasible to purchase large testing stations. However, if thousands of articles are to be inspected on a regular schedule, a highly automated system will reduce the cost and increase accuracy.

Surface Conditions of the Specimen The roughness of the surface will help or hinder the test. A rough, porous surface will cause the penetrant to be drawn into the small openings of the pores and obscure the indication of an actual defect. Extremely rough surfaces on castings may need to be inspected using some other method of NDT.

Configuration of the Specimen A specimen that has many machined angles and small machined holes will be hard to inspect. It is mandatory that the inspector be able to see into cavities to be inspected. If this is not possible, another type of NDT should be used. Remember also that any penetrant or developer left in small orifices may cause accelerated wear of the part when put into service. Specimen configuration determines the ease of post-test cleaning. As the geometric complexity of the part increases, a more seeable indication is required.

Availability of Water, Electricity, Compressed Air, and Test Area Some processes demand that a large area and utilities be used. Many of the older testing systems took up a lot of space. These factors must be considered when selecting a process and penetrant. If water is not available, a water-washable penetrant process would not be a wise choice.

9–3 SELECTION OF EQUIPMENT AND MATERIALS

When choosing the process and penetrant, selecting the equipment and materials is also a factor. The equipment requirements vary within a process depending on several factors. One factor is the inherent requirements of the test procedure. Composition and size of the specimen may require certain types of equipment, such as a large tank for dipping the part. The frequency of the tests may dictate the use of certain types of equipment. Types of suspected discontinuity also play a part in what type of equipment should be used.

9–4 GENERAL PROCEDURES FOR DYE PENETRANT

There are seven basic steps to be accomplished in all penetrant inspections. Individual test procedures dictate the specific steps, but these steps are included in all procedures in one form or another.

- Precleaning
- Penetrant application
- Dwell time
- Excess penetrant removal
- Developer application (sometimes optional in fluorescent testing)

- Specimen inspection
- Postcleaning

Precleaning This is one of the most important steps in any procedure. If the surface of the specimen is not completely clean, the penetrant cannot enter the discontinuities, and no indication will be given.

The purpose of precleaning is to remove all contamination such as grease, dirt, oils, paint, and surface finishes to expose the base material of the specimen. Cleaning can be a dangerous step in the process if done improperly. It is even possible to damage the specimen if too harsh a cleaner is used. Cleaners should be just strong enough to clean what is necessary. Cleaning methods such as detergent, vapor degreasing, and ultrasonic cleaning are commonly used in PT. Take caution when using steam cleaning, chemical cleaning, acid and alkaline cleaners, and paint removers. A neutral solvent is perhaps the most common cleaner. It is safe for the part and does a good job of removing organic contaminations such as grease and oil. However, it does leave a light film of powder that may act as a blotting agent in very sensitive tests.

Contamination is commonly broken down into two categories—organic and inorganic. Organic denotes oils and greases, whereas inorganic is the baked-on carbon and finishes applied to the parts. Each cleaner is used on a special type of contamination and, therefore, one must study the cleaners before proceeding with the test.

Note that abrasive cleaning methods (such as sandblasting and wire brushes) must be used with the greatest of caution. These types of cleaning can often damage the specimen and obscure indications by closing the openings of the surface discontinuities.

Penetrant Application This may be done in several ways. Application is not classification specific; that is, each of the application methods to be discussed can be done in all the processes with each penetrant. The three commonly accepted methods of application are:

- Spraying (using spray can or in a spray booth)
- Brushing (usually used for spot checks)
- Dipping (complete immersion of the part, considered to be the most efficient method)

Dwell Time *Dwell time* is the period of time during which the penetrant is permitted to remain on the specimen.[4] It is the time in which capillary action takes place and the penetrant enters the defect.

The time may vary, depending on the type of material being inspected and the kind of suspected discontinuity.

Specimens with "tight flaws" (such as stainless steel) require a longer dwell time to allow the penetrant to enter the flaws. Commonly, dwell times range from 3 - 30 minutes. Heating the specimen accelerates the capillary action; however, the penetrant may begin to dry out and the specimen may have to be rewet.

Excess Penetrant Removal This is an important step in the process. All excess penetrant is to be removed without removing the penetrant that is in the defect. If the part is not clean enough, the remaining penetrant will cause false indications to appear. If the penetrant is washed from the defect, there will be no indication and a flaw will be missed.

Water-washable penetrant is removed with a gentle, warm water rinse. Do not use coarse sprayers, since they will flush the penetrant from the flaw. Try to aim the water at a 45° angle to the surface to minimize the possibility of flushing the penetrant from the flaw. If the penetrant is fluorescent, the rinse may take place under ultraviolet light. Stop the rinse as soon as the excess penetrant is removed.

Solvent-removable penetrant is rinsed with a suitable solvent. This type of penetrant cannot be removed with water. The remover must be volatile and leave no residue. Such a residue can act as a blotting agent and prematurely bring the penetrant out of the flaw. Do not spray the solvent directly on the surface of the specimen, since this will wash the penetrant from the flaws. Instead, spray the solvent on a rag and wipe the surface clean.

The removal of excess penetrant in the postemulsification process is more difficult than in the other processes. It requires the application of an emulsifier. The emulsifier breaks down the penetrant so that it becomes water washable.

The idea is to apply the emulsifier and to let it work long enough to emulsify the excess penetrant, but not long enough to emulsify the penetrant in the discontinuities. That way the water will wash the excess penetrant from the surface, but the penetrant in the flaws will not be washed away by the water. To do this, there is a critical time called the emulsification time—generally three to five minutes. The exact time is determined by trial and error on a specimen (standard) with a known flaw. Once the procedure is established on a standard it will not vary, and the time is monitored closely. After emulsification, a gentle water spray is used to remove the excess penetrant, as in the water washable procedure.

Figure 9–4

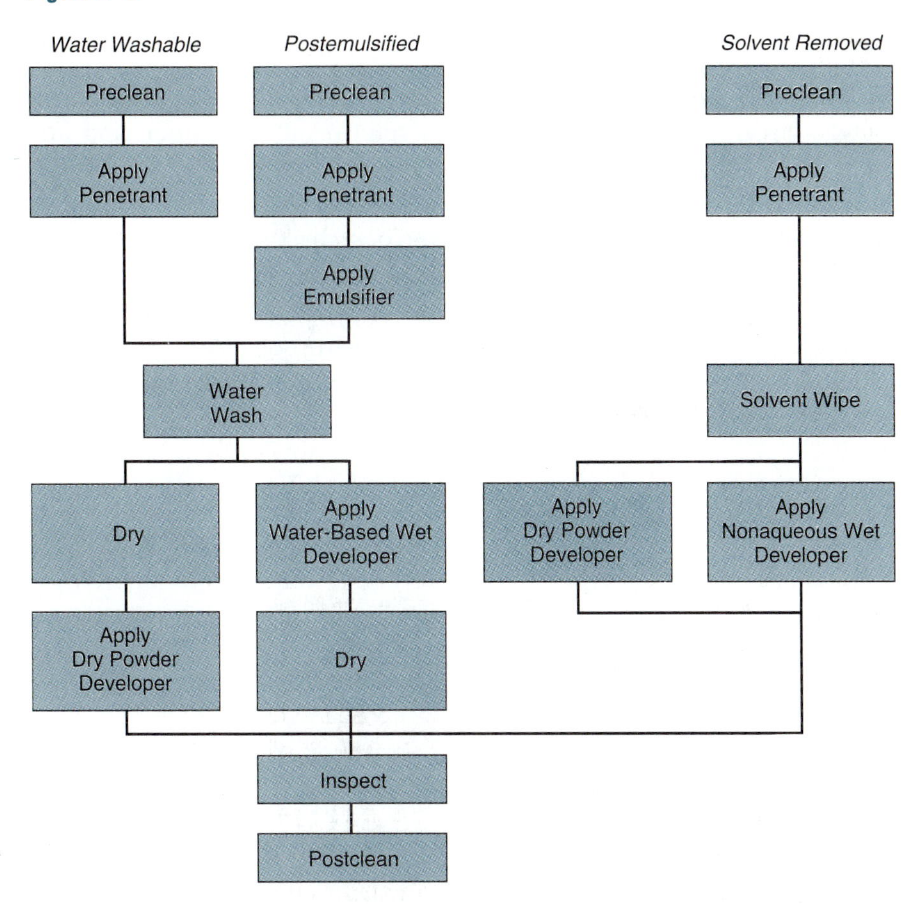

Developer Application Developers are blotting agents that draw the penetrant from the flaw to the surface. These developers come in a wet or dry form. If using a wet developer, it can be applied directly after the excess penetrant removal. As the developer dries, a powder is left on the surface that draws the penetrant back to the surface. If a dry developer is used, the specimen must be completely dried before the developer is applied. Dry developers are fine powders resembling white flour. This step may be omitted on some procedures when using the proper penetrant under the right conditions.

Specimen Inspection This step of the procedure is critical and cannot be rushed. The indication of a flaw takes the outline of the flaw on a contrasting color to the background. For visual inspection, the indication generally appears as a red mark on a white background. When using fluorescent inspection, the indication appears as a bright blue-green mark.

Postcleaning This step is often not given the time and attention it needs. Great care should be taken to remove all developer, and any penetrant missed in the excess penetrant removal step, from the specimen. Developer and penetrant not removed can attack the specimen and cause wear or corrosion. After postcleaning, you should take steps to protect the specimen from corrosion. Remember that the protective coating, such as paint and oil, and other contamination was removed. Something must be applied to retard corrosion. Figure 9-4 contains a flowchart that demonstrates the steps described in the preceding sections.

SUMMARY

Liquid dye penetrant testing is a method of nondestructive testing that is employed to find defects open to the surface on an object. The only major restriction on the material to be tested is that it be nonporous. Plastics, glass, nonporous metals, and ceramics are all good candidates for liquid dye penetrant inspection.

After cleaning, the part is dipped, brushed, or sprayed with a liquid dye that is allowed to sit on the surface. Capillary action causes the dye to be drawn into any flaws open to the surface. The excess dye is removed from the surface and a blotting agent is applied. This draws the dye back to the surface, where it outlines the flaw on a contrasting background. Any action that causes flaws to be closed to the surface will cause flaws to be missed. This may be caused by not cleaning the part completely or by doing so improperly.

This method is widely used in industry. The training required for this method is short and relatively easy. The equipment ranges in cost from a few dollars to tens of thousands of dollars. This method can be applied to automated manufacturing lines or to one piece at a time.

KEY TERMS

Blotting agent A compound that draws the penetrant out of the defect.
Capillary action The action by which the surface of a liquid, where in contact with a solid, is elevated or depressed.
Dwell time The amount of time the penetrant is allowed to sit on the surface of the part and be drawn into the flaw.
Sensitivity The ability of the test to give an indication on flaws of different sizes.

LIQUID DYE PENETRANT TEST

1. Define capillary action.
2. What are the two major categories of penetrant?
3. What is viscosity?
4. What influences sensitivity?
5. What is the possible damage to the specimen if postcleaning is done improperly?

6. What are the seven basic steps in penetrant testing?
7. How can precleaning affect the integrity of the part?
8. What are the two basic categories of contamination covering a test article?
9. What are the three ways to apply penetrant?
10. Define dwell time.
11. When removing excess penetrant, why must the technician be careful?
12. How is excess postemulsified penetrant removed?
13. What is the purpose of the emulsifiers in postemulsification penetrant?
14. Why is emulsification time important?
15. How is emulsification time determined?
16. What is a blotting agent?
17. What should be done to protect the specimen after postcleaning?

NOTES

1. *Nondestructive Testing: Liquid Penetrant* (CT-6-2), 4th ed. General Dynamics Convair Division, 1977.
2. *NDT: Liquid Penetrant.*
3. *NDT: Liquid Penetrant.*
4. *NDT: Liquid Penetrant.*

EDDY CURRENT TESTING (ET)

10

Eddy current testing is an electromagnetic surface inspection used on conductive parts. It is growing in popularity because of its quick response time and sensitivity. There is no developing period or dwell time. This type of test can be done on tubing or sheet stock. It will measure conductivity, detect discontinuities, and determine the thickness of coatings or plating. Because ET gives a continuous indication, it is useful for continuous processes and automated lines.

OBJECTIVES

After completing this chapter, the reader should be able to:

- List the advantages and limitations of eddy current testing.
- Discuss the basic principles of operation.
- Understand the vocabulary relating to eddy current testing.
- List the critical variables in test parameters.
- Discuss coil types and their arrangements.

10–1 ADVANTAGES

There are numerous advantages to ET, such as:

- Fast response time
- High sensitivity to physical and metallurgical variables
- Adaptability to automatic processes
- Computer compatibility
- No direct contact needed with specimen
- Adaptability to go/no go standards and preprogrammed inspections
- Testing equipment can be compact and self-powered[1]

10–2 LIMITATIONS

Some of the limitations of ET are:

- It can be used only on conductive materials.
- Rough surfaces do not respond well.
- Test results may vary with operator's ability to filter out noise.
- It is limited to surface and near-surface indications.
- Testing of ferromagnetic parts can be difficult.[2]

10–3 PRINCIPLE OF OPERATION

Eddy current inspection uses the theory of electromagnetic induction as a principle of operation. Because electromagnetic induction involves the use of magnetism, it may be helpful to refer back to Sections 8-1 and 8-2 (the theoretical section of the magnetic particle testing chapter).

Remember that when current is passed through a conductor, a magnetic field is established around that conductor. Also, it is possible to create a magnet by coiling a wire around a ferromagnetic material and passing a current through the wire. The field from the wire aligns the domains in the ferromagnetic material, and the material begins to exhibit magnetism. It is possible to change the polarity of the coil by reversing the direction of current flow through the conductor. Whenever electrons move through a conductor, a magnetic field will be present.

The reverse is also true. If a conductor with a closed loop is passed through a magnetic field, current will flow. Three things are all that is needed to generate electricity:

- A conductor
- A magnetic field
- Relative motion between the conductor and the magnetic field

If these three are present, current will flow. For example, when coils of wire are rotated in the magnetic fields of a generator, a current flows. If any of the three variables changes, there will be a corresponding change in current flow. If the number of conductors is increased while the RPM (relative motion) and flux density stay the same, the current will increase. Likewise, if the magnetic field is decreased with a constant RPM and number of conductors, the current will decrease.

In the case of a generator, a conductor is moved inside a stationary magnetic field. The same effect can be achieved by moving a magnetic field over a conductor. For clarity, let us go back to the generation of the magnetic field.

A simple dc circuit will be used for study. When the switch is closed (turned on), current begins to flow. The number of electrons flowing increases to its maximum. This takes only a fraction of a second, but a building effect does take place. As the number of electrons grows, so does the magnetic field that the electrons create. As long as the direction and number of electrons flowing remain constant, there will be no change in the magnetic field. However, when the switch is opened (turned off), the number of electrons is reduced to zero. Again, this takes time, and the magnetic field diminishes as the number of electrons diminishes. Although the conductor in the circuit has remained stationary, the magnetic field has moved.

Every time the switch is opened or closed, there will be a corresponding buildup and collapse of the magnetic field. If the switch is opened and closed fast enough, the magnetic field will appear to pulsate. If a coil of wire is introduced into this circuit, it will appear to be a pulsating magnet.

If another coil of wire is put close to the coil of wire in the circuit and the switch is closed, the magnetic field moving across the conductors in the new coil will cause a pulse of current. When the switch is opened, the collapse of the magnetic field will cause another pulse of current to flow in the opposite direction.

Study this situation and note that all three of the requirements to generate electricity are present:

- Conductor (the second coil of wire)
- A magnetic field (from the coil in the circuit)
- Relative motion between the conductor and the magnetic field (the buildup and collapse of the field)

This is called *electromagnetic induction.* It is commonly referred to as transformer action. Increasing the frequency with which the switch is opened and closed should increase the current flow in the second (secondary) coil. This is absolutely true. Also, increasing the strength of the magnetic field should increase the current flowing in the secondary. Right again!

Now apply this theory to nondestructive testing. Remove the secondary coil and replace it with a conductive specimen. As the poles of the magnet cut the conductor, current flows in the direction shown in Figure 10-1. As the magnetic field reverses, the direction of current flow also reverses.

It becomes relatively easy to see how changes in the specimen cause changes in the current flow in the specimen. If there is a discontinuity in the specimen, it causes the electrons to flow further around the discontinuity (Figure 10-2). This appears as an increase in resistance. Likewise, if another conductive specimen is substituted, the difference in conductivity will appear as a change in resistance.

It is not possible actually to monitor the changes in current flow in the specimen. However, it is possible to monitor changes in the primary coil that occur because of changes in the specimen. Power is drawn off the primary coil by the conductive specimen. The more conductive the specimen, the more power is drawn off the coil.[3]

Figure 10–1

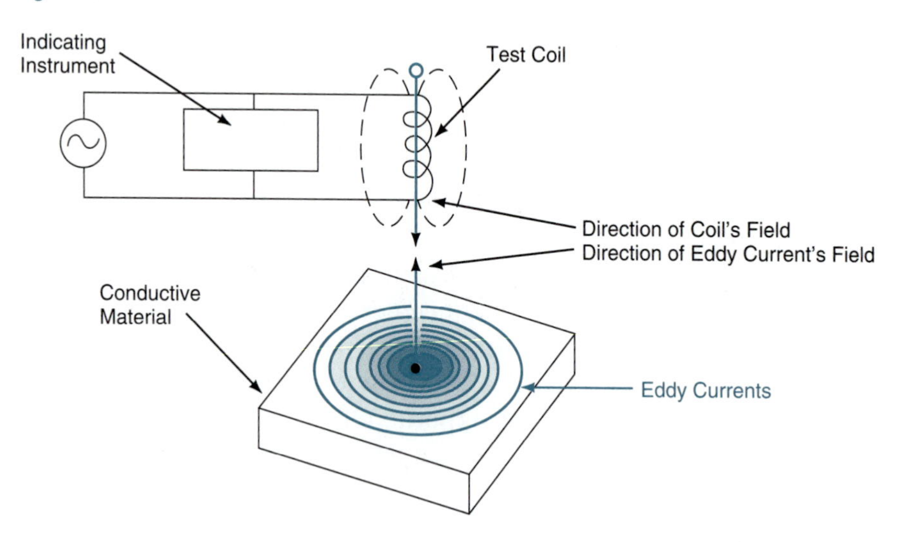

For example, if the specimen is copper, the magnetic flux cutting the copper causes a great number of electrons to flow. If aluminum is substituted for the copper, fewer electrons will flow. This changes the impedance of the primary coil. *Impedance* is the opposition to current flow in an ac circuit. Although the primary coil and the specimen are not connected electrically, they are connected magnetically. Changes in the specimen's current flow cause changes in the magnetic fields of the primary coil, and that in turn causes changes in the current flow in the

Figure 10–2

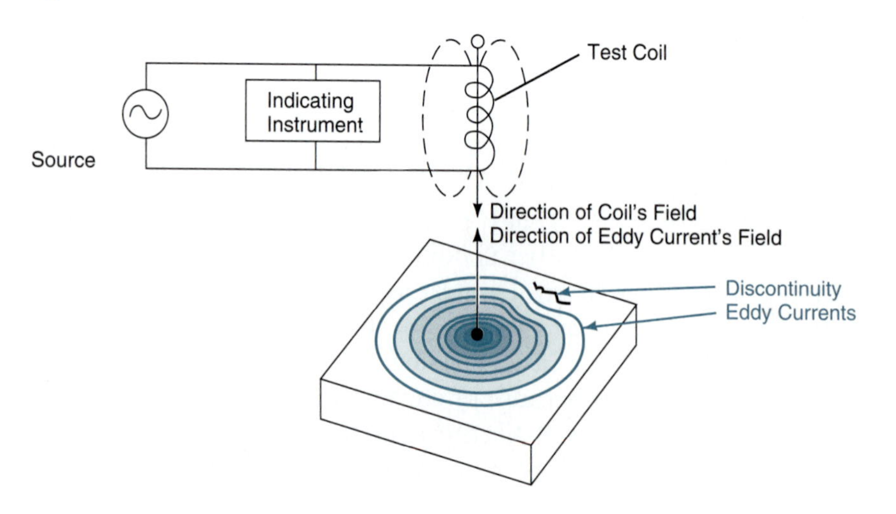

coil. If you want further information on impedance and its causes, there are many good books for reference. It is sufficient for our purpose to say that the quality professional monitors changes in the current flow of the coil, which relate to the specimen characteristics.

The depth of penetration is very small, and it can be affected by several variables. These variables will become evident when the cause of the change is explained. When the magnetic lines of flux cut the conductive specimen, currents flow (eddy currents). Also remember that when current flows through a conductor, a magnetic field is established around that conductor. The same laws hold true in the specimen. As the eddy current flows, small flux fields are set up inside the specimen. The polarity of the small flux field in the specimen is opposite that of the larger flux field that caused current to flow in the specimen in the first place. Generally, the higher the number of eddy currents, the less depth of penetration. Consequently, the flux field from the little currents becomes additive and pushes the flux field of the primary up toward the surface.

Therefore, anything that causes a larger number of electrons to flow in the specimen will cause less depth of penetration (refer to Faraday's law). Changing any of the three variables that generate electricity will also vary the depth of penetration. Increasing the number of turns (magnetism) in the coil would increase current flow, causing a decrease in the depth of penetration. However, the number of turns in the coil is fixed and generally cannot be changed. The only way to change this variable is to change coils.

Increasing the conductivity of the specimen causes more electrons to flow at a given frequency. That causes a stronger opposing magnetic field and forces the depth of penetration to decrease. If the conductivity and coil are kept constant (but the frequency is increased), that is the same as increasing the relative motion. More current flow causes a greater opposing magnetic field, which again pushes the primary magnetic field toward the surface. To summarize:

- Increasing the frequency on a given conductivity decreases depth of penetration.
- Increasing the conductivity at a given frequency decreases the depth of penetration.

The chart in Figure 10-3 shows the standard depth of penetration versus frequency for different types of materials.[4]

Frequency can be increased until *"skin effect"* is reached. This is where there is hardly any penetration at all and an extreme number of currents are flowing on the surface. As the eddy currents are forced

Figure 10–3

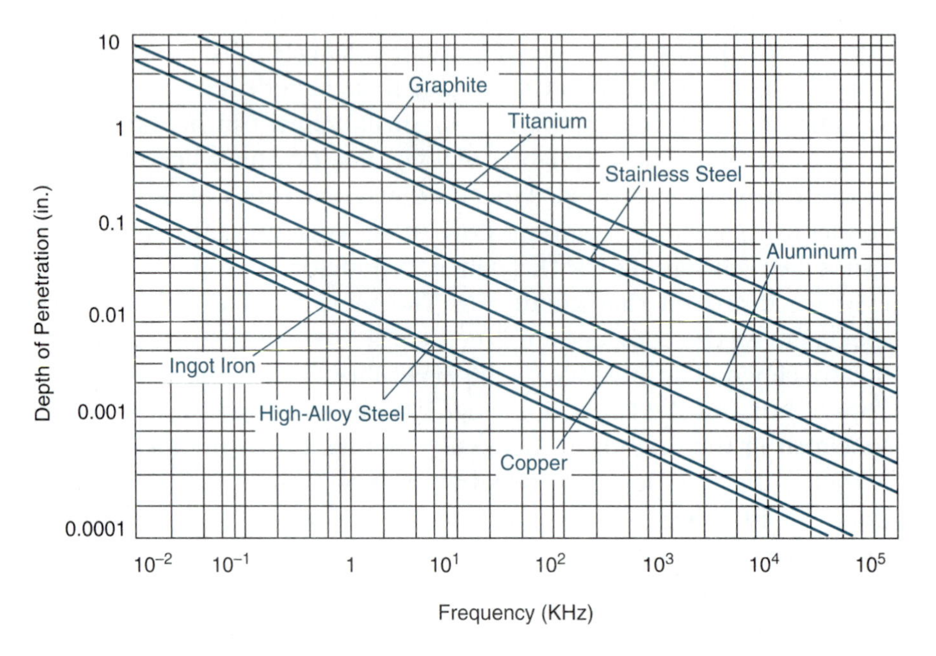

toward the surface, they become more concentrated. These magnetic fields are extremely powerful and exert a great influence on the primary coil. There is a much greater influence than if fewer electrons are flowing. This causes another characteristic: Less depth of penetration causes increased sensitivity. As a general rule, increasing the frequency will increase the sensitivity of the test. When the magnetic coil is moved further away from the test specimen, the number of eddy currents generated is reduced. This is called "lift-off." It becomes a factor when testing the thickness of nonconductive coatings (such as paint) on conductive material.

When the test coil comes near the edge of the test specimen, the flux field is again distorted. This distortion causes an apparent increase in resistance. The indication is false, however, and should be discounted by the technician. This "edge effect" makes it difficult to inspect the outer perimeter of the specimen.

10–4 TYPES OF COILS

There are three basic types of coils used in eddy current testing. There are many possible arrangements of these coils, but each of these arrangements uses one, or a combination of, these three types of coils:

Figure 10–4

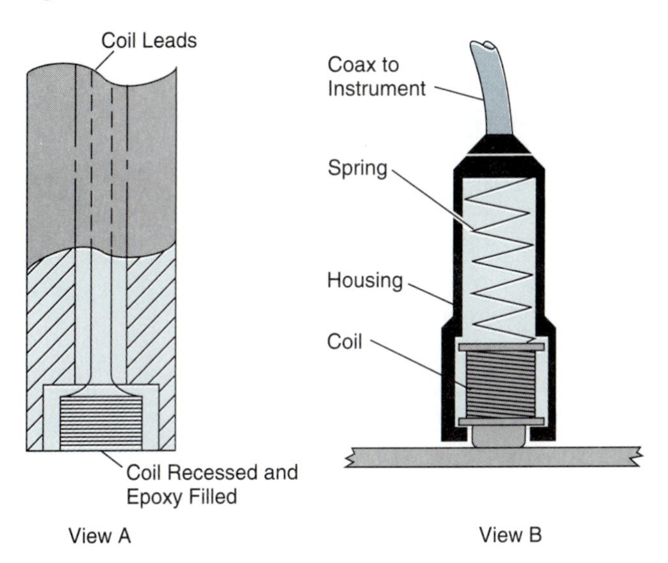

- Surface coil (probe)
- Encircling coil (through coil)
- Internal coil (bobbin coil)

A *surface coil* is often called a probe. It is usually built in such a way as to orient the centerline axis of the coil 90° to the surface of the material. This orientation of the coil offers the most useful flux pattern. Figure 10-4 shows the basic design of the surface probe. Figure 10-5 shows an adaptation used to inspect the inner surface of a hole.

There are many other physical variations of this probe: probes with a gap between the surface contacts, probes that keep the contact flush on the surface, and even probes that spin inside a hole. Companies that make these probes take great pride in being able to engineer any type of probe needed. One should expect to pay a handsome price for a specially made probe.

Encircling coils are wound around the outside of the pipe, tubing, bar stock, and wire. They are used to locate outside surface discontinuities. The coils may be wide or narrow (Figure 10-6). The wide coil can locate gradual changes such as thickness changes. Narrow coils pick up rapid changes such as cracks and voids. This type of coil centers the specimen in the coil and, as it passes through the coil, inspects the specimen without ever touching it. It is an extremely sensitive device. Many of these devices are connected to processing equipment and can even isolate and remove defective material without slowing the process.

Figure 10–5

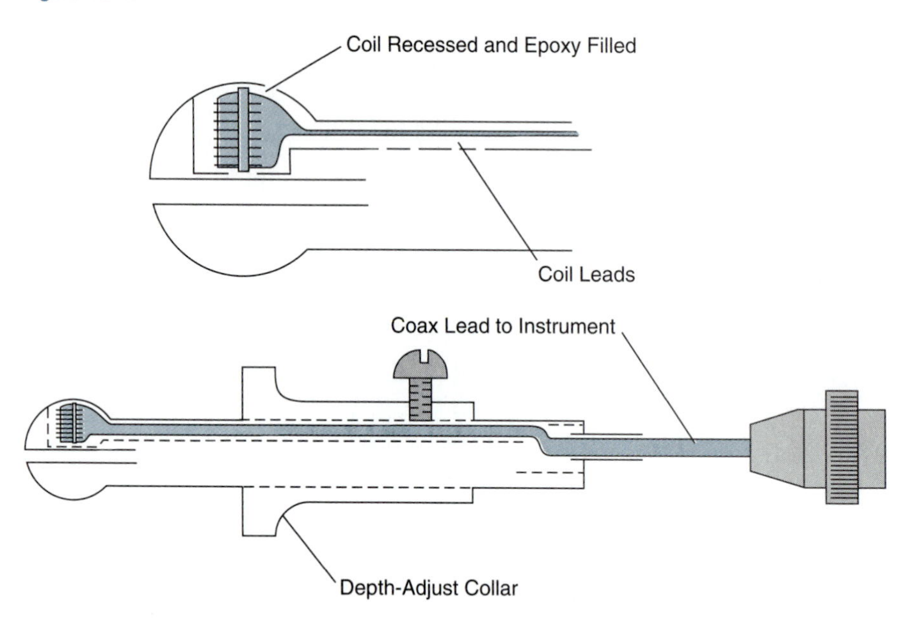

Inside-diameter coils are often called *bobbin* or *internal coils.* They are inserted into tubing or pipe to inspect the inside surface. This type of coil does not lend itself to high-speed applications. The coil has a cable attachment that requires the coil to be retracted. The amount of space between the outer surface of the bobbin coil and the inner surface of the tubing is critical. The relationship of these two diameters is called the "fill factor."

The *fill factor* is a critical parameter with both the encircling coil and the inside-diameter coil. It is expressed by the formula D1/D2.[5] In the encircling coil, D1 is the outside diameter of the tubing and D2 is the inside test coil. The bobbin coil requires a different set of values. D1

Figure 10–6

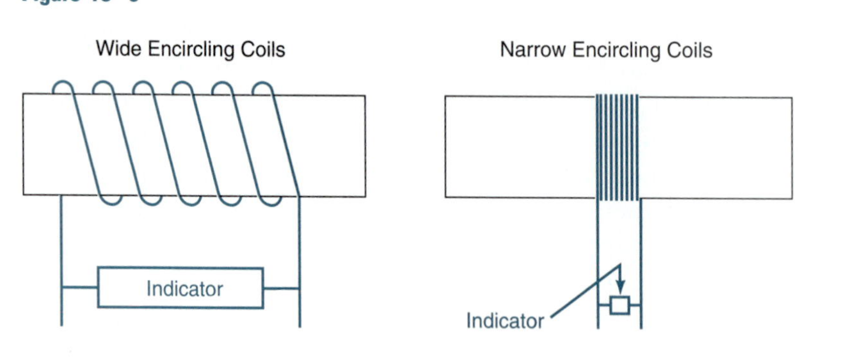

Figure 10–7

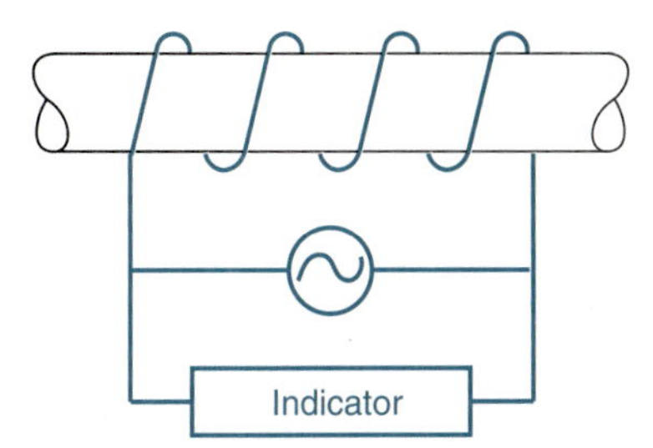

is the outside diameter of the coil, whereas D2 is the inside diameter of the tubing. In either case, the fill factor must be less than 1. If the fill factor is 1, the diameters will be the same size, and the coils will not be able to move. The fill factor should be as close to 1 as possible, yet still allow free movement of the coil.

10–5 COIL ARRANGEMENTS

Single and Double Up to this point only the single coil has been discussed. The single coil both sends the magnetic flux field into the test article and senses changes in the eddy current flow. It is the only coil used in the inspection. The double coil, on the other hand, has two separate coils. One is connected to the power supply and is used to develop the eddy current within the test article. The second is a secondary coil connected to a meter movement. This secondary coil monitors the changes in power from the primary coil caused by specimen changes. Figure 10-7 shows the single coil; Figure 10-8 shows the double coil.

Figure 10–8

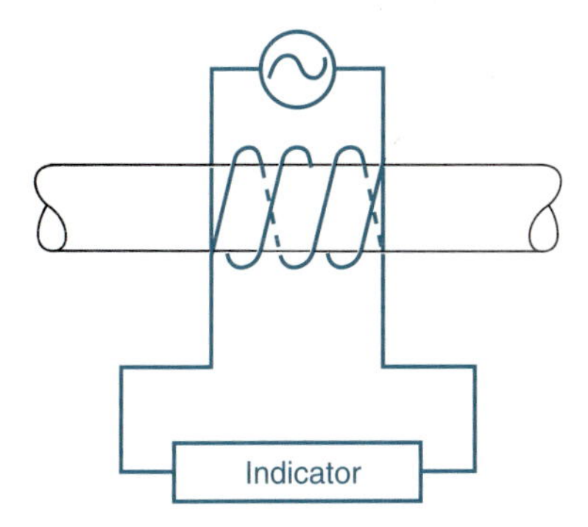

The single- and double-coil arrangements are used on all three of the coil types previously discussed. The coil arrangement is determined by the equipment used.

Absolute and Differential The absolute coil is a coil that gives one response for each discontinuity. It is a coil of wire wound in one direction. The differential coil gives two responses for each discontinuity. It does this because it is first wound in one direction, and then wound the same number of turns in the other direction. This causes an indication on the meter of a specific magnitude in one direction as the defect passes under the coil wound in the clockwise direction. As the defect passes under the coil wound in the counterclockwise direction, there is an indication on the meter of equal magnitude in the opposite direction. This makes the indication more seeable. However, it does not increase the sensitivity of the test.

It is possible to mix and match these coil arrangements to achieve the desired signal on a specific type of indicator. For example, one might find a double differential coil on a machine that could interpret those signals.

It is extremely important to have a reliable *standard* when using eddy current. As in most types of NDT, there must be an acceptable standard to compare against the test specimen. Many machines today use small internal computers to set the values of test parameters. Therefore, the machine must contain a good set of parameters. The standards are essential to the validity of the test.

In some cases the standard is used as a comparison while the test is being done (e.g., in the use of reference comparison testing [Figure 10-9]). This test is done on a test article at the same time it is being done on a standard. As long as both legs of the test circuit are balanced, there is no indication on the meter. If the legs become unbalanced, it is because of a change in the test article. It is easy to see that, if there were a flaw in the standard, it could be very costly in causing good parts to be scrapped.

It is also possible to use the test article itself as a standard. For example, if only the variation in tubing is to be recognized, a self-referencing system could be used. As long as there are no great changes in the tubing from one coil to the other, there is no imbalance. If an imbalance does occur, it is because of a fluctuation in a critical variable.

Critical variables may be the conductivity of the material, dimensions of the material, or permeability of the material. Changes in any of these areas cause a shift in the phase angle of the test equipment's

Figure 10–9

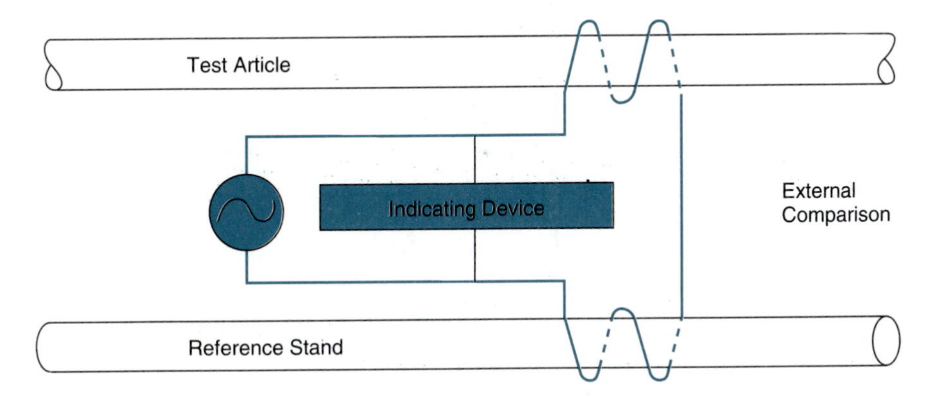

circuits. Each phase angle shift is like a fingerprint. A shift in the phase angle caused by a change in conductivity will always look the same. The key to accurate eddy current testing lies in the training of the technician. He or she must be able to identify these fingerprints and must know how to set up the equipment to maximize the desired indications while filtering out irrelevant indications (*noise*). The test can tell you everything about the specimen—but without a well-trained technician, you will not be able to sort the information out.

Summary

Eddy current testing is a method of nondestructive testing that inspects for surface and near-surface flaws. This method of testing is limited to conductive materials such as copper, stainless steel, aluminum, and tin. Ferromagnetic materials may be difficult to inspect using this method.

Various coil arrangements are used to induce a circulating current into the test specimen. These circulating currents exhibit in turn their own magnetic field, which reacts to the magnetic flux from the coil. Changes in the test specimen cause a phase angle shift in the test equipment that is recorded as a flaw. This method can detect changes in thickness in nonconductive coatings or in the conductivity of the specimen, cracks in the specimen, or thickness changes in thin-walled specimens.

This method is adaptable to continuous process manufacturing. It gives extremely fast responses and can send these signals to a computer in the process. ET equipment is generally expensive, and training for this method is extensive. The technician is critical in this method of inspection. He or she must be able to set up the equipment to filter out irrelevant information and maximize the signal of suspected flaws.

Key Terms

Conductive The ability of a material to carry an electrical current.

Differential coil A coil arrangement using wire wrapped first in one direction and then in the opposite direction.

Electromagnetic Magnetism produced by passing electrical current through a coil wrapped around a core material.

Encircling coil A coil with its axis wound parallel to the surface of the specimen and used in checking the outer surface of thin-walled pipe and tubing. This coil is sometimes called a through coil.

Fill factor An expression of the amount of space between the inner wall of tubing and an internal or encircling coil, which affects the efficiency of the test.

Impedance Total opposition to current flow in an ac circuit.

Internal coil A coil with its axis 90° to or parallel to the inner surface of tubing or thin-walled pipe; sometimes called a bobbin coil.

Noise Signals picked up from irrelevant discontinuities.

Sensitivity The ability to pick up indications from a very small flaw.

Skin effect The condition of having an extreme concentration of eddy currents near the surface with very little depth of penetration.

Standard A specimen that is flawless; used to compare with test specimens.

Surface coil A coil with its axis oriented 90° to the plane of the surface, usually hand held; sometimes called a probe.

EDDY CURRENT TEST

1. List the advantages and limitations of eddy current testing.
2. What is the principle of operation employed in eddy current testing?
3. What three things are required to generate electricity?
4. Why does a defect in the conductive material cause an indication on the eddy current machine?
5. Define *impedance.*
6. At a given frequency, as conductivity increases, what happens to the depth of penetration?
7. How does increasing the frequency affect sensitivity?
8. What is skin effect?
9. What is edge effect?
10. List the three types of coil.
11. What application would employ an encircling coil?
12. What is the purpose of the secondary coil in a double-coil system?
13. What is the differential coil?
14. Why is a reference standard important?
15. What are the critical variables monitored in eddy current testing?

NOTES

1. *Nondestructive Testing: Eddy Current Testing* (CT-6-5), 2d ed. General Dynamic Convair Division, 1979.
2. *NDT: ET.*
3. *NDT: ET.*
4. *NDT: ET.*
5. *NDT: ET.*

ULTRASONIC INSPECTION

11

OBJECTIVES

After completing Chapter 11, the reader should be able to:

- Define piezoelectric effect.
- Distinguish between immersion and contact testing.
- Name the parts of the test unit.
- Discuss the purpose of a couplant.
- Discuss equipment calibration.
- Discuss the basic principle of ultrasonic testing.

The use of sound to identify the presence of objects and to measure distance is not a new concept. From the time of tall ships it has been common practice during times of poor visibility to make a noise (shooting a cannon, ringing a bell, or sounding a fog horn) and listen for an echo to determine whether the ship was close to land. The time between the noise and its echo was measured, and the distance from land could be calculated. This distance was found using a simple formula: the time it took to hear the echo, multiplied by a constant feet per second for sound to travel through air. This number was then divided by two, since the sound had to travel to the land and back to the boat.

In this example, all that the sailors had to know to calculate the distance was the time it took for the sound to travel to land and echo back to the ship, and the approximate speed of sound in air. If the speed of sound in air is 1000 feet per second and it took six seconds for the sound to return to the point of origin, then the ship was 3000 feet from land. The sound took three seconds to travel the 3000 feet to the land and the echo took three seconds to travel the same distance back to the ship. This method enabled navigators to use sound echoes to measure distance and discover the presence of land.

In *ultrasonic* testing the same principle is used to measure thicknesses of material and to find defects hidden inside a part. This is accomplished by using sound transmitted into a part and measuring the time it takes for the sound to reflect off the back side of the part or to reflect off a discontinuity hidden within the part. The sound source used in ultrasonic testing (UT) has all the same properties as the sound that people can hear. However, it is transmitted at a very high frequency (higher than a person can hear), and the return echo is captured and the result displayed on an oscilloscope screen (Figure 11 - 1).

Figure 11–1

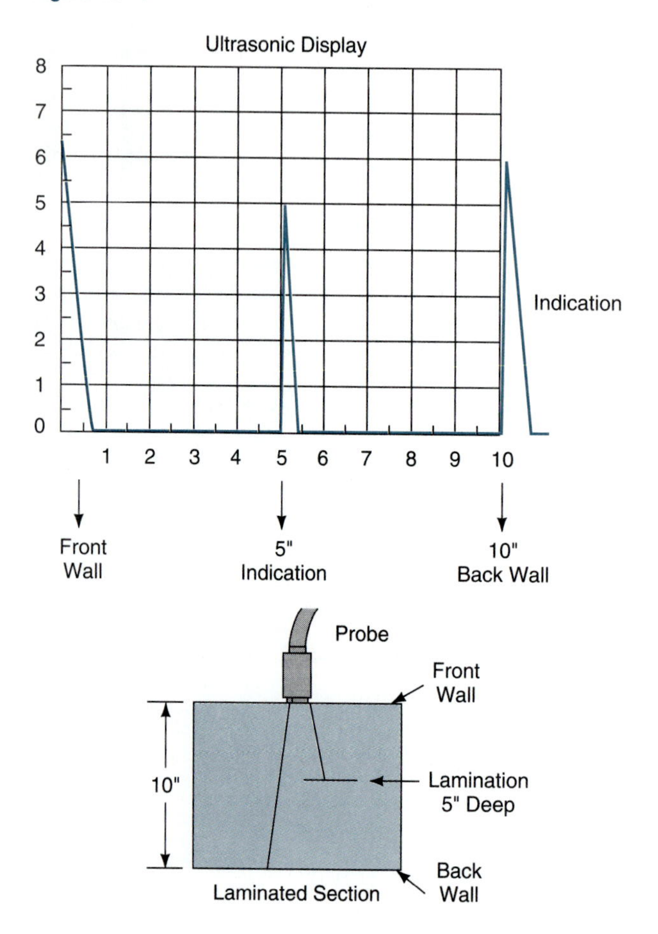

This figure shows a part that is 10 in. thick with a crack-like lamination 5 in. deep inside it. If the horizontal scale measures in inches, the spike at the 5 in. mark is indicating that the sound hit something 5 in. deep in the specimen that caused the sound to reflect (echo) back to the UT machine. The spike (or pip) at the 10-in. mark indicates the sound hit the back wall and reflected back. The principle is exactly the same as a ship blowing a horn to find land, except that the equipment is designed to be much more precise.

This chapter discusses the basic physics that are used to perform UT, followed by the inspection methods. It provides the reader with an understanding of the basics of ultrasonic inspection. For a quality professional to gain a working knowledge of UT, practice is required, along with an in-depth understanding of this type of NDT.

Figure 11–2

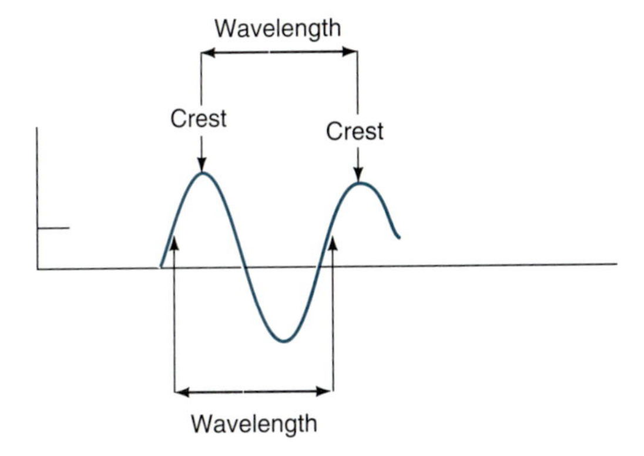

11–1 PRINCIPLES OF SOUND TRANSMISSION

Sound is a form of energy that can be transmitted through (or passed along) a *medium.*[1] The medium might be air, water, steel, glass, or anything that has density.

Density is defined as mass per unit. This definition refers to how tightly the *molecules* from which the part is made are held together. A part with a high mass has molecules that are held closely together. The denser the material, the better the sound energy is transmitted.[2]

One other physical property of a medium that affects sound transmission is *elasticity.* Elasticity is defined as how much force is present in the medium holding the molecules together. If a high force is present, the part is elastic. If a low force is present, the part is inelastic. The more elastic the part, the better sound is transmitted.

Of the two properties (density and elasticity), elasticity is the more important when it comes to sound transmission. For example, sound moves at a rate of 2.56×10^5 in. per second through 2117 aluminum, whereas it travels at a speed of 2.32×10^5 in. per second for steel. This means that aluminum, although less dense, is more elastic. On the other hand, if several media had the same elasticity, the medium that had the highest density would transmit sound faster.

The speed (or velocity) that sound is transmitted through a medium also depends on the *wavelength* of the sound wave. Sound is transmitted through a part in such a way that it can be shown as a normal sine wave. The wavelength is a measurement from crest to crest (Figure 11–2).

Wavelength is determined by the frequency of the sound and by the velocity of the sound. As was discussed earlier, the velocity of the sound is determined by the medium the sound is passing through. The wavelength is affected by many factors. It becomes important to be able to determine the wavelength when performing UT because the shorter the wavelength, the smaller the defect that can be located. One problem exists when using short wavelengths—the sound will not penetrate as well. When performing UT, the general rule of thumb is to use a wavelength that is twice the size of the maximum allowable defect of interest. The reflection from the defect will show on the screen, and the sound will still penetrate the medium well.

Determining the wavelength is not difficult. Standard charts have been developed that give the velocity of sound through different materials, and the part of the UT equipment that produces the sound (the transducer) comes in different frequencies. The wavelength can be calculated by dividing the velocity of sound through a material by the frequency of the sound. This information is discussed later in this chapter and will be used to determine the proper transducer for a given inspection.

11–2 EQUIPMENT

The equipment used to perform an ultrasonic inspection can be divided into the following three groups:

- The transducer (the unit that makes the sound wave and receives the returned echo)
- The transmitter/receiver (the machine that generates the impulse and reads the returned sound wave)
- The display (the screen) that shows the indications

In this section we provide generic information about typical equipment in use today.

Transducer The transducer is the part of the UT equipment that sends the sound into the specimens and listens for the return sound wave. These units are also known as search units or *probes.* The probe (transducer) is designed to convert electrical impulses sent from the transmitter/receiver into mechanical energy (sound). This sound is induced into the part being tested. When the sound hits something in the part, some of the energy is returned to the probe in the form of an echo. The probe converts this mechanical energy (the echo) back into an electrical impulse that is sent to the receiving side of the UT electronics.

Figure 11–3 Longitudinal Wave Contact-Type Search Unit

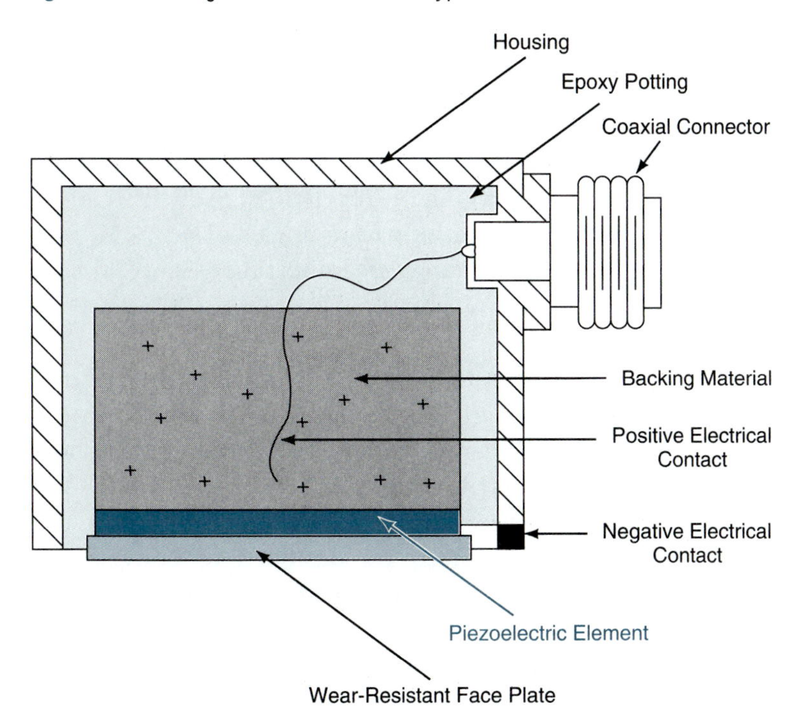

To convert electrical impulses into mechanical energy and mechanical energy into electrical energy, a *piezoelectric* material is used. Piezoelectricity is sometimes called pressure electricity.[3] It is the unique characteristic of some materials to expand in physical size when an electrical voltage is applied. When the voltage is removed, the material returns to its normal size.

This expansion in the presence of voltage is how electrical energy is converted into mechanical energy. These materials have a second unique characteristic. If the material is squeezed or put under pressure, it will produce an electrical impulse. This is how the mechanical energy (the echo) is converted into an electrical impulse. Materials such as lithium sulfide, quartz, and polarized ceramics possess this characteristic.

The typical probe is constructed as shown in Figure 11-3. The transducer has a housing that, in many cases, is the negative contact for the piezoelectric element. It contains an electrical conductive backing material that is in contact with the piezoelectric element and the positive electrical contact. All these elements are held in place with epoxy potting. A wear-resistant face plate is attached to the piezoelectric element. This is to prevent the element from wearing as the probe is slid over the surface of a part being inspected. All probes have some type of electrical connector (uhf, bnc, lemo, microdot, etc.), which transfers the electrical impulse to and from the transducer.

Transducers come in different physical sizes (from around 0.250 in. in diameter and larger) and different frequencies (from 0.5 MHz to over 25 MHz). Some are designed to send and receive sound; some are designed only to send and others only to receive. Transducers can be designed to send sound waves into the part at angles of 30°, 45°, 60°, and 70°. These probes are used for angle-beam testing (discussed later in this chapter).

Transmitter/Receiver The transmitter/receiver consists of electronics that are designed to generate an electrical impulse of the proper frequency to allow the transducer to generate the sound wave and induce it into the part. When the sound echo is received by the transducer, an electrical impulse is sent to the electronics in the transmitter/receiver. These electronics are designed to convert the electrical signal received from the transducer to a signal that is displayed on the cathode ray tube (CRT).

For the electronics to properly display the information that is being received from the transducer, the machine (transmitter/receiver) has several controls. These controls are used to adjust the screen display. They can be grouped into three categories:

- Display controls
- Pulse controls
- Timer (pulser/receiver) controls

Display Controls The primary purpose of this set of controls is to adjust the display positioning and viewability on the screen. Properly adjusted, the display line is easy to see and positioned properly on the screen. The display controls that affect the screen presentations are the following:

- Vertical and horizontal controls
- Intensity control
- Focus control
- Astigmatism control
- Power and scale illumination control
- Repetition rate control

The *vertical and horizontal controls* allow the inspector to move the baseline indication on the display grid. This adjustment is needed to

Figure 11–4

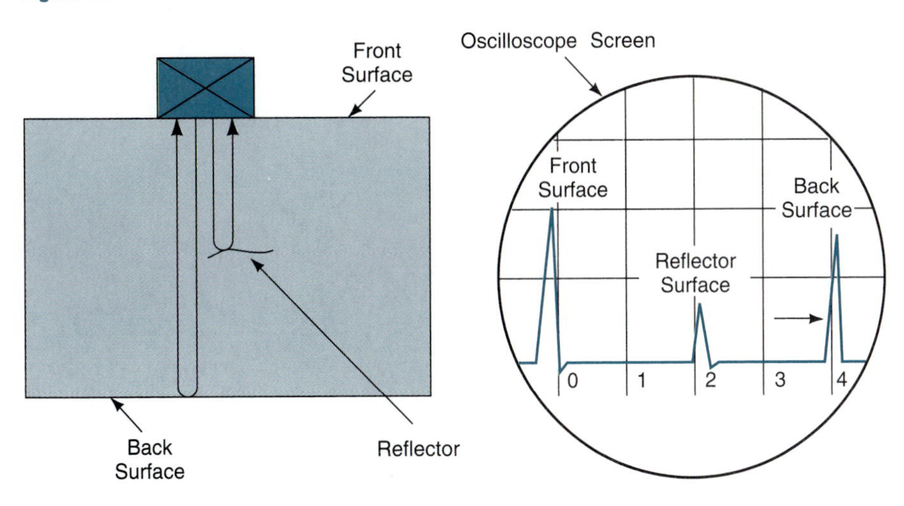

locate the display line horizontally and vertically on the grid. The *intensity control* adjusts the brightness of the baseline compared to the background. This provides proper contrast between the line and the background. The *focus control* adjusts the sharpness of the baseline. The *astigmatism control* corrects the distortion that may occur as the electron beam is redirected across the face of the CRT. The distortion is caused by the transit time of the electron beam across the screen, which could result in the baseline becoming distorted. The *power and scale illumination control* turns the power on and off and also adjusts the contrast of the edges of the display. This adjustment provides contrast between the display line and the display background. The *repetition rate control* governs the cycle time of the transmit and receive impulses.

Pulse Controls The pulse controls have an effect on the pips or spikes that indicate that a return echo was received by the transducer. These are caused by the initial pulse, the front surface of the specimen, the back wall of the specimen, and any reflector in between. All of these cause a reflection or echo to be returned to the transducer (Figure 11-4).

The controls consist of the following:

- Sweep delay
- Sweep length controls

The *sweep delay control* moves the baseline horizontally on the screen. This adjustment allows the inspector to align the pip indicating

Figure 11–5

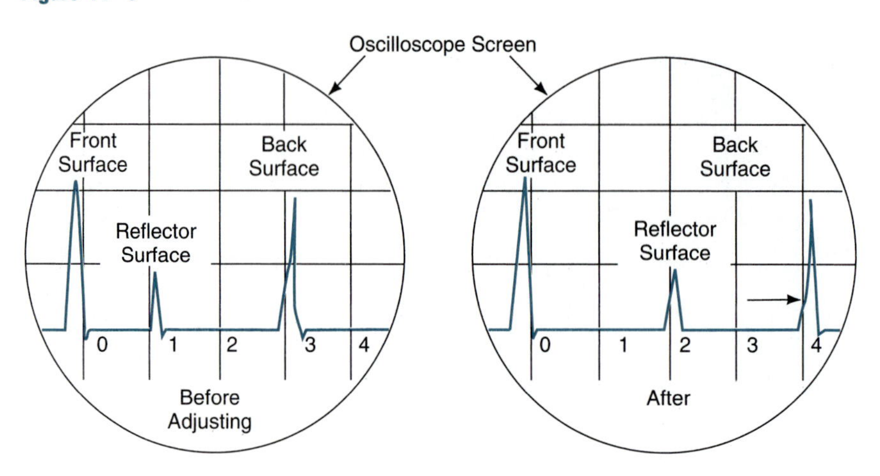

the front surface with the zero mark on the screen grid (Figure 11-5). The *sweep length control* allows the inspector to stretch or shorten the baseline to move the pip indicating the back wall to the proper grid location. Figure 11-5 shows two screen displays. The first display indicates the sweep delay has been used to place the front surface pip at the zero location. The back surface indication is at the three-inch mark. For this example, suppose the specimen that is being tested is actually 4 in. from the front surface to the back wall. The sweep length control would be adjusted to lengthen the baseline so the back wall pip would show at the proper division.

Timer (Pulser/Receiver) Controls The last set of controls is the pulser/receiver controls. These controls allow the inspector to fine tune the electronics with the selected transducer. They also allow the inspector to select the type of inspection that will be performed. These controls consist of the following:

- Pulse length control
- Pulse tuning control
- Reject control
- Mode select control
- Frequency control
- Sensitivity (dB gain) control

The *pulse length control* sets the duration of the electrical pulse being applied to the transducer. The *pulse tuning control* adjusts for slight electrical differences between the transducer and the cable connecting it to the electronics. The *reject control* is used to eliminate very low amplitude spikes that are usually caused by the grain structure in the metal. They are not associated with flaws and are referred to as "grass" on the screen. The reject control adjusts the screen so these indications do not appear. The *mode select control* is used to select the type of technique used for testing. Typically, the methods include pulse echo, through transmission, or pitch catch. The *frequency control* sets the electronic frequency to the same frequency as the transducer. The *sensitivity* or *dB gain control* adjusts the amplification of the signals that are sent and received from the receiver/amplifier.

The abbreviation *dB* stands for *decibel,* a measurement that describes the logarithmic ratio between the height of two echoes. Although the effect of changing the gain can be mathematically proved by a formula, for the purpose of this text we will use the general rule that adding 6 dB doubles the height of an indication. Likewise, if 6 dB are removed from an indication, the height will drop by 50 percent. In other words, if the distance of a reflecting surface from the transducer is doubled, the indication will be reduced by 6 dB and the indication height will be reduced by one-half.

The purpose of these controls is to set the screen display so that the screen indications fairly represent the part being tested. Adjusting the equipment to a part of known size and indications is called *calibration.* Calibration is the process by which the signal and the indications are adjusted to show accurately the indications of a block with known defects. Calibration is discussed in greater detail later in this chapter.

Display Most displays used in UT are much like a television screen. The display is known as a *cathode ray tube* (*CRT*). The CRT is made up of a glass tube that has been evacuated of air. In other words, it is a vacuum tube with a screen coated with *phosphor* on one end. The other end contains an *electron gun,* which emits a stream of *electrons* that are accelerated and directed across the tube to hit the screen. When the electrons hit the phosphor-coated screen, the phosphor emits light that can be seen by the human eye. The electron beam is directed across the screen to display the indications by horizontal and vertical deflector plates. Figure 11-6 is a diagram of a typical CRT.

Most CRTs have a grid over the face of the screen. This grid allows the indication (baseline) to be properly interpreted. The grid is usually divided into 10 sections; some CRTs subdivide each of the 10 divisions (Figure 11-7).

Figure 11-6

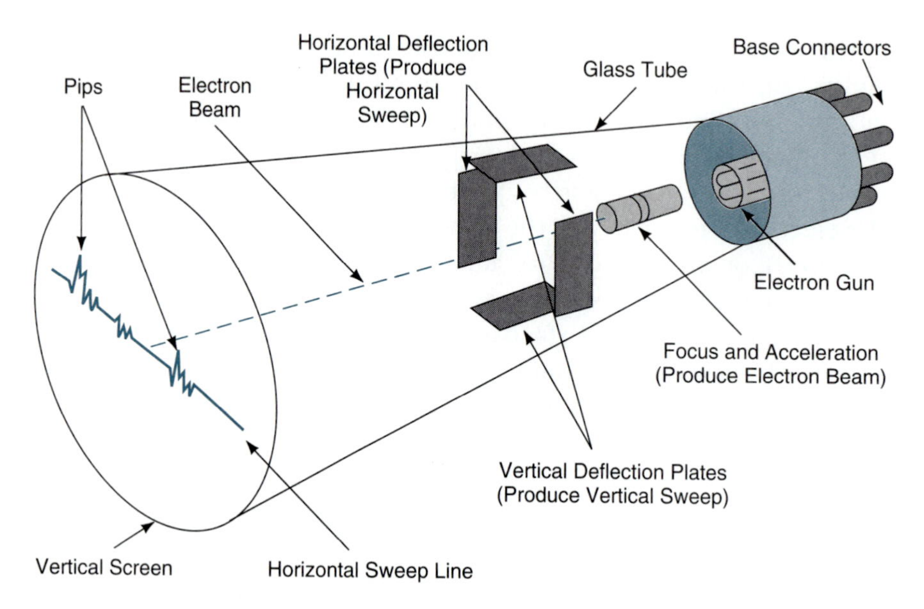

CRT displays are of two types. The first type is a linear display. This type of display has a linear relationship between the signal voltage and the echo's signal height. As discussed earlier in this chapter, the echo height will decrease by 50 percent if the distance is doubled.[4] With this type of display, each repeated back wall will be 50 percent as high as the previous indication, as shown in Figure 11-8.

Figure 11-7

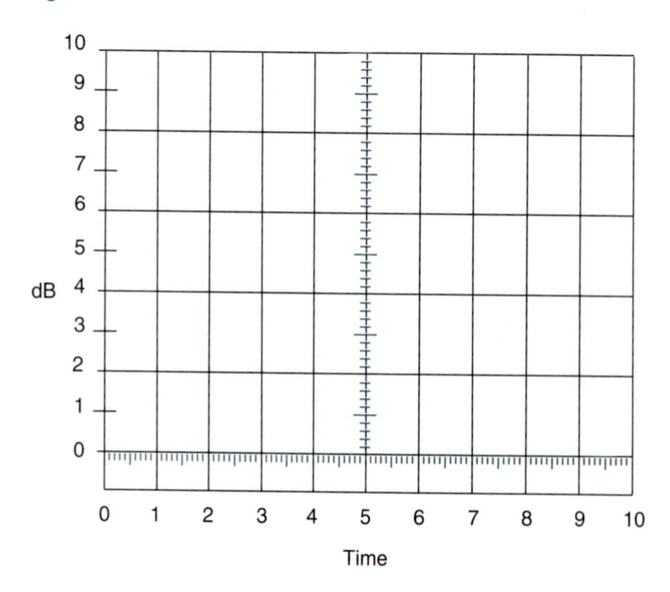

Figure 11–8

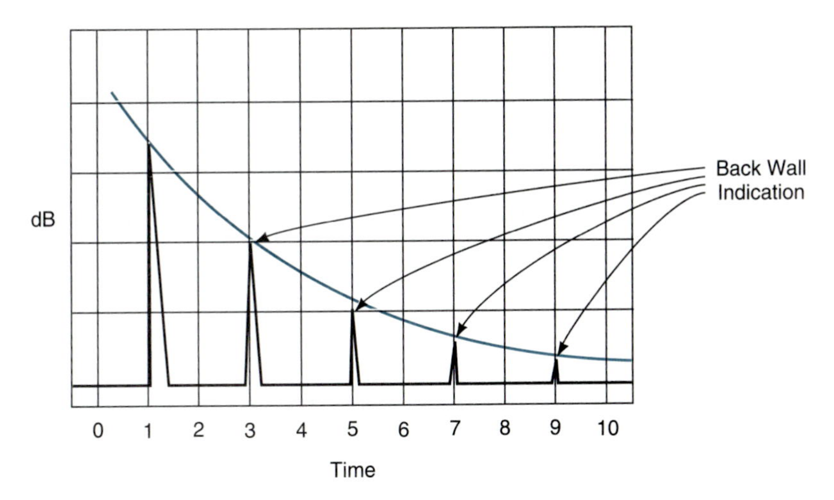

This denotes that each back wall delivers 50 percent of the voltage compared to the previous indication. If the first back-wall indication delivered 10 volts, this display shows that the second back-wall indication delivered 5, the third delivered 2.5, and so on. On this type of display, the amount of voltage returned to the transducer is proportional to the height of the indication. The amount of power the echo returns is directly proportional to the type of reflector, the size of the reflector, and the distance the reflector is from the transducer.

The second type of display is a logarithmic display. This type of display differs from the linear display by the way repeated back walls are indicated. There is a logarithmic relationship between the signal voltage and the echo height. This means that the echo height is reduced by a given distance every time the signal voltage is reduced by 50 percent. Each repeated back wall is reduced by the same amount.

The difference between the two displays is that, in the linear display, the difference between two back-wall indications is the difference in the signal voltage. In the linear display, the difference between two back walls is 6 dB. The type of display used is at the discretion of the inspector. The inspector should be familiar with the different displays, because the same indication on the two different displays looks different. Inspectors need training to interpret each type of display.

11–3 SCREEN PRESENTATIONS

The screen presentations in UT equipment range from electronic spikes on a screen (A-scan) to the type that provides a picture resembling an

Figure 11–9 A-Scan

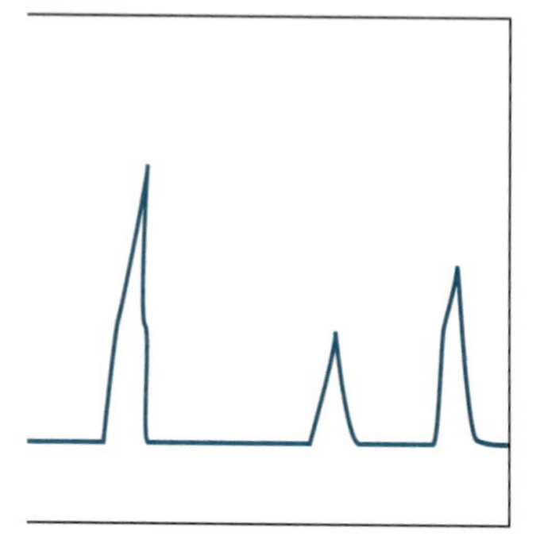

X-ray (C-scan). The displays are called A-scan, B-scan, and C-scan, and the resolution increases from A to C.

The A-scan presentation is displayed as a spike (pip) on a CRT display. This type gives a time versus amplitude display that is either a linear or logarithmic display (Figure 11 - 9).

The B-scan produces a cross-sectional view of the part being tested. It requires a special CRT and is used primarily in medical applications (Figure 11 - 10).

The C-scan gives a picture much like an X-ray, showing the outline and the discontinuity of the part being tested. In most cases the C-scan produces a permanent record on paper or on magnetic tape.

For through transmission, or pitch/catch, transducers are placed on each side of the part. The pitch transducer sends the sound into the part while the catch transducer listens for the sound (Figure 11 - 12). As the sound is picked up by the catch transducer, the amplitude of the sound is shown on the screen. If a discontinuity is located, some or all of the sound is reflected back to the pitch transducer and less is passed

Figure 11–10 B-Scan

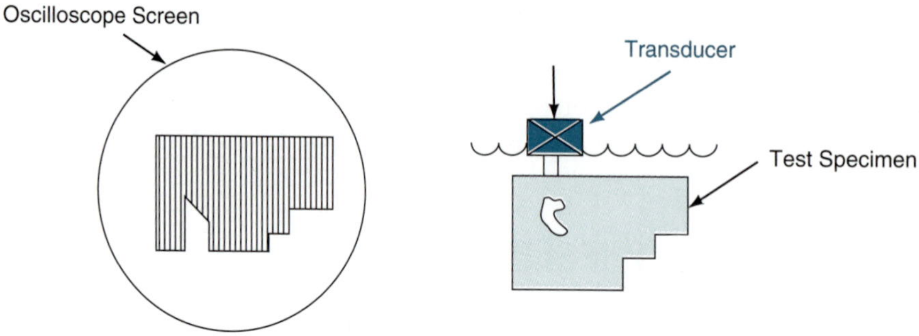

Figure 11–11 (a) C-Scan Principle of Operations (b) Typical C-Scan Recording

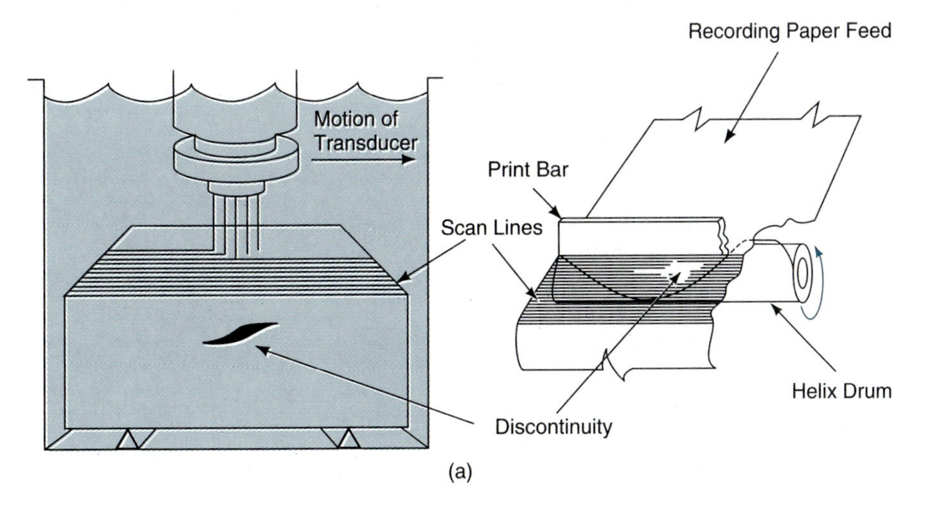

(a)

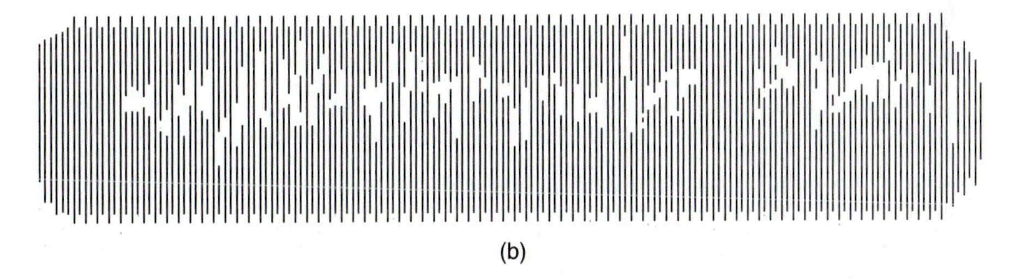

(b)

on to the catch transducer. This reduces the amplitude of the sound being received. This is indicated by a drop in the indicated amplitude on the screen.

A second form of pitch/catch is when both the sending and the receiving transducers are on the same surface (Figure 11–13). The sound generated by the sending transducer is sent into the part, and the echo is received by a receiving transducer some distance away from the first. This type of testing has many of the features of through

Figure 11–12

Transmit

Receive

Through Transmission

Figure 11–13 Straight-Beam Pulse/Echo Contact Using a Pitch/Catch or Dual-Element Transducer

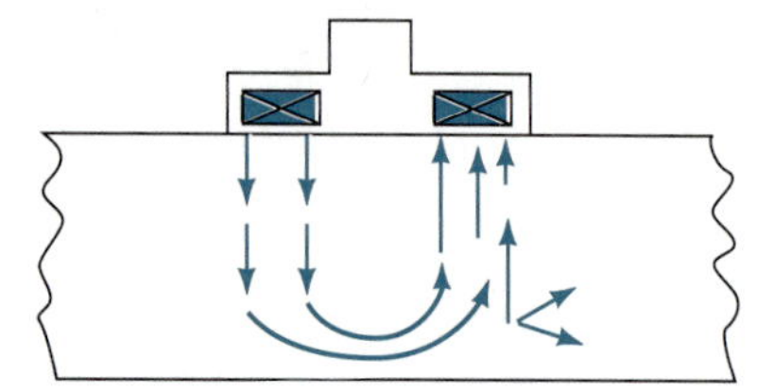

transmission, but gives the same type of indication as in the pulse/echo technique. The amount of reflected sound is indicated on the screen.

11–4 TESTING TECHNIQUES

Ultrasonic testing of a part consists of many steps that require the close attention of a trained inspector. The result of a properly performed UT inspection is a highly accurate, detailed description of the internal condition of a part. The testing methods are generally a type of contact testing or immersion testing. In contact testing, the transducer is, as the name implies, in direct contact with the part. In immersion testing the transducer is held some distance from the part and the sound is transmitted through water to and from the part. The method of choice depends on the part's shape and type and the availability of test equipment. Each of the methods is highly accurate when properly conducted.

Contact Method In contact testing, the transducer is in direct contact with the part being tested. The techniques used are straight-beam, angle-beam, or surface-wave testing. Each of the three testing methods has unique and specific uses. Each is discussed in the following section.

Straight-Beam Testing In straight-beam testing the sound energy is transmitted in a straight line into (pulse/echo) or through the part (through transmission or pitch/catch). In the pulse/echo technique, sound is induced into the part by a transducer. As the sound hits a discontinuity and the back wall of the part, the energy is reflected or echoed back to the transducer. The echo is picked up by the transducer and the screen indicates the echo pattern.

Figure 11–14 Angle-Beam Testing of a Weld

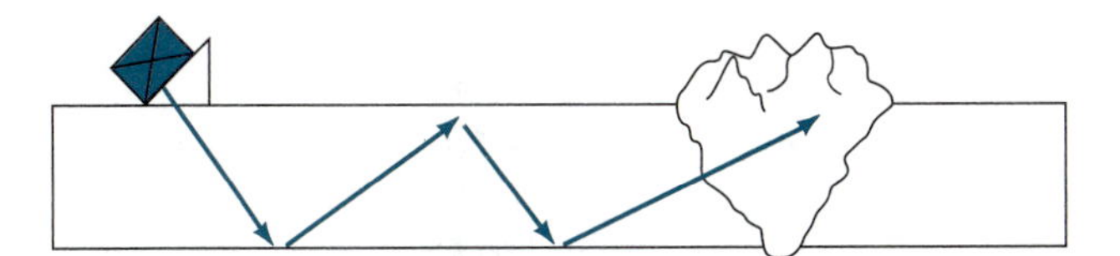

Angle-Beam Technique For parts such as weldments, where it is difficult or impossible for the transducer to be positioned over the area to be tested, the sound can be bounced into the desired test area using the angle-beam technique. Sound is induced into the part at an angle that allows the sound to travel in a zig-zag direction, reflecting from side to side until a discontinuity is hit, at which time the echo reverses direction and returns to the transducer.

The indication is the same as for a straight beam, with the exception that the angles must be calculated to determine the path the sound traveled.[5] The sound will have traveled further to get to the discontinuity than the discontinuity's distance from the transducer (Figure 11–14).

This form of testing is popular for testing welds and parts with rough or irregular surfaces. Determining the exact location of the defect becomes an exercise in math to determine the path the sound has traveled within the part. Once the path is plotted, the exact location of the defect can be determined.

Figure 11–15 Surface-Wave Technique

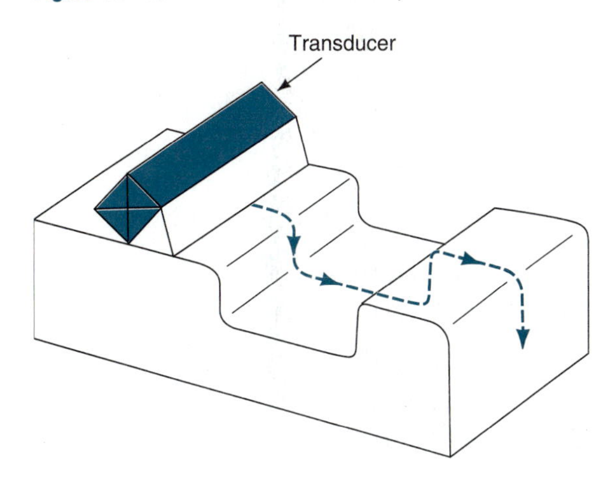

Figure 11–16

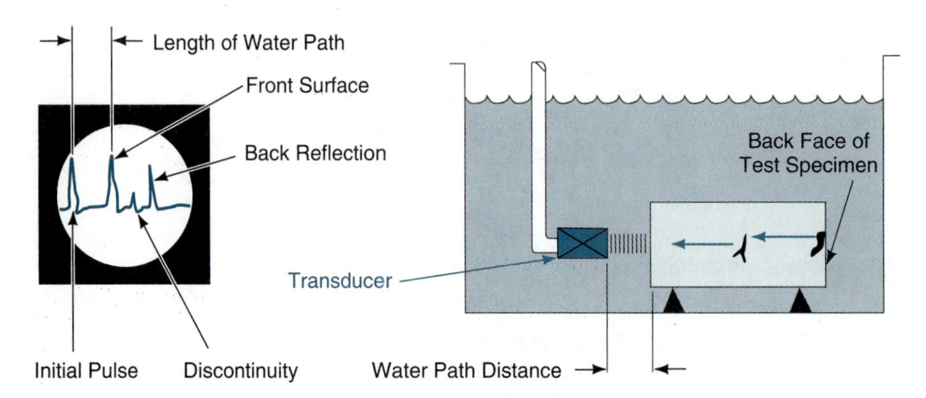

Surface-Wave Method In many applications a part must be tested for surface discontinuity. For such discontinuities to be identified, the sound must be transmitted along the surface of the part instead of being transmitted into the part (Figure 11 - 15). This *surface wave* (also called a Raleigh wave) is induced by a special transducer to travel along the surface of the part.

The surface wave shows defects in the same way as straight-beam and angle-beam testing. This method of testing is highly effective, although it requires a high level of skill to produce consistent and accurate results.

Immersion Method Immersion testing is a technique by which the sound waves produced by the transducer are carried to the part and the echo is returned from the part to the transducer through water. This means that, in some cases, the part is submerged in a tank of water along with the transducer, and the sound is sent to the part through the water (Figure 11 - 16).

In other cases the sound is passed through a column of water (like water coming out of a hose) and returned to the transducer through the same column of water. This equipment is known as a squirter or bubbler system (Figure 11 - 17).

The third method is known as the roller method. The transducer is contained in a roller (wheel) that is full of water. The wheel is rolled along the part, and the sound passes through the water to the part and is returned to the transducer through the wheel (Figure 11 - 18).

Figure 11–17 Bubbler Technique

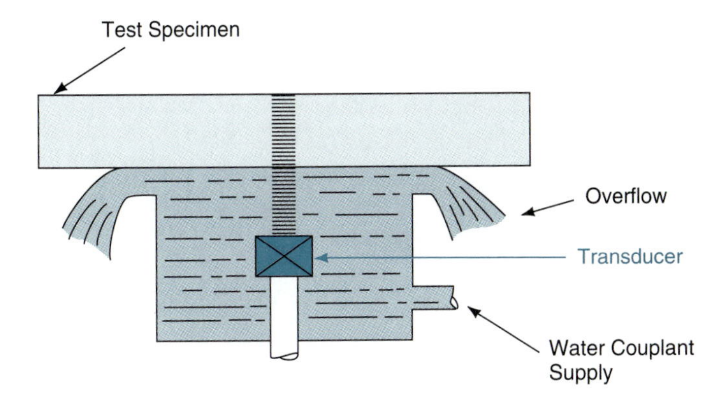

11–5 TESTING PROCESS

The testing process starts with selecting the method to be used for the test. The proper equipment must be selected and calibrated, and then the test can be accomplished. Each of these steps will be explored.

Selecting the Method As previously discussed in this chapter, two basic methods for testing can be used:

- Contact method
- Immersion method

Both the contact method and the immersion method can use either straight-beam, angle-beam, or surface-wave techniques. For this reason the technician must first select between contact testing and immersion testing. This choice is usually dictated by the part. Obviously, the part size, in comparison to the test equipment, could dictate that contact testing must be used. If the part is larger than the immersion tank or would not fit in the squirter fixture, then contact testing will be used. Contact testing is more commonly used because the equipment is less expensive and is highly portable. This type of testing method is accurate and flexible. Indeed, the main advantage is its flexibility. It can be used almost anywhere a surface of the part can be accessed.

Immersion testing is commonly used for production testing of parts that fit into an immersion tank or fit the fixtures of a squirter setup. The advantages of this type of testing are that the coupling between the transducer and the part is much more complete using water, and that this method is conducive to automated testing. With immersion testing

Figure 11–18 Wheel-Transducer Technique

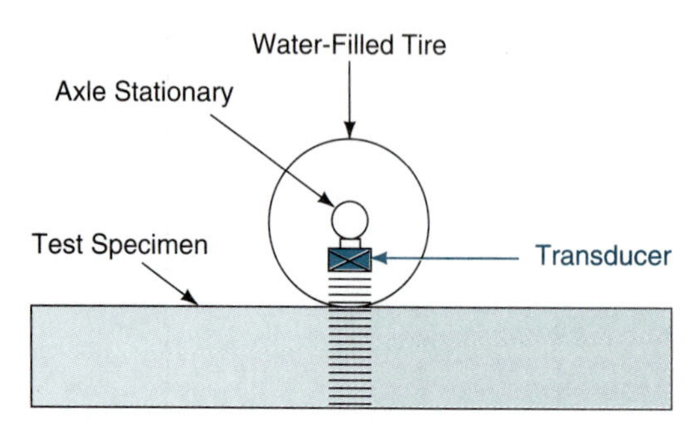

the part is commonly placed in the fixtures and the test is automatically conducted and monitored by computer. This type of testing is fast and efficient compared to contact testing.

Calibrating the Equipment Once the method of testing to be used has been determined, the next step is to calibrate the equipment to a known standard. This step is universal for all UT equipment. The ultrasonic equipment in use today has many adjustments that need to be set for the test to give reliable results. The method of ensuring that the equipment is properly set is to adjust the equipment to give the proper results for a known standard. The standard is a metal block that is designed to have specific places where indications will show under testing.

These calibration blocks are used by placing the transducer on the block where it will give a known indication and adjusting the equipment to give the known indication. As an example, a block might have an overall thickness of 1 in. The equipment should be adjusted to indicate the same thickness as the known block. Some blocks have a flat-bottomed hole that is 2.5 in. below the surface. The equipment should be adjusted to indicate that depth when it is calibrated. This is a critical step that should be accomplished with precision. If this step is not accomplished, the readings cannot be trusted. Figure 11 - 19 shows some of the types of calibration blocks in common use.

Testing the Part Once the equipment has been adjusted, the part can be tested. The test procedures are usually outlined by written specifications. In many cases, these are controlled by the company requesting the inspection. Military, nuclear, or industry standards are the most

Figure 11–19

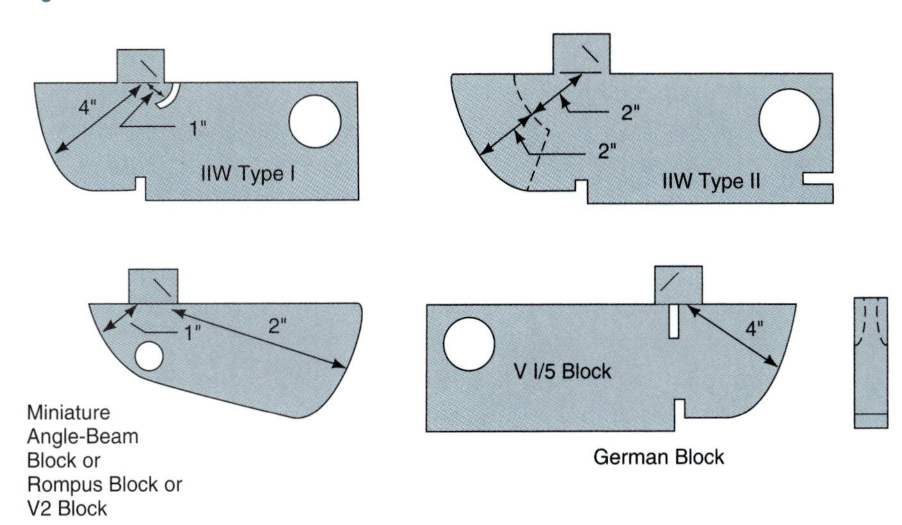

common. These usually dictate the method of test, the calibration equipment to use, and how to document the results. The appropriate standard should be followed when testing any part.

The specific steps of the testing process are precise and unique to the method chosen and the type of material being tested. One important step in any testing process is to make sure that all air is removed from between the transducer and the part. In immersion testing this is accomplished by placing the part and the transducer in water. In contact testing, a gel is placed between the part and the transducer. This gel (or glycerine) is used to couple the transducer and the part together. This material transfers the sound from the transducer to the part without interference. Use of a couplant is important in any type of ultrasonic testing.

Scanning the part is a step that is common among all UT methods. The common methods that are employed are hand manipulation and automatic scanning. Either method of scanning can produce precise results. Equipment is designed for one of the two methods. Many of the automatic scanning testers are computer controlled and are designed to be programmed. This programming is like a computer numerical control (CNC) machine, in that specific movements are preprogrammed. Complicated parts can be placed in a fixture for testing. The inspector simply accesses and runs the program for that part. In some equipment, the computer can analyze the results and accept or reject the part.

SUMMARY

Ultrasonic testing is one NDT method that requires skill and knowledge to accomplish properly. It can be used to detect many types of discontinuities in a wide range of materials. The basic concept behind this method is that sound energy can be induced into a part and the returned echoes can be identified. As the sound travels through the part, changes in the material's structure (e.g., a crack, an inclusion, or a material-type change) cause a portion of the sound energy to be reflected. This reflection can be used to identify the presence of a defect.

The equipment to accomplish this consists of a transmitter and a receiver that are connected to a transducer. This equipment creates the impulse of energy that is induced into the part and displays the returned sound wave as it is received by the transducer.

This method of testing can be used over a wide range of materials. The range of testing methods that can be used includes direct contact, angle-beam, squirter, and immersion. Each of these methods can be highly efficient and effective. However, these methods require a high degree of knowledge and skill for the results to be accepted as accurate.

This chapter should be thought of as an introduction to the concept of ultrasonic testing; it presents only the rudimentary concepts behind this testing method. In-depth study is required, along with practice in a controlled environment, before a technician can gain sufficient knowledge to accomplish this testing method. Once mastered, this technique is highly accurate. Technicians with this knowledge and skill are in demand in many industries.

KEY TERMS

Absolute 1. Not relative, independent; e.g., absolute zero temperature, as distinct from zero on an arbitrary scale, as the Celsius scale. 2. One way (as opposed to round trip), used with reference to distance measurement.

Absorption Changing part of a sound beam's mechanical energy into heat energy, evidenced by a slight increase in the temperature of specimen molecular particles. This is one cause of material loss attenuation.

Acoustic impedance A measure of the work sound does to pass through a medium, equal in magnitude to the product of the sound velocity (C) multiplied by the medium's density (ρ).

Acoustic interface The boundary between two media of different acoustic impedances.
Acoustic zero Can be considered for practical purposes to be the point on the CRT display that represents the specimen entry surface.
Amplifier An electronic device that increases signal strength fed into it by obtaining power from a source other than the input signal.
Amplitude Usually used in the combined term *echo amplitude,* signifying echo-pulse peak height, seen in linear form on the CRT screen above a reference line (usually the baseline).
Attenuation The loss of sound pressure in a traveling wave front, caused by the reflection and/or absorption of some of the wave's sound pressure by the grain structure and/or porosity of the medium
Back echo or back-wall echo The echo representing the specimen side opposite the side to which the transducer is coupled. This echo represents specimen thickness at this point.
Calibration The graduations (markings that indicate scale) of an instrument that allow measurements in definite units. The arbitrary 0–10 CRT scale may be calibrated in units of distance, converting the scale into time base information and, thus, into distance information.
Cathode ray tube (CRT) A vacuum tube that allows the direct observation of cathode ray behavior. It consists essentially of an electron gun producing an electron beam that, after passing between horizontal and vertical deflection plates, falls upon a luminescent screen; beam position can be observed by luminescence produced upon the screen. Electric potentials applied to the deflection plates are used to control beam position, and its movement across the screen, in any desired manner.
Clock interval The clock period time elapsing between each clock pulse.
Coaxial cable A cable consisting of a central conducting wire together with a concentric cylindrical conductor, the space between the two being filled with a dielectric substance, e.g., polyethylene, air, etc. The outer conductor is normally connected to ground. A coaxial cable's main use is to transmit high-frequency power from one place to another with a maximum loss of energy.
Code specification The document that prescribes approved procedures to be followed in a test.
Couplant A material (usually a liquid) used between the transducer and the test specimen to eliminate air from this space and thus ensure sound-wave passage into and out of the specimen.
Critical defect Either the largest tolerable defect or the smallest intolerable defect. The critical defect size is usually given by a specification.
Decibel One tenth of a bel. A unit that compares levels of power. Two power levels, P1 and P2, are said to differ by n decibels when: $n = 10 \log_{10} p2/p1$. This unit is often used to express sound intensities. In this case, P2 is sound intensity under consideration and p1 is a reference level intensity. In the case of displayed voltages on a cathode ray tube screen, the relationship becomes: $n = 20 \log_{10} v2/v1$.
Defect level The number of decibels of calibrated gain that must be added to the defect echo to bring its peak to the reference line on the CRT.

Detectability The ability to detect a given-size defect.

Dual element probe A probe containing two piezoelectric crystals, one of which transmits only and one that receives only.

Elasticity A material's ability to return to original form and dimension when forces acting upon it are removed. If the forces are sufficiently large for deformation to cause a break in the material's molecular structure, elasticity is lost and the "elastic limit" is said to have been reached.

Electrical zero 1. The point in time when a transmitter fires the initial pulse to the transducer and receiver. 2. The point on a cathode ray tube screen where an electron beam leaves the baseline, caused by an initial pulse signal coming from the transmitter.

Electron An elementary particle having a rest mass of 9.1091×10^{-31} kg, approximately 1/1836 that of a hydrogen atom, and bearing a negative electric charge of 1.06021×10^{-19} coulombs. The electron radius is 2.81777×10^{-15} meters. All atoms have electrons.

Electron gun A cathode ray tube electron source, consisting of a cathode emitter, an anode with an aperture through which an electron beam can pass, and one or more focusing and control electrodes.

Frequency The number of cycles per second undergone or produced by an oscillating body.

Gain A term used in electronics with reference to an increase in signal power, usually expressed as the ratio of output power (for example, of an amplifier) to input power, in decibels.

Hertz The derived frequency unit of a periodic phenomenon with a one-second period, equal to one cycle per second. Symbol: Hz. 1 Kilohertz (kHz) = 10^3 cycles per second; 1 megahertz (MHz) = 10^6 cycles per second. Named after Heinrich Hertz (1857–1894).

Incidence, angle of The angle of a sound beam striking an acoustic interface, normal (perpendicular) to the surface at that point. Usually designated by the Greek symbol α (alpha).

Interference (specifically of wave motions) Vector addition or combination of waves (also, superposition).

IP Abbreviation of the term *initial pulse,* which is the electrical pulse sent out by the transmitter to the receiver and the transducer. IP generally refers to this trailing edge caused by the "ringing" (or continuing vibration) of the transducer crystal.

Longitudinal wave Wave propagation characterized by particle movement parallel to the direction of wave propagation.

Medium A substance through which a force acts or an effect is transmitted; surrounding or enveloping substance; environment.

Mode conversion Changing a portion of a sound beam's energy into an opposite-mode wave, caused by reflection and/or refraction at incident angles other than 0°.

Molecule The smallest portion of a substance capable of existing independently and retaining properties of the original substance.

Normal probe A transducer that sends sound into a test specimen perpendicular to the entry surface.

Oscillator A device for producing sonic or ultrasonic pressure waves in a medium.

Parallel Extending in exactly the same direction so that there is neither divergence nor convergence; being an equal distance apart at all points.

Penetration The ability of the test system to detect a given-size defect at a given distance.

Perpendicular At right angles; a straight line making an angle of 90° with another line or plane.

Phase Points in the path of a wave motion are said to be points of equal phase if the displacements at those points, at any instant, are exactly similar, i.e., of the same magnitude and variation.

Phosphor A substance capable of "luminescence" (light emission from a body from any cause other than high temperature): storing energy (particularly from ionizing radiation) and later releasing it as light.

Piezoelectric crystals A family of crystals that possesses the characteristic ability to produce: a) a voltage differential across their faces when deformed by an externally applied mechanical force, and b) a change in their own physical configuration (dimensions) when an external voltage is applied to them.

Probe The transducer, or search unit.

Pulse A wave disturbance of short duration.

Pulse repetition rate The frequency with which a clock circuit sends out its trigger pulses to the sweep generator and transmitter, usually quoted in terms of pulses per second.

Radio frequency An oscillation frequency that falls within the range used in radio, i.e., 10 kHz to 100,000 MHz.

Range The total distance (specimen depth) being displayed at any one time across the CRT screen.

Ray A line giving wave direction of advance at any point. This direction corresponds to that of the radius of curvature.

Reference echo The echo from a reference reflector.

Reference level The number of decibels of calibrated gain that must be added to the reference-echo signal to bring its peak to the reference line on the CRT.

Reference reflector A known-size reflector at a known distance, such as a flat-bottomed hole.

Refraction Sound beam bending when passing through an acoustic interface at an incident angle other than 0°. The bending is caused by the difference in wave speed on either side of the interface, so refraction is accompanied by a wavelength change.

Refraction, angle of The angle between a refracted sound beam and the perpendicular.

Resolution Test system ability to distinguish defects at slightly different depths.

Sensitivity The ability of the test system to detect a given-size defect at a given distance.

Shock wave A particularly sudden and intense wave disturbance of short duration.

Single-element probe A probe containing only one piezoelectric crystal, which is used both to transmit only and to receive only.

Sonic Of or relating to frequencies within human audible range between 20 and 20,000 cycles/sec.

Sound path distance The distance from the transducer beam index to the reflector located in the specimen, measured along the actual path sound traveled.
Subsonic Of or relating to frequencies below the human audible range; below 20 cycles/sec.
Supersonic Of or relating to movement through some medium at speeds greater than the speed of sound in that medium.
Surface wave Wave propagation characterized by an elliptical movement of particles (molecules) on a specimen surface penetrating the specimen to a depth of one wavelength.
Trace The illuminated line on a cathode ray tube screen, caused by the luminescence of the phosphor layer by an electron beam.
Transverse wave Wave propagation characterized by particle movement perpendicular to the direction of wave propagation.
Ultrasonic Of or relating to frequencies above human audible range, above 20,000 cycles/sec.
Ultrasonics The study of pressure waves, which are similar to sound waves but which have frequencies above the human audible limit, above 20 kHz.
Volt The derived unit of electrode potential, defined as the difference in potential between two points on a conducting wire carrying a constant current of one ampere when power dissipated between these points is one watt. Named after Allesandro Volta (1745–1827).
Voltmeter An instrument for measuring potential difference between two points.
Wave A periodic disturbance in a medium (or a vacuum, as in the case of electromagnetic waves), which may involve the elastic displacement of material particles or a periodic change in some physical quantity such as temperature, pressure, electric potential, or electromagnetic field strength.
Wave form The shape of a wave, illustrated graphically by plotting the values of the periodic quantity against time.
Wave front The locus of adjacent points in the path of a wave motion that possess the same phase.
Wave motion The propagation of a periodic disturbance carrying energy. At any point along the path of a wave motion, a periodic displacement or vibration about a mean (average) position takes place. This may take the form of a displacement of electromagnetic vectors. The locus of these displacements at any instant is called a "wave." The wave motion moves forward a distance equal to its wavelength in the time taken for the displacement, at any point, to undergo a complete cycle about its mean position.
Wavelength The distance between like points on successive wave fronts, i.e., the distance between any two successive particles of an oscillating medium that are in the same phase. It is denoted by the Greek letter λ (lambda).
Wetting agent A substance that lowers the surface tension of a liquid.

ULTRASONIC TEST

1. Define the term *piezoelectric* and describe how it is used in ultrasonic testing.
2. Describe immersion testing and contact testing. List the advantages of immersion testing.
3. List the major components of the test unit.
4. Discuss in a short paragraph the purpose of a couplant.
5. Discuss in a short paragraph the term *equipment calibration.*

NOTES

1. *Nondestructive Testing: Ultrasonic Testing* (CT-6-4), 2d ed. General Dynamics Convair Division, 1967.
2. *NDT: UT.*
3. *NDT: UT.*
4. *NDT: UT.*
5. *NDT: UT.*

RADIATION SAFETY

12

12–1 PHYSICS OF RADIATION

Industrial radiography is the use of penetrating electromagnetic radiation to expose an indicating medium. Radiography (RT) is one of the most important, and most versatile, of all the nondestructive test methods used by modern industry. It provides a permanent, visible record of internal conditions.

Radiography uses ionizing electromagnetic radiation to penetrate a specimen and expose an indicating medium. The concept is simple (Figure 12-1). However, to get the proper result, the technician must be familiar with the intricacies of the method.

RT is widely used in industry and requires extensive training for technicians. Of all the nondestructive testing methods, RT poses the greatest health hazard to the technician and the public.

The health hazards associated with radiography are real. However, often the hazards do not measure up to the misconceptions. The fear of radiation in the general public is so great that the very mention of radiation may bias a conversation. The bias may go so far as to disqualify radiography as a test method. For this reason, we spend a great deal of time in this chapter discussing the physics of radiation. When one begins to apply the physics to safety and testing, he or she will have a better grasp of the realities of the situation.

OBJECTIVES

After completing Chapter 12, the reader should be able to:

- Distinguish between the two types of radiation.
- Distinguish between an isotope and an ion.
- Discuss the origin of alpha, beta, gamma, and X radiation.
- Identify the two types of electromagnetic radiation used in radiography.
- Discuss the basic operation of the X-ray machine.
- Identify terms associated with dose and dose rate.
- Identify regulatory authorities associated with industrial radiography.
- Identify and discuss the factors in reducing radiation exposure.
- List the dose limits outlined by the regulating authority.
- Name the two types of radiation detection devices used in industrial radiography.
- Discuss the basic operation of radiation detection instruments.

Figure 12–1

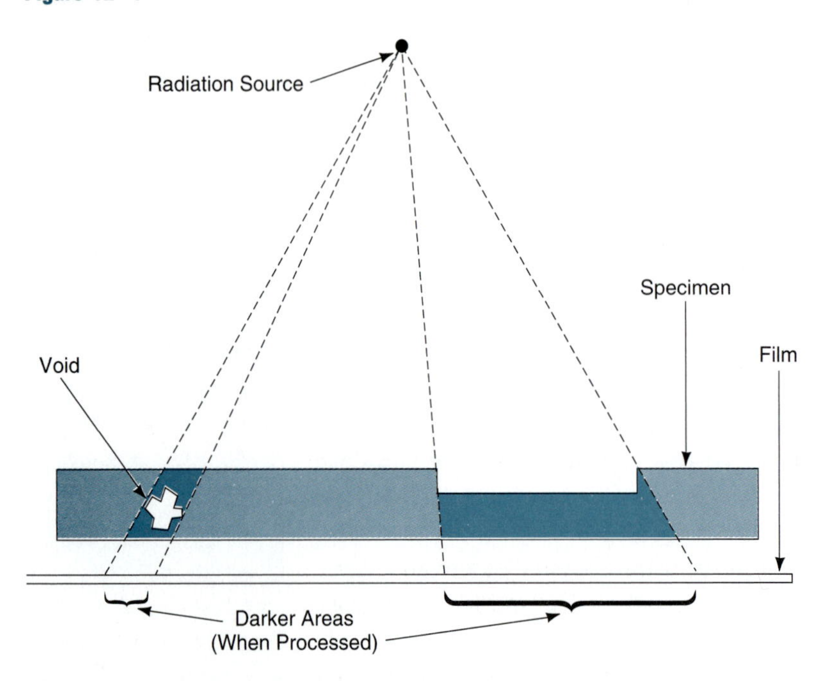

12–2 TYPES OF RADIATION

There are two types of radiation of interest in this study: particulate radiation and electromagnetic radiation. Both are what is termed *ionizing*. This means that both are capable of ripping electrons away from an atom.[1] Only electromagnetic radiation is used in industrial radiography. This is very important to remember.

Electromagnetic radiation is a wave-type radiation much like visible light. The wavelength of useful radiation is about 1/10,000 that of visible light.[2] This type of radiation can come from many sources: the sun, television, or a radioactive isotope, just to name a few. Particulate radiation, on the other hand, comes only from radioactive isotopes. To better understand the difference, let us look at each in a little more detail.

12–3 WHAT IS AN ISOTOPE?

An *isotope* is a material that has an imbalance in the nucleus of the atom. Specifically, there are too many neutrons for the number of protons. Isotopes are either stable or unstable. A stable isotope has an excess number of neutrons but does not emit radiation. An unstable isotope is one that is decaying, or trying to return to a stable state. *Particulate radiation* is a byproduct of the decay of a radioactive isotope. As the atom decays, pieces (particles) of the atom are ejected from the nucleus along with electromagnetic radiation.

It is possible to make any material an isotope by bombarding the atoms with neutrons. Remember that one must bombard with neutrons, not electromagnetic waves. To bombard with neutrons, there must be a nuclear reaction much like what takes place in the explosion of an atomic bomb or inside the reactor of a nuclear power plant.

For example, a helium atom is made up of one proton and one electron. When bombarded with neutrons, the helium atom first becomes deuterium—which is made up of one electron, one proton, and one neutron. Deuterium is a stable isotope that gives off no radiation. As neutron bombardment continues, however, deuterium becomes tritium. At this point the isotope is unstable, giving off radiation as it decays back to a stable condition.[3]

Because of the imbalance in the nucleus, isotopes have so much energy inside that they must cast it off. Energy is eliminated in the form of small particles and electromagnetic radiation. The particles are called alpha and beta particles, and the electromagnetic radiation is called gamma rays (Figure 12-2).

The *alpha particle* is a positively charged particle emitted from the nucleus of radioactive materials. It is made up of two neutrons and two protons; hence it is identical to the nucleus of a helium atom.[4] Comparatively speaking, it is a large particle. Its size makes it easy to stop. This type of radiation can be stopped with a sheet of paper. It is

Figure 12–2

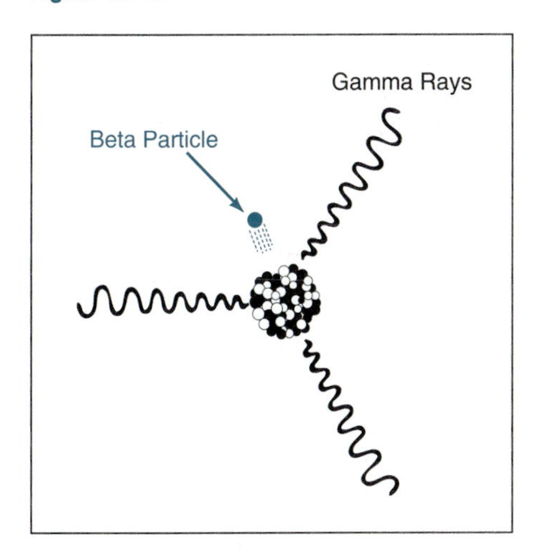

extremely ionizing. If allowed to get into the body, it can do a lot of damage.

A *beta particle* (beta ray) is an elementary particle emitted from a nucleus during radioactive decay. It has a single electrical charge and a mass equal to 1/1840 that of a proton. Beta particles are easily stopped by a thin sheet of metal. A negatively charged beta particle is physically identical to the electron. If the beta particle is positively charged, it is called a *positron.* Beta radiation may cause skin burns, and beta emitters are harmful if inhaled or ingested.[5] The most important thing to remember here is that a beta particle is an electron that is emitted from the nucleus and does not come from the shell of the atom. Beta particles are more penetrating than alpha particles. Beta particles can penetrate the skin and can cause extensive damage.

Alpha and beta particles are extremely dangerous and hard to contain. They are extremely ionizing and of no use in industrial radiography. They are considered contamination and are not a factor in the everyday world of RT. These particles are given off by radioactive isotopes only. Although radioactive isotopes are used in RT, great efforts are made to keep all particulate radiation confined. The isotope is sealed into a stainless steel capsule to prevent the particulate radiation from escaping. Only the electromagnetic (gamma) radiation is used in industrial radiography.

12–4 ELECTROMAGNETIC RADIATION

Unlike the particulate radiation, electromagnetic radiation has no mass or weight; the photons are pure energy and are extremely penetrating and ionizing. Both these characteristics are important to the usefulness of electromagnetic radiation in radiography. The penetrating power of the electromagnetic wave allows the ray to go through the part under inspection and the ionization ability of the wave exposes the film, causing an image to be formed.

Light is a form of electromagnetic radiation. Its frequency is in the range that can be sensed by the human eye. At a much higher frequency is the electromagnetic radiation used in RT. Figure 12-3 gives a graphic representation of the relationship of the different frequencies on the electromagnetic spectrum.

Note that X-rays and gamma rays occupy the same place on the electromagnetic spectrum. These are the two types of electromagnetic radiation used in RT. The major difference between the two is their

Figure 12–3

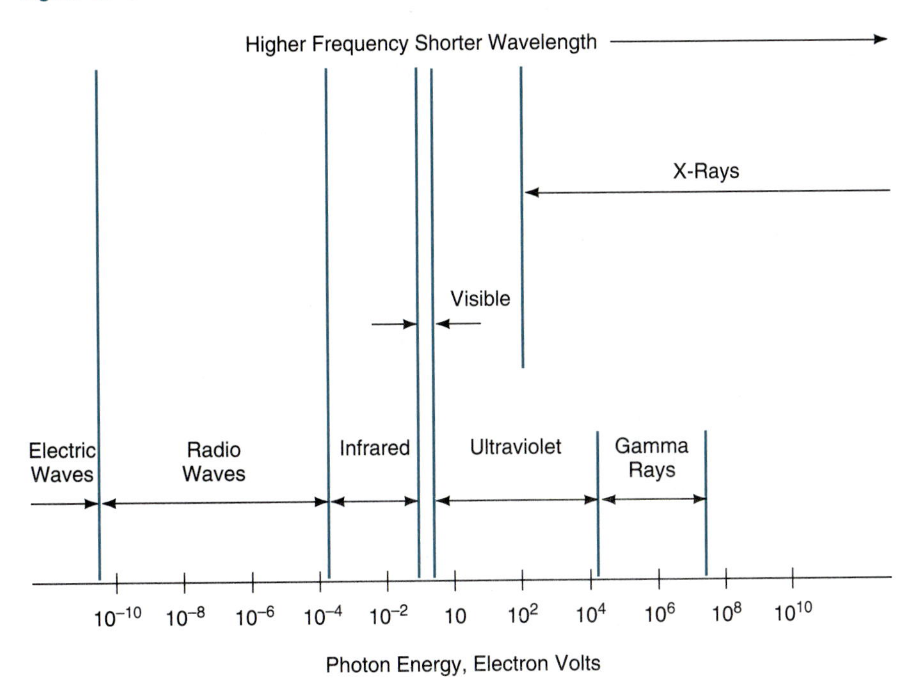

origin. X-rays come from electrical equipment; gamma rays come from the decay of a radioactive isotope. Since these types of radiation are used in RT, note how they are generated.

Gamma Rays When a material is bombarded with neutrons, it becomes *activated.* During activation the atom becomes heavier because of the mass of the added neutron. It also stores the kinetic energy from the neutron.

Gamma rays are given off when an unstable isotope decays. *Decay* is the spontaneous act of the isotope undergoing change back to the stable state. Each atom of an isotope decays in much the same way.[6] As the subatomic particles are cast off, a photon of energy is given off also. This photon is the gamma ray. One or more rays are given off as each atom decays, depending on the isotope. The same number of rays at the same frequency are given off as each atom decays.

The analogy of a shotgun can be used to demonstrate the decay of an atom. Picture the atom as a shotgun. When the trigger is pulled, the gun fires (decay begins). Pellets shoot from the muzzle; the alpha and beta particles are ejected from the nucleus of the atom. The pellets, then, are particulate radiation. There is also a concussion that goes along with the discharge of a gun. This sound is like the gamma ray being given off. Different guns give off different sounds—a 12-gauge shotgun has a different sound than a 410-gauge shotgun. Likewise, each isotope gives off its own ray pattern.

For example, as each atom of cobalt 60 (Co^{60}) decays, it gives off two primary rays. One of these rays has a wavelength of just under 0.01 angstrom, and the other slightly over 0.01 angstrom. With iridium 192 (Ir^{192}), approximately nine frequencies are given off as primary rays during decay. They range from 0.02 angstroms to just over 0.1 angstrom. Note that the frequency of Co^{60} is higher than that given off by Ir^{192} (Figure 12-4, p. 225).

The frequency (Hz) of the wave (ray) is called its *intensity.* Co^{60} always gives off rays of the same intensity; Ir^{192} always gives off rays of its intensity. These rays are always classified as gamma rays simply because they originated from an isotope.

The total number of rays per second given off by the isotope is called its "activity," and is measured in curies. *Activity* is defined as the number of disintegrations per second. Since gamma rays are emitted from the nucleus as it decays to a stable state, the number of disintegrations has a direct correlation to the number of rays. One *curie* is equal to 37,000,000 disintegrations per second. It is not uncommon to have a 100-curie source (i.e., 3,700,000,000,000 disintegrations per second).[7] As the isotope decays, there are fewer and fewer active atoms. Therefore, there are fewer and fewer rays given off. The amount of time required for the curie strength to be reduced by one-half is called the *half-life.*

In the case of Co^{60}, the half-life is 5.7 years (Table 12-1, p. 227). Ir^{192} has a half-life of 75 days. At 12.5 ci, Co^{60} isotope is of little value to a radiographer. It takes too long to expose the film with such a "light" source. However, the source is still decaying at a rate of 462,500,000 disintegrations per second (dist/sec). This is still an extremely dangerous amount of radiation. One of the largest problems facing our society today is what to do with spent isotope. One-half of its activity is still there after each half-life. Even after 57 years this is still very dangerous material.

It is customary to use a graph to determine the activity of a source. On the graph in Figure 12-5, p. 226, the Ir^{192} source was new on June 1, 1994. Simply follow the horizontal axis to the desired date, move vertically to the diagonal line and then left to read the activity of the source.

Figure 12–4 Gamma-Ray Spectrum of Cobalt 60 (Solid Lines) and Principal Gamma Rays of Iridium 192 (Dashed Lines)

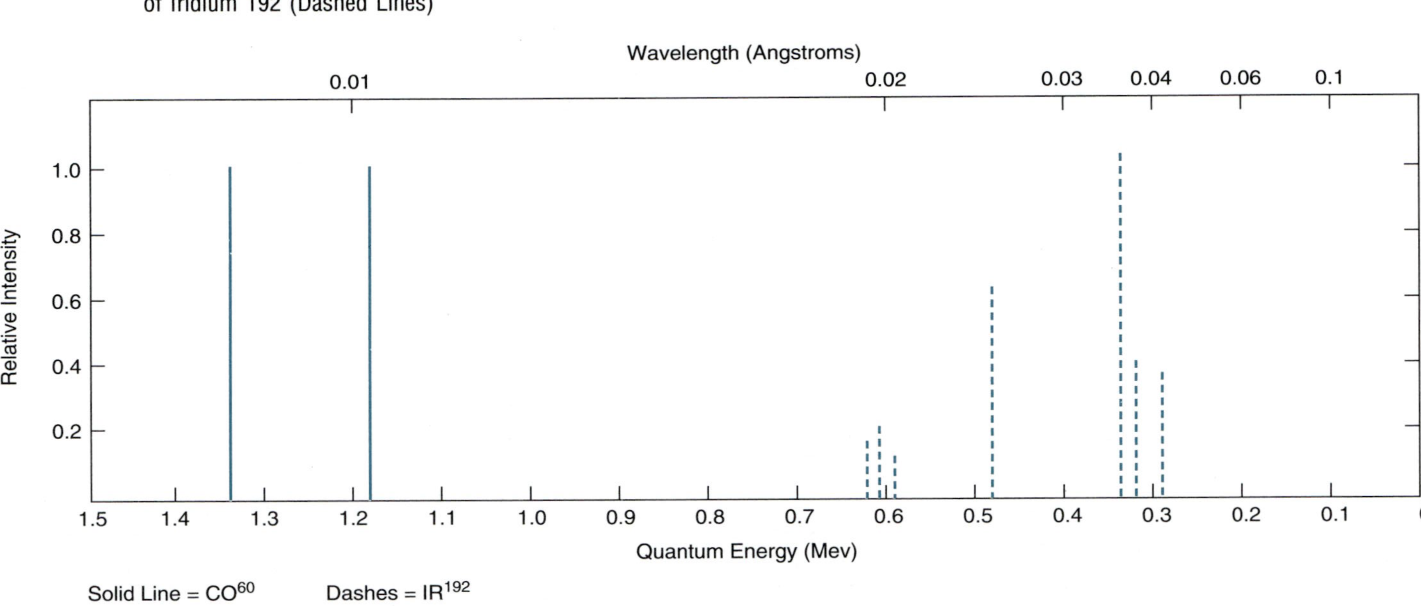

Figure 12–5

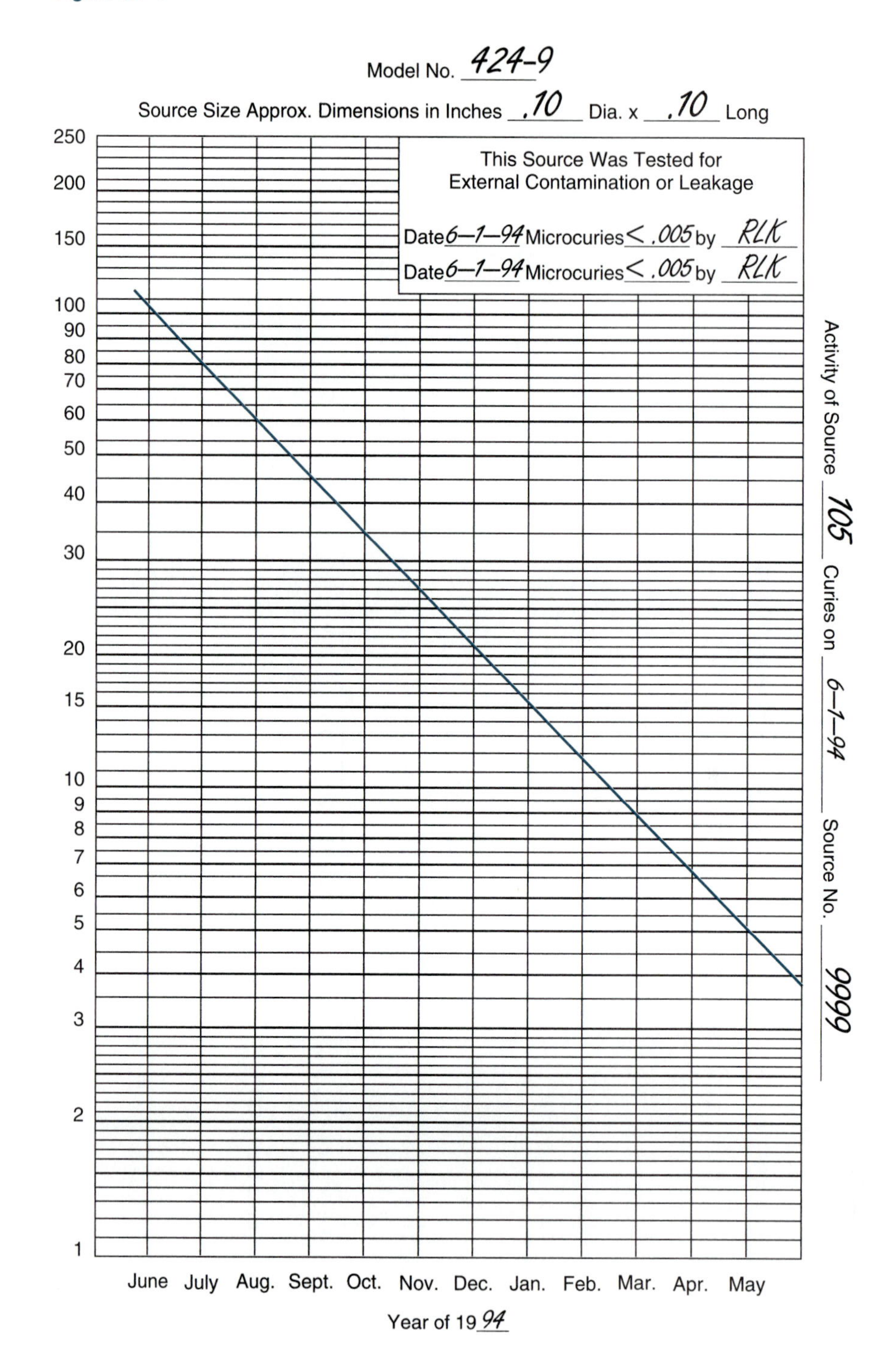

Model No. 424-9
Source Size Approx. Dimensions in Inches .10 Dia. x .10 Long
This Source Was Tested for External Contamination or Leakage
Date 6–1–94 Microcuries < .005 by RLK
Date 6–1–94 Microcuries < .005 by RLK
250
200
150
100
90
80
70
60
50
40
30
20
15
10
9
8
7
6
5
4
3
2
1
June July Aug. Sept. Oct. Nov. Dec. Jan. Feb. Mar. Apr. May
Year of 19 94
Activity of Source 105 Curies on 6–1–94 Source No. 9999

Table 12–1 Life Cycle of Co^{60}

TIME (IN YEARS)	STRENGTH (IN CURIES)	
new	100.0	
5.7	50.0	
11.4	25.0	
17.1	12.5	
22.8	6.25	
28.5	3.125	
34.2	1.56	
39.9	0.78	
45.6	0.39	
51.3	0.195	
57	0.1	(3,700,000 dist/sec)

This type of chart accompanies a source from the manufacturer. It should accompany the source throughout its life.

X-Rays *X-rays* are the product of the interaction of atoms with electrons, not of the decay of radioactive material. There is *no* radioactive material in an X-ray tube. When it is turned off, it is no more dangerous than an electric skillet.

The X-ray tube is made like a light bulb. The tube itself is evacuated of all gases. An anode and a cathode are inside this vacuum tube. The cathode is much like the filament in a light bulb. The cathode heats up to white hot. At this temperature, the electrons moving through the cathode are "boiling" around the filament. Not far away is the tungsten anode, which has a positive charge in relationship to the cathode. The voltage difference across the anode and cathode is extremely high, ranging from hundreds of kilovolts to megavolts.

The negatively charged electrons boiling around the cathode are attracted by the extreme positive charge of the anode. They shoot across the gap and strike the anode. The point where they impact the anode is called the target. The interaction of the electrons with the atoms of the anode gives off X-rays (Figure 12-6).

As the electrons from the cathode come near the atom, the strong magnetic forces of the atom's electrons and nucleus cause the electrons suddenly to slow. This sudden reduction in speed causes electro-

Figure 12–6

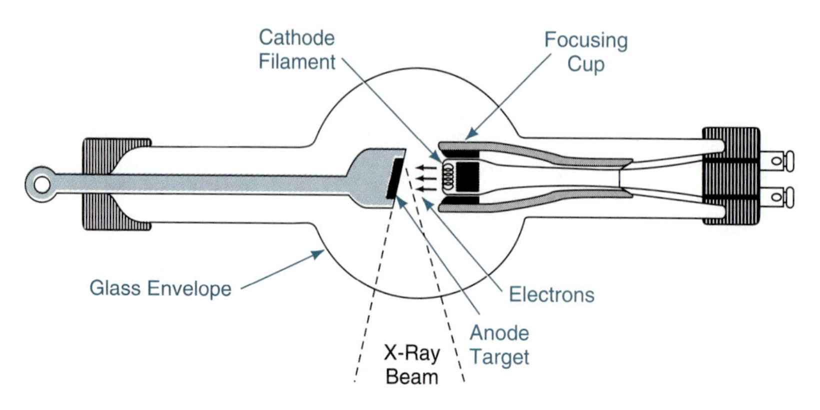

magnetic radiation to be emitted. This phenomenon was discovered in Germany, and is called *bremsstrahlung,* or "breaking rays." This sudden reduction in speed varies with the relative position of the electron to the parts of the atom on the target and the speed of the electron. The electromagnetic radiation given off during bremsstrahlung also varies.

Another analogy can be used to demonstrate the working of the X-ray machine. In this analogy, picture someone standing in front of a wall with a tennis ball in his or her hand. The person is the cathode and the tennis ball is an electron. The wall is the target of the anode. The person throws the ball at the wall; determine what the X-ray would be.

The sound of the ball striking the wall equates to the X-ray. The sudden reduction in the speed of the tennis ball gives off a sound. The sudden reduction in the speed of the electron gives off a photon of energy (the X-ray). Refer to Figure 12–4 and note that the X-ray wavelength is also on that chart. It is measured in million electron volts (Mev), which is the force of one electron powered by 1 million volts (that is, 1.1 to 1.4 Mev will give the same frequency as Co^{60}). It takes an extremely large machine to achieve such power. Such a powerful machine would still put out rays at the very low end of the spectrum. The major difference in gamma- and X-radiation equipment is that the isotope gives a few frequencies of wave, depending on the kind of isotope used, and the X-ray machine gives off the full spectrum of frequencies up to the maximum power setting. Gamma and X-radiation are similar—only their sources are different.

12–5 RADIATION SAFETY REGULATIONS

There are many regulations governing the use of radiation for any reason in the U.S. The use of radioactive isotopes is governed by each state, in the case of "agreement" states. It is governed by the Nuclear Regulatory Commission (NRC) and Code of Federal Regulations (CFR) 10 in the case of "nonagreement" states.

An *agreement state* is one that has signed an agreement with the NRC allowing the state to regulate certain activities using radioactive materials, such as gamma radiography using Ir^{192} or Co^{60} sources.[8]

A *nonagreement state* is one in which the NRC regulates the use of radioactive materials.[9] In any state the process required to gain a license to use and store radioactive isotopes is lengthy and costly. The licensee must have secure areas for storage, radiation detection equipment, and an in-depth and ongoing education program for technicians, and must undergo an exhaustive investigation just to qualify for the license. There are also regular audits by the government and stiff penalties for violations.

In nonagreement states the NRC is responsible for regulating radioactive materials. X-ray radiography, accelerator radiography, and radiography using radium 226 (Ra^{226}) is regulated by the U.S. Occupational Safety and Health Administration (OSHA).[10]

The U.S. Department of Transportation (DOT) has the responsibility to regulate the transportation of radioactive materials between states. The NRC and agreement states also regulate the transportation of radioactive materials within a state's borders.[11]

Consequently, most companies do not use in-house isotopes. Many companies contract to a third-party inspection company to do RT inspection if an isotope is required. The contracted company then assumes the responsibility and much of the liability. There are times when only isotope testing can meet the need—for example, in a remote area without electrical power, where a crew is building a bridge or welding a pipeline.

There are no licensing requirements, in most cases, for the use of X-ray equipment—one factor that makes the use of X-ray testing so attractive to industry. The liability and cost for technician training are still very high, but less than for the use of isotopes.

If a manufacturer or builder is considering the use of radiation inspections, a company representative should contact the appropriate authorities and get the information for licensing and certification. This is a very important step in considering this type of inspection.

12–6 HARMFUL EFFECTS

There are two types of harmful effects from radiation—*prompt effects,* which occur within a few days or weeks of exposure, and *delayed effects,* which take many years to develop.

Prompt effects are injuries such as radiation burns and radiation sickness. Exposures of 2000 to 3000 rem (roentgen equivalent man) in a localized area (such as a hand) result in burns resembling chemical burns. In this case, the dose is so massive that the cells die quickly and are not able to reproduce themselves. With exposures this high, the damage is often permanent and can result in amputation of the exposed area.

Radiation sickness results when the trunk of the body receives in excess of 100 rem. In this case the cells are not killed outright by the radiation, but they are not able to reproduce. A person gets sick because the valve in the stomach (the duodenum) is not able to regenerate itself. Because of its location, the duodenum is continuously eaten away by the acids in the stomach. When exposed to radiation, the cells are not able to reproduce, and the valve is damaged by the acids it is to control. This allows the contents of the small intestine to back up into the stomach, causing nausea and vomiting.

Radiation affects a cell's ability to reproduce by disturbing the DNA molecules in the nucleus of the cell. As the photon passes through the cell, it interacts with the atoms that come in contact with it. It knocks electrons away from their atoms (ionization) and changes the chemical structure of the nucleus. The water in the nucleus is momentarily changed to bleach when hit by a photon. This change does not last very long before the cell finds its electrical balance again. The greater the number of photons that react in a given time period, however, the more harmful the result. The bleach does not have time to return to water, and the concentration becomes more and more toxic to the DNA molecules in the cell. When the bleach reaches a certain level, the DNA is damaged beyond repair, and the cell will mutate. This can be the first step in cancer.

It is hard to determine the cause of a particular cancer because there are so many factors that can prompt a cell to mutate. However, it is known that cell reproduction is dramatically affected by radiation. Cells that reproduce the fastest are affected the most. Small children and pregnant women are therefore not allowed into radiation areas. The rapid rate at which the cells of fetuses and small children divide makes the possibility for birth defects or cancer very great.

The amount of radiation and the time over which the radiation is absorbed are critical to the extent of the effect. The higher the dose or the shorter the period of time, the greater the damage. The same dose over a longer period of time will not have as drastic an effect. To expand on this subject, let us study the terms associated with exposure.

The *roentgen* is a unit of radiation dose. It was named in honor of Wilhelm Roentgen, who discovered X-rays in 1895. The roentgen is the amount of radiation required to give one cubic centimeter (1 cm^3) of air one electrostatic unit of charge at standard temperature and pressure. Remember that radiation is ionizing. Therefore, it rips electrons away from the molecules in the air. This gives the molecules in the air a relative positive or negative charge. This charge is termed an *electrostatic* charge. If one wants to know the charge in the air, the roentgen is the appropriate choice. However, if we want to know the ionizing effect on tissue, we have to use a different unit of measure.

The *roentgen equivalent man (rem)* is the unit of measure for radiation exposure to a human. It gives the effect of 1 roentgen on a human. When talking about dosage, the term *millirem* (mrem), which equals 1/1000 rem, is generally used.

A person living in the U.S. receives approximately 200 mrem per year. This radiation comes from the natural background. Cosmic radiation, medical tests, and the earth itself expose many people to various levels of radiation.

The time through which the radiation is absorbed is called the dose rate. It can be expressed in rem per hour (rem/hr), but mrem per hour is a more common measurement.

12–7 TIME, SHIELDING, AND DISTANCE

The total dose depends on the amount of radiation being absorbed and the amount of time during which it is absorbed. There are several things that can be done to minimize the radiation dose. The amount of radiation can be reduced, and/or the amount of time spent in the radiation area can be shortened. There are two ways to reduce the amount of radiation—(1) put something between the person and the source, and (2) move away from the source.

Time The less time spent in a radiation area, the less radiation will be received. This can be expressed mathematically as

$$\text{Dose} = \text{Dose rate} \times \text{time}$$

The dose is expressed in millirem, the dose rate in millirem per hour, and the time in hours or minutes.

Table 12–2 Half-Value Layer

MATERIAL	IRIDIUM 192 (IN.)	COBALT 60 (IN.)
Lead	0.19	0.49
Concrete	1.75	2.38

If one spends 3 hours in an area of 5 mrem/hr, he or she will receive 15 mrem. If one spends 20 minutes in that area, he or she will receive 1.66 mrem.

Shielding Putting something between a person and a radiation source is called "shielding." If there is a dense material between the person and the source, it will absorb some of the rays being emitted from the source. This, in effect, reduces the amount of radiation hitting the person. The amount of material it takes to reduce the radiation by one-half is called a *half-value layer.* The thickness of a half-value layer varies with the density of the material and the Hz of the ray. For example, one half-value layer of concrete would be thicker than one half-value layer of lead, because lead is more dense than concrete. Also, a half-value layer of lead for Co^{60} is thicker than a lead half-value layer for Ir^{192}, because the Hz of cobalt is higher than that of iridium (Table 12-2). A half-value layer is always that amount of material required to reduce the radiation dose by one-half (Table 12-3).

If one is in an area of 100 mrem/hr, it would take 5 half-value layers to reduce the dose rate to approximately 3 mrem/hr. If using an Ir^{192} source, it would require approximately 1 inch of lead.

Distance Distancing a person from a radiation source will also reduce the exposure. If the distance is doubled, the dose is reduced to

Table 12–3

STRENGTH (CI)	MATERIAL THICKNESS (IN.)	STRENGTH (CI)
100	0.19	50
50	0.19	25
25	0.19	12.5
12.5	0.19	6.25
6.25	0.19	3.13

one-fourth. This can also be expressed mathematically. It is called the *inverse square law.*

$$\frac{I2}{I1} = \frac{(D1)^2}{(D2)}$$

I1 = known dose rate
I2 = unknown dose rate
D1 = known distance
D2 = unknown distance

If a person is receiving 200 mrem/hr at 100 ft, how much will he or she receive at 400 ft?

$$I2 = I1 \times \frac{(D1)^2}{(D2)}$$

$$I2 = 200 \text{ mrem/hr} \times \frac{(100)^2}{(400)}$$

$$I2 = \frac{200 \text{ mrem/hr} \times 1}{16}$$

$$I2 = 12.5 \text{ mrem/hr}$$

12-8 DOSE LIMITS

CFR 10, Part 20 lists the dose limits set by the NRC. A person can receive 3 rem of whole-body radiation per quarter year if the person does not average more than 5 rem per year.[12] A record of each worker's radiation history must be kept by employers. There is a formula to compute the maximum amount of whole-body radiation that a person of a certain age may receive over his or her lifetime. This accumulation is called the *bank account.* The formula is 5(N − 18). N equals the present age in years, 18 is the youngest age at which a person may enter a radiation area, and 5 is the average yearly exposure.

What, then, is the maximum amount of radiation a person can take by the age of 30?

$$5(30 - 18) = 60 \text{ rem}$$

A 30-year-old person may receive no more than 60 rem as a lifetime dose. However, in addition, he or she may receive no more than 3 rem per quarter in any given quarter.

These numbers are for whole-body doses; it is known that different parts of the body react differently to radiation. There are some additional dose limits for extremities. A person may receive 18.75 rem per quarter year to his or her hands, forearms, feet, and ankles.[13] To help radiation workers maintain these minimums, the NRC has set classifications for radiation areas:

Unrestricted Area An area where a person will receive no more than 2 mrem per hour.[14]

Restricted Area An area where the licensee must set barricades to restrict public access.[15]

Radiation Area An area where a person would receive in excess of 5 mrem per hour.[16]

High-Radiation Area An area where a person would receive in excess of 100 mrem per hour.[17]

12–9 RADIATION DETECTION DEVICES

There are two types of detection devices used in industrial radiography. The dose rate is given by a hand-held survey meter, and the dose is recorded on two dosimeters.[18]

The *survey meter* is a gas-filled ion chamber. The chamber is sensitive to the ionizing effects of radiation. The wall of the chamber is negatively charged. There is an electrode that is positively charged in the center of the chamber. The chamber is filled with an inert gas that normally has no electrical charge. As a photon of gamma or X-ray passes through the chamber, it strikes gas molecules, breaking them apart. These pieces of the molecule are called ion pairs. One ion is positively charged, and the other ion is negatively charged. The negatively charged ion moves to the positive electrode, and the positive ion moves to the negatively charged wall. This electron movement appears to the meter to be current flow, and that causes the meter to move.

The higher the radiation, the more photons that interact with the gas, which causes more "current" to flow through the meter, thus, the higher the dose the meter reads. The meter is marked in millirem per hour. The purpose of the survey meter is to enable the radiographer to know the dose rate in the area. It gives immediate indication. The radiographer is required to have a survey meter whenever he or she is in a radiation area.

There are two types of *dosimeters* that a radiographer must carry. The dosimeters read the total dose received by the radiographer. The first type is the self-reading pocket dosimeter. It can be read at any time to see the total amount of radiation taken over a time period. Radiographers are required to read the pocket dosimeter at least once a day.

The pocket dosimeter is a miniature, sealed ion chamber. Inside the chamber is a thin fiber that is insulated away from the rest of the chamber. It receives a charge to cause an electrical difference. This electrical charge causes a repulsion and attraction, which in turn causes the fiber to move. The fiber is located beneath an eyepiece that has a scale painted on it. The repulsion is reduced by ionization, and the fiber moves across the eyepiece and changes the reading. Again, the radiographer is required to carry a pocket dosimeter at all times when in a radiation area.

The second type of dosimeter to be carried is the delayed reading dosimeter. This may be a thermoluminescent dosimeter (TLD) or a film badge. These dosimeters must be read by a third party. The readings from these dosimeters are sent back to the licensee to be recorded in the radiographer's "bank account."

The thermoluminescent dosimeter uses a small crystal to capture the energy of a photon as it passes through. It stores the energy until it is heated. It then releases the energy in the form of visible light. The intensity of the light given off is directly proportional to the amount of radiation received by the radiographer. This type of delayed reading dosimeter service costs a little more than the film badge, but the TLD is sent in to be read every quarter (three months).

The film badge is sent in to be read every month. It consists of a small piece of radiography film inside a protective plastic case. As the radiographer is exposed to radiation, the film is exposed. As the dose goes up, the exposure goes up. This causes the film to turn dark when developed. The darker the film, the higher the exposure.

To summarize, there are two types of monitoring devices required—one measures dose rate (survey meter) and the other measures dose (dosimeter). There are two types of dosimeters—the self-reading (pocket) dosimeter and the delayed reading dosimeter. There are two types of delayed reading dosimeters—the film badge and the TLD.

Summary

There are two basic types of radiation: particulate and electromagnetic. Only electromagnetic is used in RT. Particulate radiation is made up of alpha particles and beta particles. This type of radiation is very dangerous because it is hard to contain. Therefore, it is not used in RT. Electromagnetic radiation is highly penetrating and is therefore well suited to RT.

There are two sources of electromagnetic radiation. Both are widely used in RT. Gamma rays are emitted from the decay of an activated isotope. An isotope gives the same intensity of ray throughout its life, making one isotope different from all others in penetrating power.

All intensities from isotopes are contained on the spectrum of X-rays. X-rays are emitted from the interaction of electrons and atoms in electrical equipment. X-ray machines contain no active material. It is possible to vary the intensity of the ray, and thereby affect penetration.

The hazards from radiation fall into two categories: prompt effects and delayed effects. Prompt effects are those that become visible within a few days of exposure. They appear as burns, sickness, hair loss, and other symptoms. Delayed effects are much harder to isolate. They may show up as cancer many years after exposure. To minimize the hazards involved with RT, technicians are required to carry two types of radiation-detection devices: one that reads radiation immediately and one that makes a delayed reading. The dose taken by each radiographer is recorded by the employer, and records are kept over the radiographer's entire career. These records are called the employee's "bank account." By staying within the limitations imposed by safety standards, the radiographer reduces the possibility of delayed effects. Accidents involving isotopes are very serious. Government agencies are responsible for overseeing the outcome.

Key Terms

Activate The act of bombarding a nucleus with neutrons until it becomes unstable.
Activity The number of rays given off per second by an isotope; measured in curies.

Agreement state A state that has signed an agreement with the NRC allowing the state to regulate certain activities using radioactive materials.
Alpha particle A positively charged subatomic particle emitted from the nucleus of radioactive materials. It consists of two neutrons and two protons.
Bank account The formula that reveals the maximum amount of whole-body radiation that a person of a certain age may safely receive over his or her lifetime.
Beta particle A negatively charged subatomic particle emitted from the nucleus of radioactive materials. It has the mass of an electron but comes from the nucleus.
Bremsstrahlung German for "breaking rays." The sudden reduction in speed of an electron as it nears the anode in an X-ray tube.
Curie The activity of 37,000,000 disintegrations per second.
Decay The spontaneous act of an isotope undergoing change back to the stable state.
Delayed effects Harmful effects from radiation exposure that occur many years after the fact.
Dosimeter A personnel-monitoring device that reads total radiation dose.
Electromagnetic radiation Radiation consisting of photons of energy such as light.
Gamma rays Electromagnetic radiation emitted from the nucleus of an atom as it decays.
Half-life The amount of time required for the curie strength to be reduced by one-half.
Half-value layer The amount of material required to reduce the radiation level by one-half.
High-radiation area An area where a person would receive in excess of 100 mrem per hour.
Intensity The frequency (Hz) of electromagnetic radiation.
Inverse square law The mathematical expression of dose versus distance.
Ionizing radiation Radiation that is capable of ripping electrons away from atoms.
Isotope A material that has an imbalance in the nucleus of the atom.
Nonagreement state A state in which the NRC regulates the use of radioactive materials.
Particulate radiation Ionizing radiation consisting of subatomic particles (alpha, beta, and positrons).
Positron A positively charged beta particle.
Prompt effects Harmful effects from radiation exposure that occur within a few days or weeks of the exposure.
Radiation area An area where a person would receive in excess of 5 mrem per hour.
Rem Roentgen equivalent man. The unit of measure of radiation exposure to a human.
Restricted area An area that the licensee barricades in order to restrict public access.
Roentgen The amount of radiation required to give 1 cm^3 of air one electrostatic unit of charge at standard temperature and pressure.
Survey meter Personnel-monitoring equipment that reads dose rate.
Unrestricted area An area where a person will receive no more than 2 mrem per hour.
X-Rays Electromagnetic radiation emitted from the interaction of electrons with atoms in electrical equipment.

RADIATION SAFETY TEST

1. There are two types of radiation discussed in this chapter. Only one is used in industrial radiography. Name the two types and discuss in a short paragraph their differences.
2. In a short paragraph, discuss the physical differences between an ion and an isotope.
3. What is a positively charged electron emitted from the nucleus of an atom?
4. Define *decay* (of an isotope).
5. Define *bremsstrahlung.*
6. Where is the target on the X-ray tube?
7. Who regulates interstate transportation of radioactive materials?
8. List the types of radiation effects on the body and discuss them briefly.
9. How many half-value layers would be required to reduce the effect of a 50-ci source of Co^{60} to 12.5 ci?
10. Name the two types of radiation detection devices and give their purpose.
11. Match the terms to the definitions on the right.

Half-life	Amount of radiation that will give the effect of one roentgen on human tissue.
Curie	Amount of radiation required to give 1 cm^3 of dry air one electrostatic unit of charge at standard temperature and pressure.
Roentgen	Unit of activity equal to 37 billion disintegrations per second.
Dose rate	Amount of time required for the curie strength of an isotope to be reduced by 50 percent.
Rem	Time over which a given amount of radiation is absorbed is expressed as __________.

NOTES

1. S. McGuiar and C. Peabody, *Working Safely in Gamma Radiography* (Washington, DC: U.S. Nuclear Regulatory Commission, 1982).
2. C. Sigl and R. Quinn, *Radiography in Modern Industry,* 4th ed. (Rochester, NY: Eastman Kodak Company, 1980).
3. Sigl and Quinn.
4. American Society for Quality Control, *SNT-TC-1A,* 1978.
5. *Nondestructive Testing: Radiographic Testing* (CT-6-6), 20 ed. General Dynamics Convair Division, 1983.
6. *NDT: RT.*
7. McGuiar and Peabody.

8. McGuiar and Peabody.
9. McGuiar and Peabody.
10. McGuiar and Peabody.
11. McGuiar and Peabody.
12. McGuiar and Peabody.
13. McGuiar and Peabody.
14. McGuiar and Peabody.
15. McGuiar and Peabody.
16. McGuiar and Peabody.
17. McGuiar and Peabody.
18. McGuiar and Peabody.

RADIOGRAPHIC TESTING (RT)

13

A radiograph is produced when X- or gamma rays are passed through a specimen, exposing a film. Some of the rays are absorbed by the specimen and some pass through to the film, forming a latent image. The image of the radiograph appears to the eye as a photographic negative. The more rays that hit the film, the higher the density (darkness).

For example, if a specimen has a void inside it caused by a bubble while it was in a molten state, there will be less material to react with the rays. There will be fewer rays absorbed, and the bubble will appear as a dark spot on the film.

Basically, only three things are needed to produce a radiograph[1] (see Figure 13-1):

- A source of electromagnetic radiation (gamma ray or X-ray)
- A specimen to be radiographed
- An indicating medium such as film

Many techniques are used to enhance the process and make a better picture. This chapter expands on these three areas.

OBJECTIVES

After completing Chapter 13, the reader should be able to:

- List the three things needed to make a radiograph.
- List and discuss the three major criteria for selecting a source.
- Discuss the major elements of the gamma ray camera.
- List and discuss the controls of the X-ray camera.
- Describe the basic construction of radiographic film.
- Differentiate between enhancing screens.
- Define terms associated with radiography.
- Describe the equipment used to determine radiographic quality.

13–1 SOURCE OF ELECTROMAGNETIC RADIATION

Much time was spent in Chapter 12 on the physics of radiation. You should now know the difference between X-ray, gamma ray, and particulate radiation. If this is not clear, take time now to reread those sections in the preceding chapter. This section deals with the selection, setup, and manipulation of a source to give various effects.

Figure 13–1

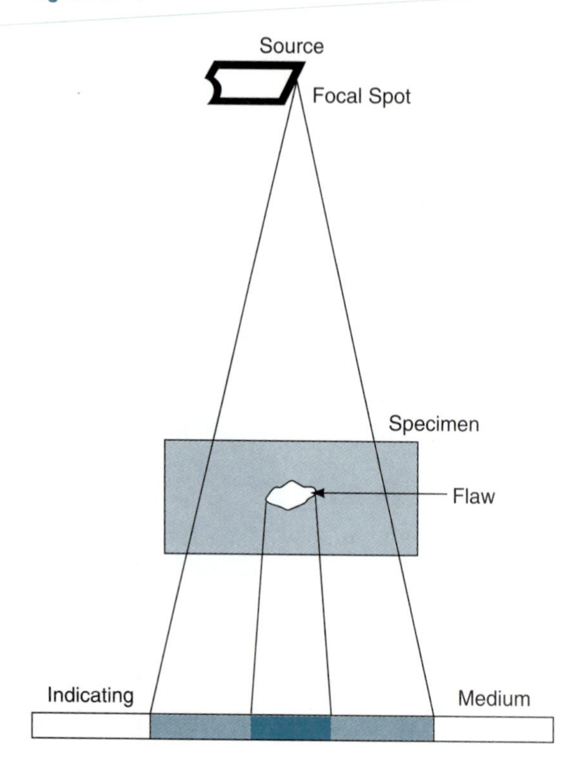

13–2 SELECTING A GAMMA RAY SOURCE

There are three major criteria in selecting a gamma ray source: penetrating power, activity, and half-life. These three factors should be weighed when selecting an isotope to be used to expose a specimen. Each type of isotope puts out frequencies peculiar to itself. Also, the higher the frequency, the more penetration by the ray. Therefore, to radiograph a thick, dense specimen, you should choose an isotope that emits a high-frequency wave.

The activity of the specimen is also important. The number of rays emitted is one of the determining factors in the amount of time it takes to make an exposure. Too much activity makes the quality of the radiograph poor; too little activity and the exposure time is too long.

When the expense of a source is considered, the activity and the half-life figure directly into the cost. However, these also determine how often the source must be replaced. It is costly to dispose of these

Table 13–1

RADIOACTIVE ELEMENT	HALF-LIFE	ENERGY OF GAMMA RAY (MeV)	APPLICATION
Thulium 170	127 days	0.084 and 0.54	Plastic, wood Light alloys
Iridium 192	70 days	0.137 to 0.65	1.5–2.5 in. steel
Cesium 137	33 years	0.66	1–3.5 in. steel
Cobalt 60	5.3 years	1.17 and 1.33	2.5–9 in. steel

materials—a source cannot simply be thrown in the trash can. It is a hazardous material and, as such, the disposal must be reported to the government.[2]

Density and geometry of the test specimen also play an important role in which isotope is chosen and how it is used. The density of the material is compared to that of steel to help the technician figure the time distance to be used with a given isotope (Table 13-1). For example, 2024 aluminum alloy is 0.35 the density of steel. The technician would look up the exposure for 0.35 in. of steel for every inch of aluminum to be radiographed.

13–3 SETTING UP THE CAMERA

Radiographic cameras are quite different from a typical photographic camera. The radiographic camera is a holding device to control the isotope when not in use. There are several types of cameras used in radiographic testing. The most popular is the S-curve camera. The S-curve camera is a feed-through camera that allows the isotope to be ejected and retracted by the technician to control the exposure. The camera has a hand crank connected to one side and a guide tube and tripod connected to the other side. As the technician cranks the source out of the camera, it follows the guide tube to the tripod. The tripod has been carefully placed in position to give the desired exposure. The technician monitors the amount of time that the isotope is in the "out" position. At a predetermined time, the technician retracts the isotope to the shielded position and the exposure is finished.

The radiographic camera is open at both ends (Figure 13-2). Gamma rays cannot escape because they travel in a straight line. When

Figure 13–2

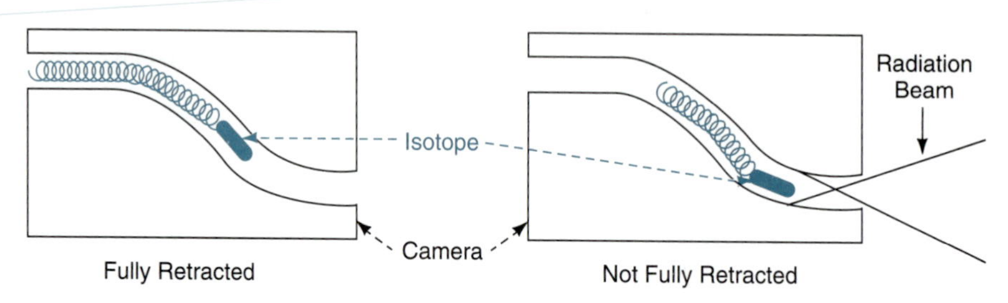

the isotope is fully retracted, the rays would have to turn a corner to get out. Since that is impossible, no rays escape. However, there is a danger with this type of camera. If the technician does not retract the isotope fully, a shaft of rays may be emitted from the opening of the camera. The technician must always approach the camera from behind with a survey meter and survey the camera from back to front for any sign of high radiation.

The hand crank is placed as far away from the camera as possible, and set up as soon as possible, to reduce the dose of radiation to the technician. If time permits, the technician should move to a safer, shielded position after cranking out the source. The tripod is placed in such a position as to give the proper exposure. In the case of gamma rays, the technician is limited in the variables that can be changed to improve the exposure. The intensity and activity of the rays are set by the type of source chosen and by the curie strength. The only variables open to the technician are source-to-film distance, film type, and exposure.

Source-to-film distance affects the sharpness of the image. As the distance increases, so does the sharpness. However, as the distance increases, so does the exposure time. The technician must decide on the quality desired and the time required.

Exposure time is the key to the *density* (darkness) of the exposure. The longer the shot, the darker it becomes. Most standards call out required densities. To be of acceptable quality, radiographs must be of that density in the area of interest.

The technician may desire to use a faster film. This type of film is made up of larger grains, which require less radiation to expose it. However, the increased grain size reduces the sharpness of the image. Often, the first shot is a trial shot to allow the technician to make corrections for better exposure.

13–4 X-RAY CAMERA CONSTRUCTION

As explained in Chapter 12, electrons crossing the vacuum from the cathode and striking the anode cause the emission of electromagnetic rays. This is an extreme simplification of the function of the X-ray machine. Radiography with X-rays gives more options and challenges to the technician.

As with gamma radiography, frequency and number of rays determine the time and quality of the image. With gamma radiography, the quantity and quality of the rays are fixed. With the X-ray tube, both of these parameters become variables.

The number of rays emitted from the X-ray camera is proportional to the number of electrons striking the target of the anode. The number of electrons striking the target is governed by controlling the current flow through the cathode. As more current flows through the filament, more electrons boil around it. This causes more electrons to be attracted by the anode. As more electrons interact with the tungsten target, more rays are produced. The milliamp (ma) control on the control panel governs the current flow through the cathode, which controls the number of rays.

The quality (frequency) of rays is controlled by how hard the electrons hit the target. This is governed by the potential difference between the anode and cathode—from tens of thousands to millions of volts.[3] The harder the electrons' impact on the target, the higher the Hertz (Hz) of the rays given off. The higher the Hz, the more penetrating is the power of the rays. Controlling this parameter is done with kilovolt (kv) control.

Consequently, if the technician wants to radiograph a thick, dense specimen, the kv adjustment would be increased. If at a given kv, time is to be reduced, the ma would be increased. Of course, the other variable, distance, also increases or decreases the time of exposure.

When a technician first shoots with a camera, it is a good practice to radiograph a "step wedge."[4] A step wedge is a piece of metal machined to different thicknesses in an increasing sequence. The shot is made at a predetermined setting. The thickness of the metal and the time of exposure are known. After the radiograph is developed, the technician can easily determine the variables for future shots.

When the technician wants to shoot a specimen of a certain material, the original step wedge would be used for the preliminary settings. Then, the specimen must be put under the camera in such a way that geometric complexities are minimized. The suspected *defect orientation* should be as close to parallel with the beam as possible. Remember that the radiographic film is sensitive to the thickness of the specimen. In a situation where there is a tight crack that runs perpendicular to the beam, the reduction in thickness would not be great enough to be detected. Often, perpendicular cracks are missed in this kind of inspection.

The geometric complexity and density of the specimen also affect the quality of the exposure. Subject contrast in the relative changes in darkness is caused by the geometric changes of the specimen. A geometrically complex part causes scatter radiation. The low-spectrum energy bouncing from the walls of the part causes the film to be exposed in a manner that causes the image to be less distinct.

Once the technician has reviewed these factors, it is time for the first shot. Using the setup numbers from the step wedge, the first shot is made and developed, and the technician evaluates the film. Then, if necessary, a correction shot is made. The exposure is altered using a chart called an H&D curve or characteristic curve. This chart uses the thickness of the material and the relative exposure that gave the first density to project a new density at another relative exposure. The technician may also use nomograms (graphs with intersecting axes that show nonlinear relationships) and other charts to calculate correction shots. Once the calculations are made on a shot for a specimen, the shot may be repeated on like specimens thereafter.

13–5 INDICATING MEDIA

There are many types of indicating media in use in the industry today. The most widely used is the radiography film. It consists of a cellulose base with a thin coating of silver bromide crystals in emulsion. When radiation hits the silver bromide, a physical change takes place, and it turns to black metallic silver. When a photon of radiation hits a crystal, it knocks an electron loose, and that electron bangs into another crystal. The more radiation that hits the film, the more bromide crystals are turned black, and the film becomes darker (denser).

The size of the silver bromide crystals can be varied to give desired results. Large crystals require less radiation to give an image. This is termed *fast film* because it does not take as much radiation to expose the film. The size of the crystals on this type of film, however, will cause the image to lack definition. It appears that one can see the grains of exposed silver. This appearance is called "graininess." The quality of this radiograph is not as high as with fine-grained film. Fast films can often be used with no enhancing screens: on exposures of especially dense, thick material, an intensifying screen may be used to speed up the exposure.

It takes longer to expose a film with a fine grain, but the image quality is much better. The small grain size tends to make sharp, crisp lines that are usually much easier to interpret. This type of film is used with an intensifying screen.

Intensifying screens increase the ionization of the bromide crystals. Only 1 percent of the radiation striking the film actually exposes the film.[5] The rest passes through the film without hitting a crystal and has no effect on the image. Intensifying screens are used to increase the efficiency of the radiographic process. As stated earlier, ionization of the bromide is the key to exposure.

There are two types of screens—lead and fluorescent. The lead screen is a thin sheet of lead, generally 0.005 to 0.020 in. thick. These sheets are placed on both sides of the film in the *cassette*. The sheet must be touching the film to work properly. Lead is a very dense material, which causes many rays to strike its molecules. As the rays knock electrons free from the lead, the electrons cross the space to the film. The free electrons cause ionization of the film much as a photon would, thereby increasing the efficiency of the radiographic process. This is the function of a lead screen.

The fluorescent screen is made of a material, such as calcium tungstate, that emits light in the presence of radiation. The film is sensitive to light rays, just as it is to gamma and X-rays. As the screen begins to fluoresce, the film is exposed and the latent image is formed. The fluorescent screen causes a loss of sharpness, but often it is the only way to make an exposure.

Both of these screens reduce exposure time, but that is not their primary purpose. Their primary purpose is to increase the efficiency of the X- and gamma rays.

Contrast is the relative difference in darkness (density) of the radiograph, a comparison of the lightest area to the darkest area. In other words, it is what makes defects stand out. There are two factors that affect contrast—specimen contrast and film contrast. *Film contrast* is the film's ability to go from light to dark, based on its manufacture. *Specimen contrast* is the difference in darkness caused by the geometry of the test specimen.

Sharpness of the radiograph is the distance it takes to get from one area of contrast to another. This is the crispness of the lines. Radio-

Figure 13–3

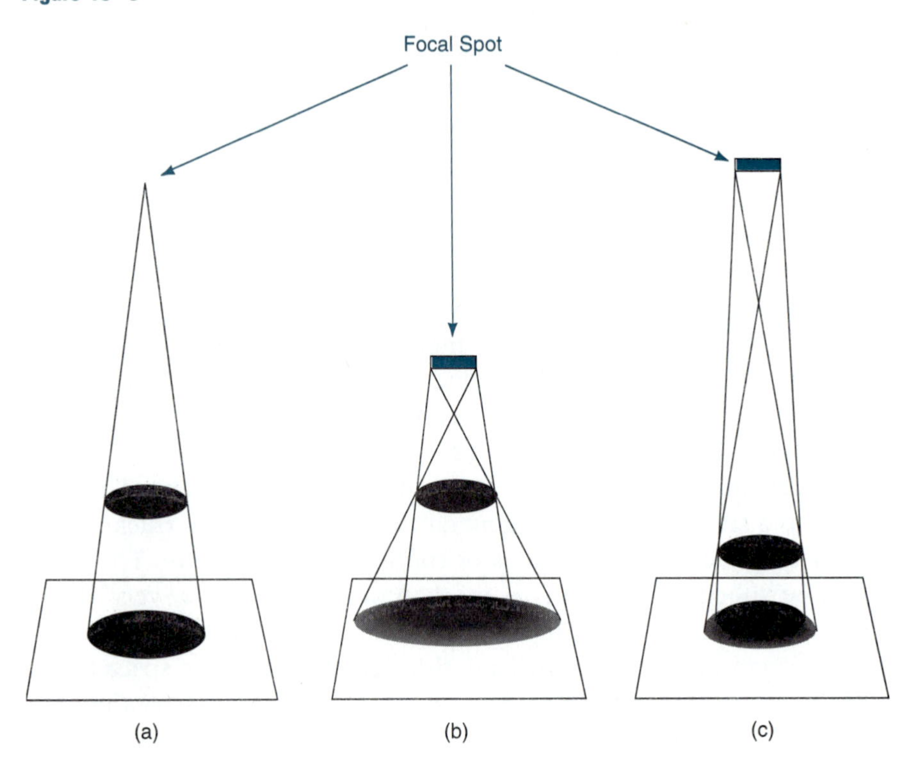

graphs can have the same contrast, but different sharpness. The major factors in this area are film grain, specimen configuration, setup, and source focal spot size.

As mentioned earlier, the graininess of the film plays an important part in sharpness. If the grains are large, it is harder to form a crisp, well-defined line.

If specimen configuration is such that there are critical geometries located at a distance from the film, that will reduce the sharpness of the image. In these cases, changing film speed or setup may help.

Distance of the source to the film also affects the sharpness. The farther away the source, the more crisp the radiograph. The shadow around the image can be affected a great deal by the distance and the angle of the source (Figure 13-3). The size of the source focal spot can be offset to some extent by increasing the distance of the source to the film.

The image on the radiograph is affected by many factors. The chart in Figure 13-4 helps to put these factors into perspective.

Figure 13–4 Factors Affecting Image Quality

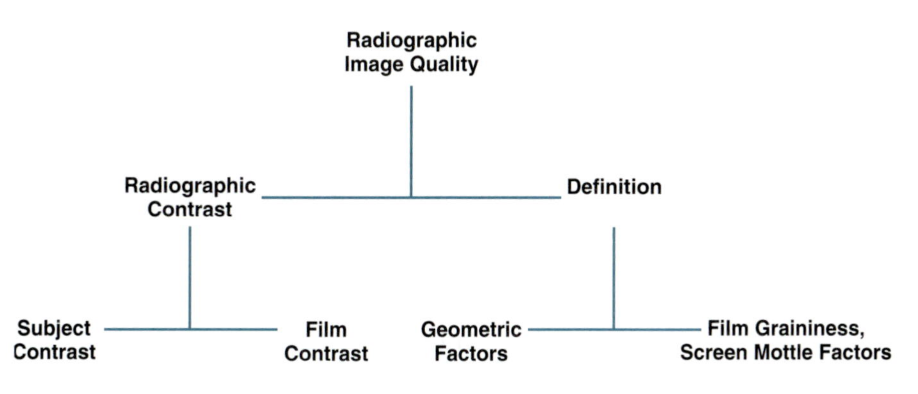

Subject Contrast	Film Contrast	Geometric Factors	Film Graininess, Screen Mottle Factors
Affected by:	**Affected by:**	**Affected by:**	**Affected by:**
A Absorption differences in specimen (thickness, composition, density)	**A** Type of film	**A** Focal-spot size	**A** Type of film
B Radiation wavelength	**B** Degree of development (type of developer; time and temperature of development; activity of developer; degree of agitation)	**B** Source-film distance	**B** Type of screen
C Scattered radiation	**C** Density	**C** Specimen-film distance	**C** Radiation wavelength
Reduced by:	**D** Type of screens (fluorescent vs. lead or none)	**D** Abruptness of thickness changes in specimen	**D** Development
1 Masks and diaphragms		**E** Screen-film contact	
2 Filters		**F** Motion of specimen	
3 Lead screens			
4 Potter-Bucky diaphragm			

13–6 DETERMINING RADIOGRAPH QUALITY

It is impossible to determine whether the quality of the radiograph itself is good just because something does not show up. There must be a known factor in each radiograph for the express reason of judging the quality of the radiograph. These indicators are generally called *pene-*

trameters. Penetrameters are of various thicknesses, with physical characteristics of predetermined size. This allows the reader to judge whether the radiograph is capable of detecting the defect.

The code to which the part is built calls out the penetrameter size and shape. There are two basic penetrameters—the hole type and the wire type. The hole type is made of material like that being radiographed. It is generally 2 percent of the thickness of the part. Three holes are drilled into the penetrameter. The holes may vary in size (determined by the thickness of the penetrameter), or they may all be the same size. The person reading the radiograph must be able to see holes of a certain size, or all the holes, depending on the code. In the past, penetrameters of this kind have been altered to make it easier to see the holes. Of course, this violates the code.

Wire penetrameters were developed in Germany and are becoming more and more popular in the U.S. This type of penetrameter uses wires of increasing size to gauge the quality of the radiograph. The technician interpreting the radiograph can determine its quality by how many wires are visible. To determine what penetrameters are required on any shot, the technician must know to what specification the work is being performed.

SUMMARY

Industrial radiography is a method of nondestructive inspection that detects subsurface defects. This method is not limited by specimen material.

Images are made on an indicating medium when electromagnetic radiation passes through a specimen and strikes the silver bromide crystals in the film, turning them black. The greater the number of photons striking the film, the darker the image. The definition of the image is affected by the intensity of the ray, the thickness of the specimen, film type, and time of the exposure. When using gamma rays, time, distance, and film type will be major variables. When using X-rays, the variables of quality and quantity of ray are added. With both, intensifying screens are used to increase the efficiency of the process. These may be lead screens or fluorescent types.

RT is most widely used where one piece is shot at a time. However, there are many applications where radiation is used on an assembly line in a real-time environment. Long and in-depth training is required for this method. The testing equipment is expensive and, in the case of isotopes, requires licensing. This method offers permanent records of test results.

KEY TERMS

Cassette Film holder used during exposure.
Contrast Relative difference between dark and light areas of the radiograph.
Density The darkness of the radiograph.
Defect orientation The position of the suspected flaw relative to the angle of the ray.
Fast film Larger-grain films requiring less exposure to be developed.
Film contrast Differences in the darkness of the radiograph caused by film type.
Penetrameters Quality indicators used on radiographic films.
Sharpness The distance it takes to get from one area of contrast to another on the radiograph.
Specimen contrast Difference in darkness of the radiograph caused by the geometry of the specimen.

RADIOGRAPHY TEST

1. List the three things needed to produce a radiograph.
2. In a short paragraph, discuss the criteria for the selection of a gamma ray source.
3. Describe the use of the "S" in the S-curve camera.
4. What control on the X-ray machine increases the depth of penetration of the ray?
5. In unexposed film, what is the material that causes the radiograph to turn dark?
6. Name and describe the two types of enhancing screens.
7. What equipment was developed in Germany to determine the quality of the radiograph?

NOTES

1. *Nondestructive Testing: Radiographic Testing* (CT-6-6), 2d ed. General Dynamics Convair Division, 1983.
2. *NDT: RT.*
3. *NDT: RT.*
4. *NDT: RT.*
5. C. Sigl and R. Quinn, *Radiography in Modern Industry,* 4th ed. (Rochester, NY: Eastman Kodak, 1980).

PART III: STATISTICS AND PROBABILITY

PART OBJECTIVES

This section introduces the basics of the following:

- Distributions of data
- Plotting and graphing data
- Measures of central tendency
- Measures of dispersion
- Probabilities
- Normal distributions
- Poisson distributions
- Binomial distributions
- Pareto diagrams

Working with numbers is no different from working with many other types of information. For example, if information (data) were collected about cars that are in a parking lot, the raw data would not reveal much. The raw data would have information such as the make, year, and color of each car that was in the lot. All this information, by itself, would be of little value because the data were not grouped so that one could see a pattern. The best way to look for patterns is to group data together and look at the number of cars in each group. If this same approach is used with numbers instead of with cars, a statistical analysis is performed. As you progress through these next chapters, we will refer occasionally to the example of cars in a parking lot. This should help you visualize what is being accomplished at each step. This section covers only ways of grouping data to look for patterns or to make predictions about the data.

Statistics is a high-powered word describing how numbers are used to delineate patterns, make decisions, and predict future trends. The car example is simple statistics—cars are counted and grouped in ways that will make the information usable. From the information on the cars in a parking lot, a store manager of an auto parts store could determine how many Ford, General Motors, Chrysler, and foreign cars were in the lot. This information could be used to determine how much of the inventory should be allotted for each manufacturer. The parts demand can be predicted by the fact that all cars will need parts sometime. If 60 percent of the cars in the lot were General Motors (GM) products, then around 60 percent of the parts purchased would be for GM cars. Future needs can be predicted by using simple statistics.

The use of statistics is a vital management tool for manufacturing, production, quality control, and accounting. Use of statistics allows management to make decisions based on quantitative fact rather than on long-held misconceptions. The general idea in business today is that if something cannot be measured, it cannot be effectively managed. To provide quantitative facts, the quality control technician must understand and be able to apply statistics to many situations. Two definitions must be introduced:

Statistics The mathematics of the collection, organization, and interpretation of numerical data.

Data Information, especially information organized for analysis or used as the basis for a decision.

Statistical methods can be applied whenever a source of variation exists. That means statistics can be applied almost anywhere, because no matter where one looks, there is evidence that *everything varies*. That statement, which may seem bold, is a fact of life. The variation might be large (differences in the sizes of lakes in the United States, for example) or small (the difference in the thickness of two sheets of the same newspaper, for instance), but the fact is that everything varies. The unique thing about this variation is that everything varies in a predictable way. The predictable variation allows statistics to be applied to many aspects of life. For example, write a name five times and compare the writing. No two of the names were written exactly the same—everything varies. In most cases, each of the names will look remarkably similar. Although everything varies to some extent, groups of things tend to be predictable. This predictability can be used to make decisions.

The word *statistics* seems to strike terror into the hearts of many people. But if one can add, subtract, multiply, divide, and find a square and a square root (using a calculator), one can master the majority of statistical operations used by most corporations. Although the terms are new and technical, they describe many math functions that people use on a daily basis.

Data can be divided into two types of information: written or numeric. Numeric data is used in this section.

Numeric data can be divided into two types: variable and attribute. Attribute data are found by counting a characteristic that is either present or absent. These types of data are presented in the form of whole numbers or in percentages (the number of red cars in a parking lot, the percentage of defective parts in a lot of product, or the number of defects in a single unit of product). Variable data can assume any value between any whole numbers. These types of data are found by measuring a unit of product (the length of a pen at 5.7 in. or the weight of a book at 1.25 lbs). Both types of data are used in quality control for making decisions.

CHAPTERS IN THIS PART

DISTRIBUTIONS

14

The term *distribution* can be described as how individual pieces of data compare to each other when viewed as a group. If the cars in a parking lot are looked at for the distribution, many criteria could be used. The color distribution of the cars could be determined. The distribution of the model years could also be evaluated. From one group of data, many areas can be evaluated by looking at how the data is distributed.

A second term that needs to be defined is *frequency.* Frequency—"how often"—is found by counting like data. If the raw data were the cars in a parking lot, the cars could be grouped together by some common element (such as color) and the number of cars in each group counted. The number in each group is termed the frequency of that group. No matter which way the information is presented, it can be grouped and counted to form patterns.

Distributions describe the way that data are arranged (distributed) in relation to the other points of data. The distribution can show (1) all the points of data in ascending order (*ungrouped distribution*); (2) the different points of data and number of times the data are repeated (*ungrouped frequency distribution*); or (3) data put together into groups, and a count taken of the number of times that the data are repeated (*grouped frequency distribution*).

Suppose the distances in miles that 25 people lived from a shopping mall were gathered. The list might look like the following:

12, 19, 21, 6, 19, 9, 13, 8, 9, 13, 15, 13, 12, 10,
13, 8, 14, 14, 16, 13, 7, 13, 11, 19, 13

The data in their raw form do not give a picture of the distribution. To find the distribution, the data should be arranged into one of the

OBJECTIVES

After completing Chapter 14, the reader should be able to:

- Understand the terms associated with distributions.
- Distinguish among the three methods of grouping data.
- Name the parts of a grouped frequency distribution.
- Discuss the procedure for constructing each type of distribution.
- Demonstrate the ability to construct each type of distribution from data given.
- Discuss the use of each distribution.

three forms described earlier. In the remainder of this section, we discuss these three distributions.

14–1 UNGROUPED DISTRIBUTION

The ungrouped distribution is the simplest to use. To form this distribution, put the information into a logical order.[1] In the example of the cars, the logical order may be that all the cars made by a given company (Ford, General Motors, etc.) are listed on a sheet of paper. The point that makes this distribution somewhat difficult to work with is that all the data are listed, no matter how often repeated. If, by chance, five cars are of the same make, model, year, and color, all will be noted one after another on the list.

In an ungrouped distribution, the data are arranged in a logical order, most often ascending. All the data are listed, no matter how often they occur in the raw data. For example—consider data on the distance from a mall that 25 customers live. If these distances were arranged in ascending order, the distribution could be identified. The raw data as collected appears as:

12, 19, 21, 6, 19, 9, 13, 8, 9, 13, 15, 13, 12, 10,
13, 8, 14, 14, 16, 13, 7, 13, 11, 19, 13

After the data are arranged, the ungrouped distribution is:

6, 7, 8, 8, 9, 9, 10, 11, 12, 12, 13, 13, 13, 13, 13,
13, 13, 14, 14, 15, 16, 19, 19, 19, 21

The raw data show 25 distances; when arranged in an ungrouped distribution (ascending order), the pattern shows that several individuals live at about the same distance from the mall as do others. Ungrouped distributions are nothing more than putting like data together and ranking the data in a logical order.

This information could be used by a store owner to decide how far away from the mall to advertise sales. From this distribution it is apparent that most of the customers are within a 15-mile radius from the mall.

Conclusions In the ungrouped distribution, the data are ranked in ascending order and every point of data is shown separately, even if it is identical to other points of data. This distribution is very effective for use with small amounts of data (25 or fewer). When the amount of data grows larger, methods of grouping like data are used to help make the distribution manageable. This distribution would not be effective if 1000 people were asked how far they lived from the mall.

Exercise No. 1 Form an ungrouped distribution from the following data:

2, 6, 4, 9, 5, 4, 5, 7, 6, 9, 7, 5, 6, 4, 5, 1

Exercise No. 2 Form an ungrouped distribution from the following data:

9, 5, 7, 3, 8, 2, 9, 5, 7, 3, 9, 6, 1, 5, 7, 1

14–2 UNGROUPED FREQUENCY DISTRIBUTION

The ungrouped frequency distribution takes the ungrouped distribution one step further. In the ungrouped distribution, all the data are listed, no matter how often the data are repeated. To convert that distribution into a frequency distribution, only the unique (different) points of data are listed, and the number of times the data are repeated is counted.[2]

In the case of the cars, it would be logical to place the cars in order by the model and the number of cars that are of each model. This simplifies the data, because data that are identical are grouped together and counted. The distribution shows only the unique data and how many times the data are present. If the information from the cars in the parking lot is grouped, it would show different models of cars and how many of each model are in the parking lot. This helps anyone looking at the data to make judgments about the data with less effort.

In the ungrouped frequency distribution, the unique points of data are arranged in ascending order and the number of times the data are repeated for each value is shown. This is used to simplify working with larger amounts of data. If the distances 25 customers live from a mall were put into a grouped distribution, the raw data would look like this:

12, 19, 21, 6, 19, 9, 13, 8, 9, 13, 15,
13, 12, 10, 13, 8, 14, 14, 16, 13, 7, 13, 11, 19, 13

In Figure 14-1, the different points of data are listed and the number of occurrences is shown above each number with "tic" marks. From this distribution, it is apparent that 6 is present in the data one time, whereas 8 has two occurrences, 13 has seven occurrences, and so on.

To summarize: In the ungrouped frequency distribution, the data are listed in ascending order. Only unique data points are shown, and the number of occurrences for each point of data is shown with tic marks over or beside the number. It is used to simplify working with large data points.

Figure 14–1 Ungrouped Frequency Distribution

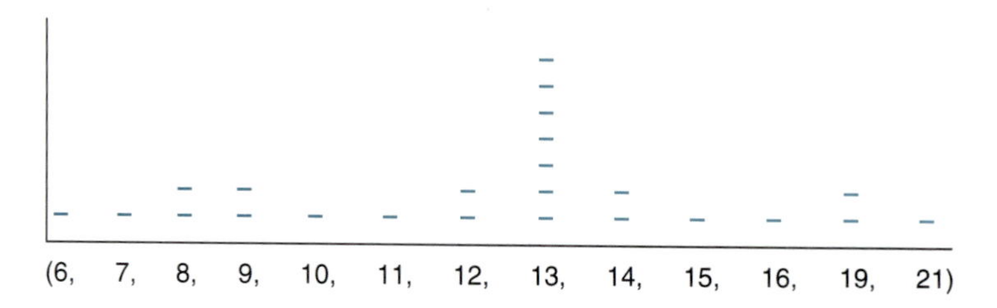

Exercise No. 1 Form an ungrouped frequency distribution from the following data:

2, 9, 6, 7, 4, 9, 5, 8, 4, 8, 5, 7, 6, 9, 7, 5, 6, 2, 7, 7, 5, 6, 4, 5, 1

Exercise No. 2 Form an ungrouped frequency distribution from the following data:

9, 5, 7, 3, 8, 7, 6, 9, 7, 5, 6, 2, 7, 7, 2, 9, 5, 7, 3, 9, 5, 1, 5, 7, 1

14–3 GROUPED FREQUENCY DISTRIBUTION

A grouped frequency distribution is used with very large amounts of data that have many unique pieces of data. Unique data points are grouped with other data points that are close in value. If the ages of 5000 people were evaluated, the distribution would range from 1 year to over 100 years. The distribution would be nearly impossible to evaluate with 5000 separate points of data. Even if a frequency distribution were used, the possibility of 100 unique points of data with this frequency would still make the distribution hard to calculate. With this large amount of data spread over many unique points of data, it would be logical to group the ages into fewer, similar groupings and find the frequency of occurrence for each group. The ages could be combined into a range of ages per group (1 year to 10 years in one group, 11 years to 20 years in a second, and so on); the total number of groups would be reduced to 10. The frequency could be found for each group and the distribution evaluated.

In the case of the cars, the information could be grouped in one of several ways. For this example, the information will be grouped by model year. With large amounts of data, this is the best way to put data into a manageable size.

The first step to make groups in which to put the cars is to look at the range of model years that is present in the data. The oldest car is a

Table 14-1

CLASS	LOWER LIMIT	UPPER LIMIT
1	1964	1967
	+ 4	+ 4
2	1968	1971
	+ 4	+ 4
3	1972	1975
	+ 4	+ 4
4	1976	1979
	+ 4	+ 4
5	1980	1983
	+ 4	+ 4
6	1984	1987

1972, whereas the newest car is a 1995. To find the range, subtract 1972 from 1995 (= 23 years). Add 1 to this figure to make the range include all the years. If all the years between 1972 and 1995 are counted, including 1972 and 1995, 24 years are included. All the cars were made in that 24-year period. The next step is to break the cars into groups (*classes*) so the distribution can be seen. The data may be divided up in several ways. The range of years is 24, so some options are 1 class of 24 years, 2 classes of 12 years, 3 classes of 8 years, 4 classes of 6 years, 6 classes of 4 years, 8 classes of 3 years, 12 classes of 2 years, or 24 classes of 1 year.

The number of classes should be large enough to show the shape of the distribution, but small enough that the width will still be apparent. General rules covering the proper number of classes to use with data are discussed later in this chapter.

For this example, the number of classes we choose is 6. Dividing the range by the number of classes ($24 \div 6 = 4$) gives the *class width.* In this case, the class width is 4 years. The data should be divided into even classes.

The class limits are found by starting with the smallest data point and adding the class width to the *lower limit.* The *upper limit* is one year below the lower limit of the next higher class. This ensures that all the cars have only one class in which they fit (see Table 14-1).

The last step is to count the number of cars that fit into each class and to make a frequency distribution of the data (Figure 14-2).

Figure 14–2 Frequency Distribution

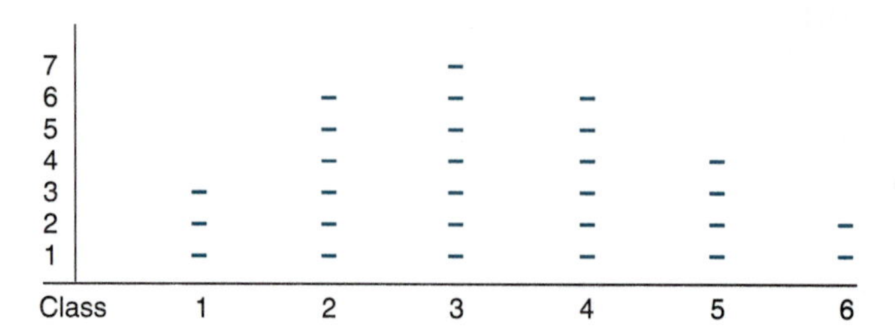

A grouped frequency distribution is like an ungrouped distribution in the way that data are first treated. All data should be ranked in ascending order. The data are divided into classes of equal width.[3] The number of classes should be more than 5 and less than 20. The larger the amount of data that are to be ranked, the larger the number of classes to be used. As a guide, Juran and Gryna (in *Quality Planning and Analysis*) developed a table for the number of observations of data versus the number of classes (Table 14-2).[4]

The number of classes (or *cells*) used does not have to follow this table to the letter, but it should be used as a guide. It is commonly accepted that one can vary one class amount either side of the recommendation and still have the data exhibit the same distribution. One last recommendation for grouped distributions—if there are fewer than 50 observations, the grouping may give indications that could lead to misinterpretations about data distribution. The following recommendations give guidance for making classes for these data. (Note: The data represent the number of miles driven by 50 people in one week.)

Table 14–2

NUMBER OF OBSERVATIONS	RECOMMENDED NUMBER OF CELLS
20 to 50	6
51 to 100	7
101 to 200	8
201 to 500	9
501 to 1000	10
over 1000	11 to 20

Data

27	68	79	91	107
43	71	80	91	108
43	71	81	93	108
44	71	82	94	116
47	73	82	94	120
49	73	84	94	120
50	74	84	96	122
54	75	86	97	123
58	76	88	103	127
65	77	88	106	128

1. Each class should be the same width.
2. The classes should not overlap and should be placed so the data will fit into only one class (mutually exclusive classes).
3. Classes should start as low or lower than the smallest data point and end as high or higher than the largest data point. All data must fit into a class (all inclusive).

The first step in setting up classes is to find the data *range*. For these data, the range is 128 − 27 = 101. Add 1 to the range, resulting in 102. This makes the range all-inclusive (include all data points). From Table 14-2, 6 classes should be used with 50 observations. The class width of 17 will be used for this example. If dividing the range by the number of classes did not give an even class width, two ways can be used to provide even class widths.

- Try to divide the data into fewer or more classes.
- Add as many numbers to the range as needed to make even class widths for the number of classes used.

In this example, the number of classes could have been 5, 6, or 7—whichever would produce even class widths. If varying the number of classes did not produce even class widths, the range could have been increased (103, 104, 105, etc.) to provide an evenly divisible range.

The process of defining class limits is now a matter of simple addition. Starting with the lower class limit (the smallest piece of data to be grouped), add the class width.

27	Lower Class Limit
+17	Class Width
44	Lower Class Limit of Second Class
+17	Class Width
61	Lower Class Limit of Third Class

Table 14-3

CLASS	LOWER LIMIT
1	27
2	44
3	61
4	78
5	95
6	112

This procedure is followed until the lower class limits are found for all classes (Table 14-3).

Upper class limits are found in much the same way. The upper class limit for the first class is one number below the lower limit for the second class. For example, 44 is the lower limit for the second class. To find the upper limit for Class 1, subtract 1 from 44. This means the first class has a lower limit of 27 and an upper limit of 43. To find the rest of the upper class limits, add the class width to the upper limit to find the next upper limit (Table 14-4).

A *class midmark* is the number that is an equal distance between the upper limit and lower limit of the same class. To find the midmark for the first class, add the lower limit to the upper limit and divide by 2 ($27 + 43 = 70 \div 2 = 35$). The midmark is 35. Add 17 (class width) to each midmark to find the rest of the midmarks just as the limits were found (Table 14-5).

A *boundary* is a number that separates classes from each other. In the example given, the first class limits are 27 and 43. The second class limits are 44 and 60. The boundary falls between 43 and 44. To locate the boundary, add 43 and 44 (= 87). If that total is divided by 2, the boundary of 43.5 is located. This boundary is used to group data into

Table 14-4

CLASS	LOWER LIMIT	UPPER LIMIT
1	27	43
2	44	60
3	61	77
4	78	94
5	95	111
6	112	128

Table 14–5

CLASS	LOWER LIMIT	MIDMARK	UPPER LIMIT
1	27	35	43
2	44	52	60
3	61	69	77
4	78	86	94
5	95	103	111
6	112	120	128

only one class. The boundaries are now located for each class as a point of separation (Table 14-6).

Table 14-6 has boundaries, class limits, and midmarks. The last step to group data in this form is to put the data into the proper class. They are grouped as in the grouped frequency distribution. The distribution is shown in Figure 14-3.

Table 14–6 Grouped Frequency Distribution

CLASS	BOUNDARIES	LIMITS	MIDMARK
	26.5		
1		27 43	35
	43.5		
2		44 60	52
	60.5		
3		61 77	69
	77.5		
4		78 94	86
	94.5		
5		95 111	103
	111.5		
6		112 128	120
	128.5		

Figure 14–3 Grouped Frequency Distribution

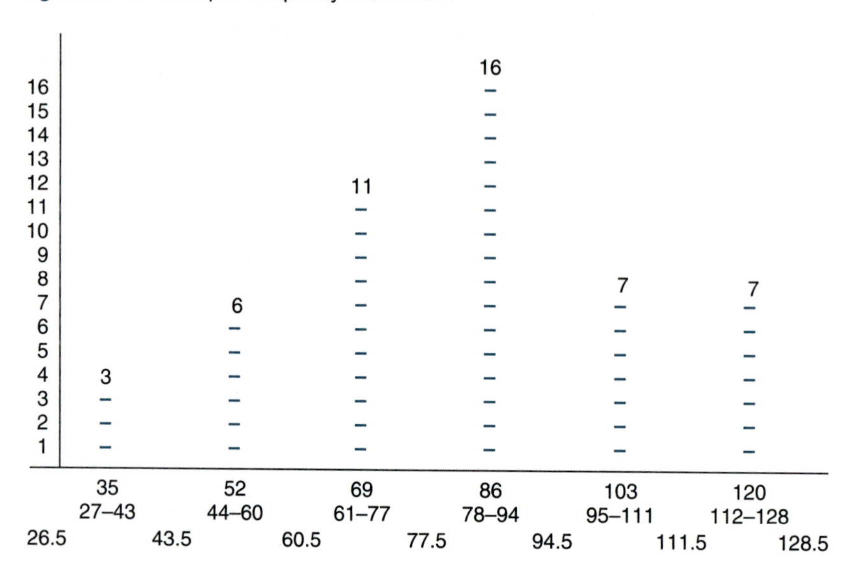

SUMMARY

The way to handle very large amounts of data is to group the data into logical classes and find the amount of data in each class. From this distribution, decisions can be made about the data.

How can these techniques assist in making judgments about data? As discussed in the introduction, everything varies, and groups of things vary in a predictable manner. The predictability of data shows in the shape of the distribution when plotted on a graph. The shape of the distribution of data that has no outside disturbance acting on it appears as in Figure 14-4.

When data are plotted and the shape of the distribution is evaluated, it can be determined whether outside forces are acting on the data. If *outside disturbances* are present, the distribution will have other than the normal (bell) curve. As we explain the significance of the normal distribution (Chapter 16), you will realize the importance of understanding methods of grouping data to show the distribution.

Distributions identify how individual pieces of data compare to each other when viewed as a group. Three of the more common types of distributions are the ungrouped, ungrouped frequency, and grouped frequency. The ungrouped distribution shows all points of data in ascending order. The ungrouped frequency distribution shows different points of data as the number of times the data are repeated. A grouped frequency distribution indicates data grouped together into classes and counts the number of times the data are repeated.

The ungrouped distribution and ungrouped frequency distribution identify each individual occurrence of a data point. The grouped frequency distribution uses classes having upper and lower limits, midmarks, and boundaries to identify occurrences within a certain range.

Figure 14–4 Bell Curve

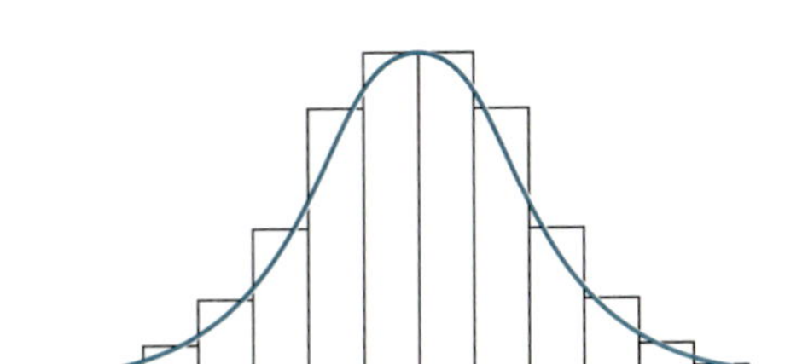

KEY TERMS

Boundary The number that separates one class from the next.
Cell Class.
Class A group of data within given parameters.
Class midmark The number that occurs halfway between the upper limit and lower limit of the same class.
Class width The distance between the upper limit of a section of grouped data and the lower limit.
Distribution The manner in which individual pieces of data compare to each other when viewed as a group.
Frequency Rate of occurrence.
Grouped frequency distribution The arrangement of data points compiled into groups and counted by the number of times the data are repeated.
Lower limit The lowest possible number that could occur within a given class.
Outside disturbance The external force acting on a system that is not inherent in the process, causing a deviation.
Range The difference between the highest and lowest values in a group of data.
Ungrouped distribution The arrangement of data points in ascending order.
Ungrouped frequency distribution The arrangement of data points indicating the number of times the data is repeated.
Upper limit The largest possible number that could occur within a given class.

DISTRIBUTIONS TEST

1. Match the terms on the right to the correct definition.

   a. The number of times the data are repeated.
   b. The groups in which to put data.
   c. The range of data divided by the number of classes gives the _____.
   d. The class lower extreme.
   e. The class upper extreme.
   f. The number that falls halfway between the upper and lower limit.
   g. The number that separates the classes from each other.

   1. Boundary
   2. Width
   3. Midmark
   4. Frequency
   5. Separation
   6. Upper limit
   7. Class
   8. Number
   9. Bottom
   10. Lower limit

2. Distinguish among ungrouped, ungrouped frequency, and grouped frequency distributions by placing an X to the left of the statement that identifies an ungrouped frequency distribution.
   a. Data put in logical order.
   b. Data put in logical order and the number of occurrences of each point listed.
   c. Data divided into groups and the number of occurrences for each group listed.
3. Name the three parts of a class of a grouped frequency distribution.
   a.
   b.
   c.
4. Write a short paragraph describing the method of constructing ungrouped, ungrouped frequency, and grouped frequency distributions.
5. a. From the following data, construct an ungrouped distribution.

   1, 9, 3, 4, 7, 3, 6, 8, 5, 6, 5, 2, 5, 3

   b. From the following data, construct an ungrouped frequency distribution.

   1, 9, 3, 4, 7, 3, 6, 8, 5, 6, 5, 2, 5, 3

   c. From the following data, construct a grouped frequency distribution with three classes. Show upper limits, lower limits, midmarks, and boundaries.

   1, 9, 3, 4, 7, 3, 6, 8, 5, 6, 5, 2, 5, 3

6. Write a short paragraph describing the circumstances where the ungrouped, ungrouped frequency, and grouped frequency distributions are used.

NOTES

1. Dale Besterfield, *Quality Control,* 2d ed. (Englewood Cliffs, NJ: Prentice-Hall, 1986), 15.
2. Joseph Juran and Frank Gryna, Jr., *Quality Planning and Analysis,* 2d ed. (New York: McGraw-Hill, 1980), 36.
3. Besterfield, 18.
4. Juran and Gryna, 37.

DISTRIBUTIONS

Exercise No. 1

Plot an ungrouped distribution from the following data.

1. 2,3,9,8,7,4,6,5,4,7,6,5,5,6
2. 10,6,5,5,1,6,7,4,3,5,6,6,4,7,8,9,3,2

3. 4,9,5,6,5,7,5,2,3,9,8,7,4,6,6,5,3,4,7,6,1,5,5,6,10
4. 5,8,7,8,7,10,6,5,5,1,6,7,4,3,5,6,6,4,7,8,9,3,2,5,7,5,6,5,9,4,7,8,8,7
5. 1,13,2,12,5,12,6,11,7,11,10,10,4,18,7,14,10,11,5,17,
8,13,9,6,12,16,7,10,11,15,9,20,8,9,19,10,8,9,9,4,18,7,14,10,11,5,17,8

Plot a grouped distribution from the following data.

6. 20,8,9,19,10,9,8,9,4,18,7,14,10,11,5,17,8,9,15,11,10,7,16,12,
6,9,13,8,17,S,11,10,14,7,18,4
7. 11,8,3,11,12,10,11,9,13,14,8,15,11,10,7,16,12,6,9,13,8,17,5,11,10,14,7,18,4
8. 4,18,7,14,10,11,5,17,8,13,9,6,12,16,7,10,11,1S,8,14,13,9,11,10,12
9. 1,13,2,12,5,12,6,11,7,11,10,10,4,18,7,14,10,11,5,17,8,13,9,6,
12,16,7,10,11,15,9,20,8,9,19,10,8,9,9,4,18,7,14,10,11,5,17,8
10. 5,8,7,8,7,10,6,5,5,1,6,7,4,3,5,6,6,4,7,8,9,3,2,5,7,5,6,5,9,4,7,8,8,7

Exercise No. 2

Plot an ungrouped and a grouped frequency distribution for the following data.

1. Inspection data of hole diameter—2.5 in., ±0.5 in.

2.7	2.5	2.6
2.2	3.0	2.4
2.4	2.6	2.7
2.6	2.3	3.0
2.7	2.5	2.8
2.9	2.8	2.8
3.1	2.9	2.6
2.7	2.8	2.7

2. Inspection of "I" beam, length 24 ft. with acceptance ±6 in.

24′3″	24′1″	23′8″
23′7″	23′10″	23′11″
23′11″	23′6″	24′0″
23′9″	23′11″	23′9″
23′10″	24′0″	23′6″
23′9″	23′8″	23′10″
23′11″	23′11″	24′0″
23′10″	24′0″	23′10″

3.

1	4	4
5	10	13
11	2	6
3	12	11
12	14	3
5	4	14
13	13	8
7	15	15
14	7	6
6	16	17

4. Measurements of the modal frequencies across a –2.25-MHz ultrasonic transducer.

2.45	2.30	2.50
2.25	2.15	2.40
2.30	2.40	2.25
2.20	2.60	2.35
2.35	2.25	2.35
2.25	2.30	2.25
2.35	2.20	2.45
2.30	2.35	2.30
2.70	2.25	2.40

5.

10	19	18	10
15	9	14	15
9	6	11	8
16	16	16	6
13	8	15	16
8	15	8	14
17	17	17	10
7	14	16	15
15	9	5	7
10	11	10	11
12	18	17	4
18	10	7	21
9	19	20	9

6. Bar stock with specifications of 2 in., ±1/16 in. was measured, with the following results:

2″	2 1/32″	2 1/16″
1 15/16″	1 31/32″	2″
2 1/32″	2 1/16″	1 31/32″
1 29/32″	2″	2″
2 1/16″	2 3/32″	1 15/16″
1 31/32″	1 31/32″	2″
2 1/16″	2 1/32″	2″
2″	2″	2″
2 1/32″	2″	1 31/32″
2″	2 1/32″	2″
1 31/32″	2″	2 3/32″

7. Sample of broken eggs per carton.

2	3	9	8	7
4	6	5	4	7
6	5	5	6	10
6	5	5	1	6
7	4	3	5	6
6	4	7	8	9
3	2	4	9	5
6	5	7	5	2
3	9	8	7	4
6	6	5	3	4
7	6	1	5	5
6	6	7	8	7
10	6	5	5	1
6	7	4	3	5
5	5	1	6	7

Exercise No. 3

Plot a grouped frequency distribution for the following data. A sample of 50 punched-hole diameters was inspected from a lot of 1000. A second sample was made the next day. Were there any changes in the process?

Day 1

0.499	0.502	0.504	0.503	0.501
0.500	0.501	0.502	0.501	0.498
0.498	0.500	0.500	0.504	0.502
0.501	0.504	0.497	0.499	0.501
0.502	0.499	0.499	0.500	0.499
0.501	0.503	0.503	0.498	0.503
0.503	0.502	0.500	0.503	0.502
0.500	0.499	0.505	0.501	0.500
0.502	0.500	0.501	0.502	0.501
0.499	0.501	0.502	0.500	0.503

Day 2

0.497	0.499	0.502	0.503	0.501
0.502	0.502	0.498	0.497	0.502
0.498	0.496	0.499	0.499	0.498
0.501	0.503	0.497	0.496	0.502
0.495	0.499	0.502	0.501	0.497
0.499	0.498	0.496	0.498	0.499
0.497	0.501	0.499	0.502	0.497
0.499	0.497	0.501	0.498	0.502
0.502	0.498	0.498	0.502	0.499
0.498	0.503	0.502	0.499	0.501

Exercise No. 4

Compare the two samples of inspection data on length of rebar taken from the same truckload. Are these samples from the same lot of material? If you think not, why would you question these data?

Sample No. 1

29′10″	29′11″	30′2″	30′4″	29′9″
30′0″	30′4″	29′7″	30′0″	30′1″
29′11″	29′10″	30′0″	30′2″	29′11″
30′3″	30′0″	30′3″	29′8″	30′6″
30′1″	29′9″	30′1″	29′11″	30′2″
29′9″	30′2″	29′10″	30′1″	30′0″

Sample No. 2

29′11″	29′9″	30′1″	30′3″	30′2″
30′2″	30′6″	30′3″	30′2″	30′4″
30′4″	29′11″	29′10″	30′0″	30′6″
29′8″	30′5″	30′4″	30′5″	30′0″
30′3″	30′1″	30′5″	29′10″	30′3″
30′1″	30′3″	30′0″	30′4″	29′11″

PLOTTING AND GRAPHING

15

The purpose of statistics is to take a large array of data and to group it into a usable form. Data being grouped may have to be arranged into distributions. From this arrangement, data can be placed in a graph to form a picture of the distribution.

In this chapter, we introduce the concepts of:

- Histograms
- Frequency polygons
- Ogives

After being grouped into one of the forms discussed in Chapter 14, the data are placed into graph form. The three methods of graphing help the technician to make sound judgments about data. Although the presentations look dissimilar, judgments made about the data are the same. Management chooses a method of presenting data, depending on which form is most practical to use. If you have ever worked with a bar graph, you can understand the basics of statistical graphs. The whole purpose of graphs is to pictorially show data that have been grouped.

OBJECTIVES

After completing Chapter 15, the reader should be able to:

- Define terms associated with graphing data.
- Distinguish among the three methods of plotting data.
- Name the parts of a histogram, frequency polygon, and ogive.
- Discuss the procedure for constructing each type of graph.
- Demonstrate the ability to construct each type of graph from given data.
- Discuss the use of each graph.
- Discuss the differences among the graphs.

15–1 HISTOGRAMS

Histogram is the proper name for a specialized bar graph. A histogram is made up of several basic components.[1]

Title: A description of the information for which the graph is made.
Vertical Scale: The number of units in each class or percentage of total data.
Horizontal Scale: The classes being graphed.

Figure 15–1 Histogram

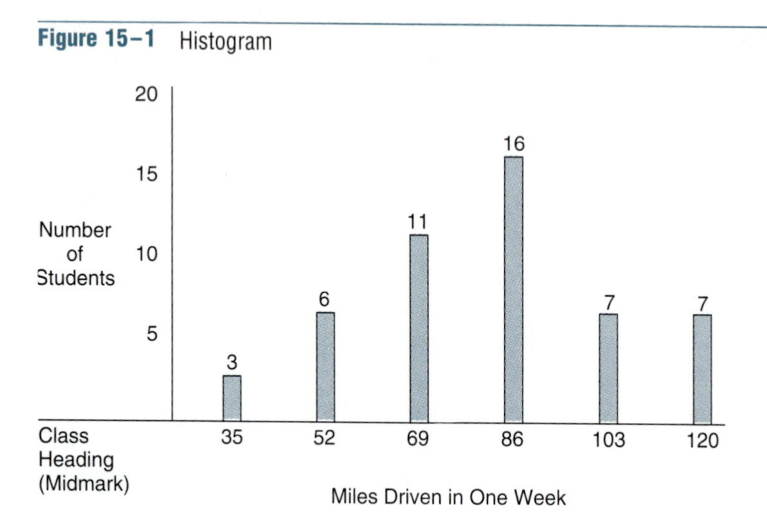

For example, using data from the previous chapter on grouped frequency distributions, a histogram would be constructed as shown in Figure 15-1. The class midmark is used as the graph's class heading. This number is a representative value for all data in the class, although all data were not that exact number. All three requirements for a histogram are presented in Figure 15-1.

Histograms can be grouped into two common types:

- Frequency (count of data in each class)
- Relative frequency (percentage each class is of all the data)

The vertical scale is the area where the two types differ. In the frequency histogram, the vertical scale shows the absolute value of occurrences for that class. That number does not change relative to the total number of data occurrences. In other words, if Class 1 has three units, that is the total number for that class. In a *relative frequency* distribution, the vertical scale depicts a percentage of the total data that a class forms, compared to all the data.

In the example in Figure 15-1, if Class 1 had three units, the percentage would be found by dividing the number of occurrences in that class by the total number of data points. If Class 1 had three occurrences and the total amount of data were 50 occurrences, the class would have a value of 3 divided by 50, or 0.06. To find the percentage, multiply the answer (0.06) by 100. The vertical scale is expressed in percent. This percentage is relative to the number of occurrences in the class and the number of occurrences in the raw data. If either number of occurrences changes, the percentage will change. That relationship marks the histogram as a relative frequency

Figure 15–2 Relative Frequency Histogram

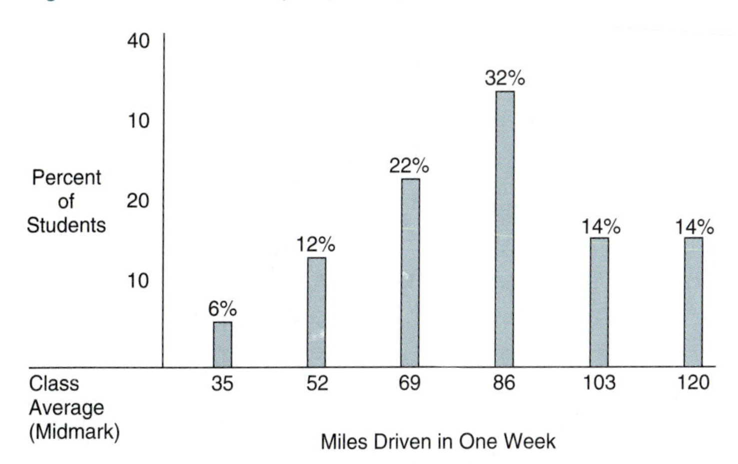

distribution. For example, the histogram from Figure 15-1 (a relative frequency histogram) is shown in Figure 15-2.

15–2 FREQUENCY POLYGON

The *frequency polygon* is a different way of plotting data. The data are treated the same way as in the histogram, but plotted with points marked at the frequency and the points connected.

Data that were graphed in the frequency histogram (Figure 15-2), plotted as a frequency polygon, are shown in Figure 15-3. The data

Figure 15–3 Frequency Polygon

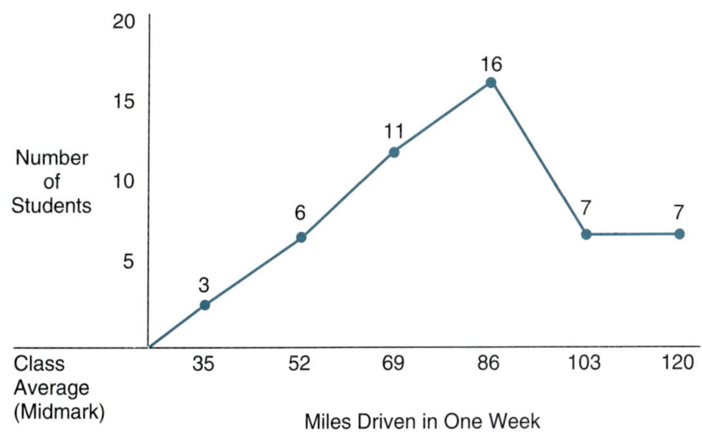

from the histogram and polygon are treated the same; only the plotting method is different.

15–3 OGIVE

The *ogive* (pronounced "o′jive") is a graph that is plotted as a cumulative frequency or a relative cumulative frequency graph. Data are grouped the same as for a frequency polygon. The difference is the way the graph is plotted. The ogive is plotted showing cumulative classes (added together). An ogive has the same requirements as a histogram.[2]

Title: Description of the information for which the graph is made.

Vertical Scale: The number of units in each class or percentage of total data.

Horizontal Scale: The classes being graphed.

The difference between a frequency polygon and an ogive is in the way that data are plotted. The ogive uses cumulative points. Consequently, Class 1 is plotted for the frequency of that class. Class 2 is plotted for the sum of Class 1 plus Class 2. Class 3 is plotted for the sum of Class 1 plus Class 2 plus Class 3. The remaining classes are plotted the same way. Each class is plotted for the frequency of that class plus all classes below it (Figure 15-4).

Figure 15–4 Ogive

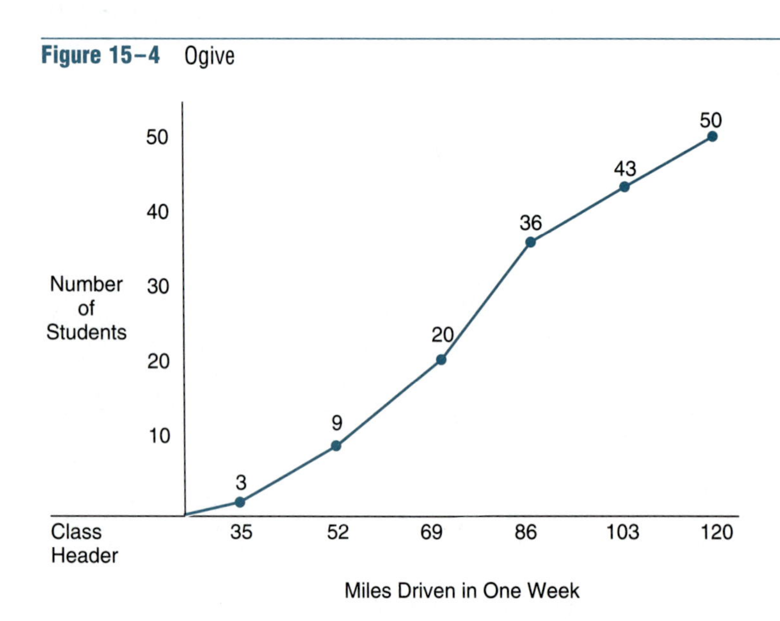

15–4 COMPARISON

The relationship between a frequency polygon and an ogive becomes apparent when they are plotted on the same graph. With both the frequency polygon and ogive plotted, it is apparent that both depict the same information. Which one is used is a matter of convenience.

The ogive can also be plotted as a relative frequency ogive. In this form, the points show the percentage of data that each cumulative class contains compared with all data (Figure 15-5).

Figure 15–5 Ogive and Frequency Distribution

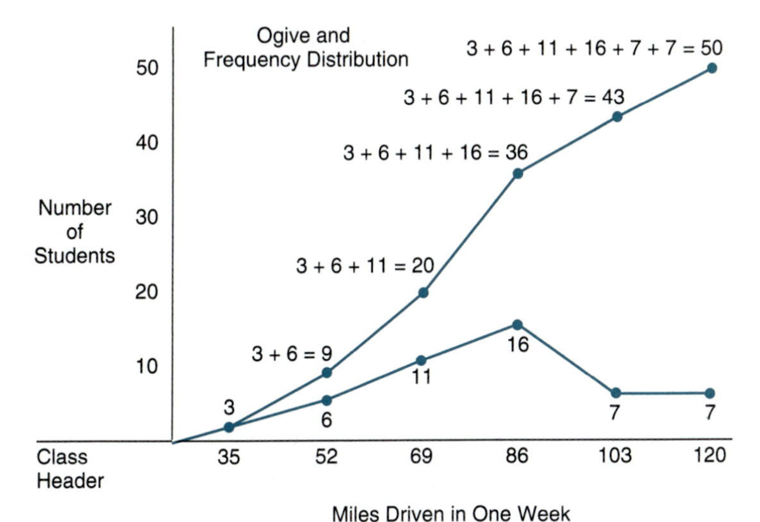

SUMMARY

The three ways that data are graphed include the histogram, frequency polygon, and ogive. Each depicts similar data, and the data are treated the same for each graph. The only difference is in the way data are represented. A histogram shows data in bar-graph form. A frequency polygon uses points and class frequency. An ogive uses points and shows the cumulative frequency of each class.

The three methods of plotting and graphing are by no means the only methods, but they are the ones in common use. Many of the more complex methods are variations of one of these three methods. The choice of which method to use depends on what information is to be gained from the graph and on what is commonly accepted in each individual company.

Many times data are plotted on one of these graphs so that the shape of the distribution can be evaluated. The shape helps determine whether outside disturbances are affecting the data. For this type of analysis, a histogram or frequency polygon is used. If the purpose of the graph is to show the effect each group (or class) has on the data as a whole, then an ogive is usually the graph of choice.

The presentation method for data is a choice that should be based on what information the presenter wants to convey to the audience reviewing the chart. The graph that conveys the maximum information in the simplest form is the best choice.

KEY TERMS

Frequency polygon A specialized graph in which the data are plotted with points marked at the frequency and the points connected.

Histogram A specialized bar graph made up of a title, vertical scale, and horizontal scale.

Ogive A specialized graph in which data are plotted as a cumulative (adding one point to the next) frequency or relative cumulative frequency.

Relative frequency A type of histogram indicating the percentage that each class composes of the entire data.

PLOTTING AND GRAPHING TEST

1. Match the terms on the right to the correct definition.
   - **a.** A description of the information for which the graph is made.
   - **b.** The scale that indicates the classes being graphed.
   - **c.** The scale that indicates the number of units in each class.
   - **d.** The number that is the representative value for all data in a class.
   - **e.** The type of histogram that shows the percentage each group makes of the whole data.
   - **f.** The type of histogram that shows the number of occurrences within each group.

   1. Horizontal
   2. Relative frequency
   3. Vertical
   4. Frequency
   5. Title
   6. Diagonal
   7. Midmark

2. Distinguish among a histogram, frequency distribution, and ogive by placing an X to the left of the description of an ogive.
   - **a.** A specialized bar graph.
   - **b.** A graph with points at the frequency of each group. The points are connected with lines.
   - **c.** A graph with cumulative points plotted for each group. The points are connected with lines.
3. Name the three parts that are common to an ogive and a frequency polygon.
4. Write a short paragraph explaining the procedure for constructing a histogram, frequency polygon, and ogive.
5. Construct, from the data given, a frequency histogram, frequency polygon, and ogive.

Number of People in Each Group

CLASS	MIDMARK	FREQUENCY
1	12	3
2	15	6
3	18	7
4	21	4

6. Write a short paragraph discussing how the histogram, frequency polygon, and ogive are used.
7. Write a short paragraph discussing the difference among the histogram, frequency polygon, and ogive.

NOTES

1. Dale Besterfield, *Quality Control,* 2d ed. (Englewood Cliffs, NJ: Prentice-Hall, 1986), 33.
2. Eugene Grant and Richard Leavenworth, *Statistical Quality Control,* 5th ed. (New York: McGraw-Hill, 1980), 38.

PLOTTING AND GRAPHING

Exercise No. 1

Plot the following histograms.

1. 2,3,9,8,7,4,6,5,4,7,6,5,5,6
2. 10,6,5,5,1,6,7,4,3,5,6,6,4,7,8,9,3,2
3. 4,9,5,6,5,7,5,2,3,9,8,7,4,6,6,5,3,4,7,6,1,5,5,6,10

Plot the following frequency polygons.

4. 5,8,7,8,7,10,6,5,5,1,6,7,4,3,5,6,6,4,7,8,9,3,2,5,7,5,6,5,9,4,7,8,8,7
5. 1,13,2,12,5,12,6,11,7,11,10,10,4,18,7,14,10,11,5,17,8,13,9,6,12,16, 7,10,11,15,9,20,8,9,19,10,8,9,9,4,18,7,14,10,11,5,17,8
6. 20,8,9,19,10,9,8,9,4,18,7,14,10,11,5,17,8,9,15,11,10,7,16,12,6,9,13,8, 17,S,11,10,14,7,18,4

Plot the following ogives.

7. 11,8,3,11,12,10,11,9,13,14,8,15,11,10,7,16,12,6,9,13,8,17,5,11,10,14,7,18,4
8. 4,18,7,14,10,11,5,17,8,13,9,6,12,16,7,10,11,1S,8,14,13,9,11,10,12

Exercise No. 2

Plot the following histogram.

1. Inspection data of hole diameter—2.5 in., ±0.5 in.

2.7	2.5	2.6
2.2	3.0	2.4
2.4	2.6	2.7
2.6	2.3	3.0
2.7	2.5	2.8
2.9	2.8	2.8
3.1	2.9	2.6
2.7	2.8	2.7

2. Inspection of "I" beam, length 24 ft. with acceptance ±6 in.

24′3″	24′1″	23′8″
23′7″	23′10″	23′11″
23′11″	23′6″	24′0″
23′9″	23′11″	23′9″
23′10″	24′0″	23′6″
23′9″	23′8″	23′10″
23′11″	23′11″	24′0″
23′10″	24′0″	23′10″

3.

1	4	4
5	10	13
11	2	6
3	12	11
12	14	3
5	4	14
13	13	8
7	15	15
14	7	6
6	16	17

4. Measurements of the modal frequencies across a –2.25-MHz ultrasonic transducer.

2.45	2.30	2.50
2.25	2.15	2.40
2.30	2.40	2.25
2.20	2.60	2.35
2.35	2.25	2.35
2.25	2.30	2.25
2.35	2.20	2.45
2.30	2.35	2.30
2.70	2.25	2.40

5.

10	19	18	10
15	9	14	15
9	6	11	8
16	16	16	6
13	8	15	16
8	15	8	14
17	17	17	10
7	14	16	15
15	9	5	7
10	11	10	11
12	18	17	4
18	10	7	21
9	19	20	9

6. Bar stock with specifications of 2 in., ± 1/16 in. was measured with the following results.

2″	2 1/32″	2 1/16″
1 15/16″	1 31/32″	2″
2 1/32″	2 1/16″	1 31/32″
1 29/32″	2″	2″
2 1/16″	2 3/32″	1 15/16″
1 31/32″	1 31/32″	2″
2 1/16″	2 1/32″	2″
2″	2″	2″
2 1/32″	2″	1 31/32″
2″	2 1/32″	2″
1 31/32″	2″	2 3/32″

Exercise No. 3

Plot a histogram from grouped frequency distributions.

A sample of 50 punched-hole diameters was inspected from a lot of 1000. A second sample was made the next day. Were there any changes in the process?

Day 1

0.499	0.502	0.504	0.503	0.501
0.500	0.501	0.502	0.501	0.498
0.498	0.500	0.500	0.504	0.502
0.501	0.504	0.497	0.499	0.501
0.502	0.499	0.499	0.500	0.499
0.501	0.503	0.503	0.498	0.503
0.503	0.502	0.500	0.503	0.502
0.500	0.499	0.505	0.501	0.500
0.502	0.500	0.501	0.502	0.501
0.499	0.501	0.502	0.500	0.503

Day 2

0.497	0.499	0.502	0.503	0.501
0.502	0.502	0.498	0.497	0.502
0.498	0.496	0.499	0.499	0.498
0.501	0.503	0.497	0.496	0.502
0.495	0.499	0.502	0.501	0.497
0.499	0.498	0.496	0.498	0.499
0.497	0.501	0.499	0.502	0.497
0.499	0.497	0.501	0.498	0.502
0.502	0.498	0.498	0.502	0.499
0.498	0.503	0.502	0.499	0.501

Exercise No. 4

Plot a histogram and a frequency polygon on the same chart for the following data.

Compare the two samples of inspection data on length of rebar taken from the same truckload. Why would you question this data?

Sample No. 1

29′10″	29′11″	30′2″	30′4″	29′9″
30′0″	30′4″	29′7″	30′0″	30′1″
29′11″	29′10″	30′0″	30′2″	29′11″
30′3″	30′0″	30′3″	29′8″	30′6″
30′1″	29′9″	30′1″	29′11″	30′2″
29′9″	30′2″	29′10″	30′1″	30′0″

Sample No. 2

29′11″	29′9″	30′1″	30′3″	30′2″
30′2″	30′6″	30′3″	30′2″	30′4″
30′4″	29′11″	29′10″	30′0″	30′6″
29′8″	30′5″	30′4″	30′5″	30′0″
30′3″	30′1″	30′5″	29′10″	30′3″
30′1″	30′3″	30′0″	30′4″	29′11″

Exercise No. 5

Plot a histogram and an ogive on the same chart from a grouped frequency distribution of the following data.

1.	0.56	0.14	0.96	0.98	0.38
	0.1	0.17	0.82	0.75	0.61
	0.88	0.72	0.66	0.43	0.75
	0.32	0.99	0.87	0.06	0.28
	0.45	0.21	0.63	0.59	0.38
	0.63	0.04	0.47	0.43	0.39
	0.09	0.70	0.57	0.53	0.04
	0.42	0.52	0.22	0.34	0.56
2.	0.55	0.22	0.26	0.35	0.24
	0.58	0.53	0.60	0.48	0.55
	0.63	0.37	0.72	0.51	0.94
	0.16	0.38	0.91	0.52	0.44
	0.83	0.55	0.65	0.95	0.39
	0.68	0.70	0.06	0.22	0.69
	0.79	0.27	0.25	0.30	0.95
	0.47	0.61	0.58	0.44	0.98

3. Two lots of shaft were received and inspected. Compare histograms from Lot A and Lot B.

Lot A

0.999	1.002	0.998	1.000	1.001
1.000	0.996	1.002	1.002	0.997
1.002	1.005	1.001	1.001	1.000
0.999	1.000	1.004	0.998	0.999
1.003	1.001	1.000	1.001	1.003
1.001	1.002	1.003	1.004	1.001

Lot B

0.998	1.000	1.002	0.998	0.999
0.999	0.997	0.999	1.000	0.997
0.996	1.001	1.003	1.004	0.999
1.000	0.999	0.998	0.999	0.998
0.999	1.002	1.000	1.001	0.999
0.998	1.001	0.999	1.005	1.002

NORMAL DISTRIBUTIONS

16

The term *normal distribution* refers to the way that data are distributed. As the amount of data that are grouped grows larger, most types of information show a normal distribution.[1] This distribution is frequently called a *bell curve* because of its shape. As an example, if the heights of 1000 people were measured at random, the distribution would look like the histogram in Figure 16-1, p. 288.

This curve shows that most of the data group around the center value of all data. The further the data are from this average, the fewer times the data are repeated. As in the example, if the heights of 1000 people were measured and graphed, the distribution would be normal, with the average—we'll pretend—being around 6 ft. The highest number of occurrences (the height that was measured most often) would be 6 ft. The further above and below that average, the fewer people there would be of those heights. The number of people who were between 6 ft. 7 in. and 6 ft. 9 in. would be fewer than those who were between 6 ft. 0 in. and 6 ft. 2 in. This would also hold true on the other end of the distribution. There would be fewer people between 5 ft. 2 in. and 5 ft. 4 in. than

Figure 16–1 Bell Curve

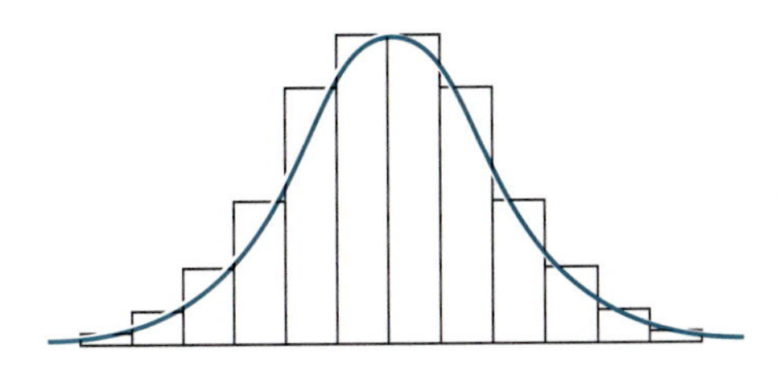

OBJECTIVES

After completing Chapter 16, the reader should be able to:

- Define terms associated with the central tendency.
- Define terms associated with dispersion.
- Distinguish between the central tendency and dispersion.
- Name the measures of the central tendency.
- Name the measures of dispersion.
- Discuss the procedure for finding each measure of the central tendency.
- Discuss the procedure for finding each measure of dispersion.
- Demonstrate the ability to find each measure of the central tendency.
- Demonstrate the ability to find each measure of dispersion.
- Discuss the use of each measure of the central tendency.
- Discuss the use of each measure of dispersion.

Figure 16–2 Normal Distribution

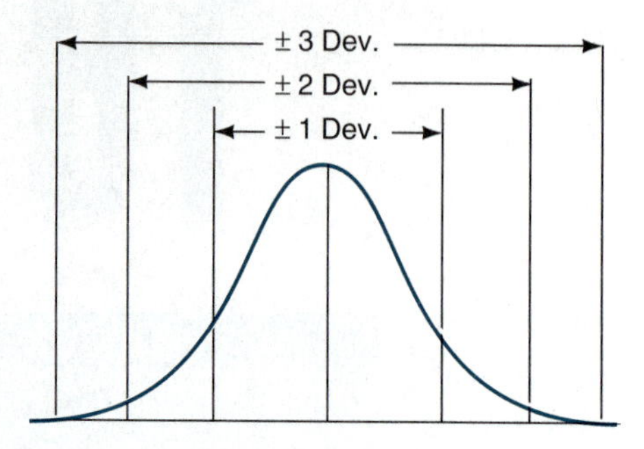

there were between 5 ft. 10 in. and 6 ft. 0 in. The higher or lower that the class (group of heights) is from the average, the fewer people there would be in that class.

If the curve were divided into six equal sections, as shown in Figure 16–2, the amount of data in each section could be counted. Since most information tends to be grouped normally, the percentages of data in each section can be predicted.

In the center of the distribution is the average of all the data. Each section of the distribution is called a deviation, because it deviates away from the average. Each section is identified by the number of deviations or sections above (+) or below (–) the average. As an example, the first deviation above the average is called +1 deviation. The first deviation below the average is called –1 deviation. If both deviations are to be referenced, they are called ±1 deviation. The same holds true for ±2 deviations and all further deviations. If data in the ±1 deviation were counted, it would contain approximately 68 percent of all data, because this distribution area contains the groups with the most data in each group. If the area of ±2 deviations were counted, it would contain approximately 95 percent of all data—the 68 percent that is in the ±1 deviation area as well as data at ±2 deviations. The area at ±3 deviations contains 99.7 percent of all data. This area contains ±1, ±2, and ±3 deviations; a total of only 0.3 percent of all data are not contained in the ±3 deviation area. These data consist of very large and very small points.[2]

This percentage is the same for all normal distributions, no matter what data are plotted. This breakdown is known as the *empirical rule* (Figure 16–3).

Figure 16–3 Normal Distribution

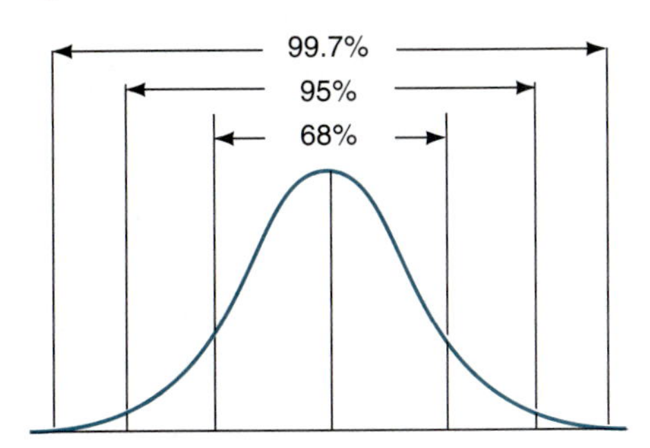

The normal curve is shaped like a bell because of two characteristics. The first is the way that data tend to group in a pattern around the center value. This is known as the *central tendency.* The second characteristic is the way the data are spread out or distributed. This is called *dispersion.*

In this chapter we discuss both the central tendency and dispersion of data. The measures of each are defined, and we explain the formulas used to find both.

16–1 CENTRAL TENDENCY

Data is a term used to describe a collection of information. For the most part, these data are in numerical form for statistical use. These are called *quantitative* data, because they describe the quantity, or amount of something, in numerical form. Quantitative data may be the miles driven each week using a company car, for example:

85 91
78 85
81 82
89 93
84

As stated in the introduction to Part III, everything varies in a predictable manner. One way that data are predictable is in their tendency to group around a center value—the central tendency. One step that will help to show this grouping around a center value is to arrange the miles in ascending order: 78, 81, 82, 84, 85, 89, 91, 93. Looking at these figures, the central tendency becomes apparent. The measures of the central tendency are the mean, the median, and the mode.

The normal distribution shape is determined by this grouping of the data around the center value. Distribution is judged for its shape, centering, and width. Centering is determined by the central tendency. Shape is determined by central tendency and dispersion. Width is determined by dispersion. The remainder of this chapter discusses central tendency and dispersion.

Mean A *mean* is the simple average of all data points. To find a mean, add data together and divide by the number of data points. The formula for finding the mean is[3]

$$\bar{x} = \sum x/n$$

where

$\bar{x}$ = average of the x values (it is read x-bar)
$\sum$ = capital sigma, meaning "the sum of"
x = any single data point
n = sample size (the number of data points)

The formula directs us to take the sum of the x values (data points) and divide that sum by n (number of data points).

For example, 78 + 81 + 82 + 84 + 85 + 85 + 89 + 91 + 93 = 768/9 = 85.33. Add the values together (= 768) and divide by the number of data points (= 9). The mean (or average) = 85.33.

Exercise No. 1 Rank the following data, showing frequency distribution and the mean.

29, 34, 12, 40, 38, 16, 19, 33, 18, 30, 33, 15

Exercise No. 2 Rank the following data, showing frequency distribution and the mean.

14, 24, 18, 31, 28, 26, 39, 43, 18, 40, 23, 25

Figure 16–4 Example No. 1 Mean

85.33

78, 81, 82, 84, 85, 85, 89, 91, 93

Median A *median* value is simply the value that is located in the middle of the data.[4] In the example, there are nine grades. The median value is simply the value that falls fifth from each end. The fifth value counting from both the front and the end is 85. The median or middle value of the data is 85.

The median is easy to find when an odd number of values is used. If an even number of values is used, there is no clear median. To find a median, divide the number of data points by 2. This will determine the two middle values by counting from each end. Add the two values together and divide by 2.

For example, suppose the number of mile counts was eight instead of nine. The data would look like this: 78, 81, 82, 84, 85, 85, 89, 91. (The 93 has been removed from the data for this example.) To find the median value, divide 8 (the number of data points) by 2. This will determine where the median is from each end. In the example, the mean is located four points from each end. Counting from the front four points, the value 84 is reached. Counting from the rear, the value 85 is reached. The exact median is the value that lies halfway between the two values. To find the median for this example, add 84 and 85 together (= 169). Divide 169 by 2 to determine the median value of 84.5.

Figure 16–5 Example No. 1 Median

				85				
78,	81,	82,	84,	85,	85,	89,	91,	93
1	2	3	4	5	4	3	2	1

Figure 16–6 Example No. 2 Median

			84.5				
78,	81,	82,	84,	85,	85,	89,	91
1	2	3	4	4	3	2	1

Exercise No. 1 Rank the following data in ascending order and mark the median.

19, 23, 25, 14, 33, 30, 29, 22, 31

Exercise No. 2 Rank the following data in ascending order and mark the median.

19, 23, 25, 15, 33, 30, 29, 22, 31, 18

Mode A *mode* is the value of the item of data that has the highest number of occurrences.[5] In the example, the value that has the highest number of occurrences is 85. For this example, 85 (which occurs twice) is the mode.

If an additional value is added to the data, the following would result. For example—78, 81, 82, 84, 85, 85, 89, 91, 91, 93. In this example, the 85 and 91 have the same number of occurrences. The data are said to be *bimodal* (having two modes). This distribution has no mode. If two or more values have the same number of occurrences, the distribution has no mode.

Figure 16–7 Example No. 1 Mode

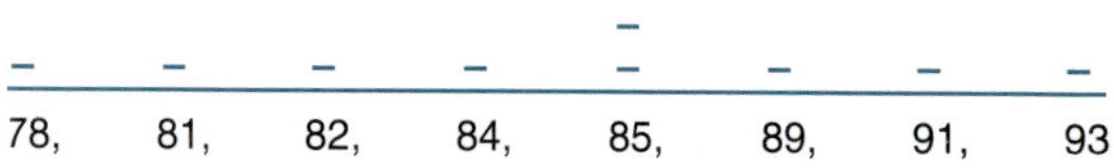

78,	81,	82,	84,	85,	89,	91,	93
–	–	–	–	– –	–	–	–

Figure 16–8 Example No. 2 Bimodal

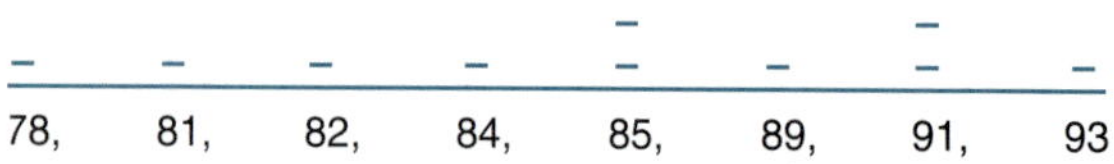

78,	81,	82,	84,	85,	89,	91,	93
–	–	–	–	– –	–	– –	–

Exercise No. 1 Rank the following data. Show the frequency distribution and mode.

27, 45, 29, 27, 33, 30, 33, 40, 38, 33, 40

Exercise No. 2 Rank the following data. Show the frequency distribution and mode.

27, 45, 29, 27, 33, 30, 33, 40, 38, 33, 40, 27

Overview In summary, the central tendency is the way that data group around a central data value. The measures of the central tendency are:

- Mean: Mathematical average
- Median: Middle value of data
- Mode: Value that occurs most

Figure 16–9 Central Tendency

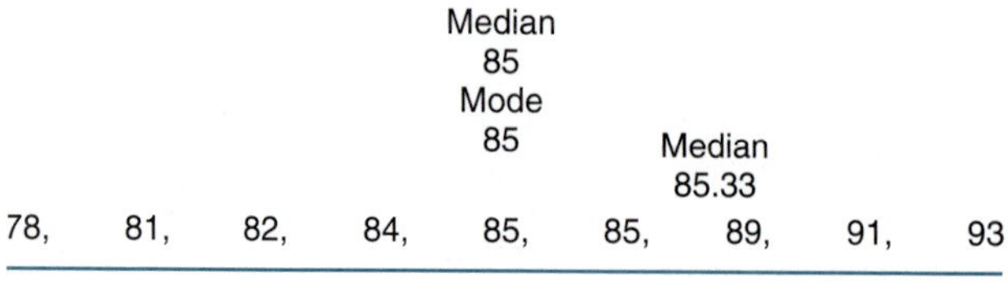

Median
85
Mode
85
Median
85.33

78, 81, 82, 84, 85, 85, 89, 91, 93

For these data, median and mode are both 85. This will not always be the case. The mean, median, and mode may be the same number or may be different numbers. The closer together the three numbers are, the higher the degree of the central tendency.

Exercise No. 1

1. Define the term *central tendency.*
2. Define the term *median.*
3. Define the term *mode.*
4. Define the term *mean*.
5. Rank the following data. Show the frequency distribution, mean, median, and mode.

14, 24, 18, 31, 28, 26, 39, 43, 18, 40, 23, 25

16–2 DISPERSION

Dispersion, you recall, refers to the way that data are spread out over the entire distribution. As stated previously, normal distributions have a shape, centering, and width. Centering is determined by the central tendency. Shape is determined by the central tendency and dispersion. Width is determined by dispersion. The measures of dispersion are range, standard deviation, and variance.

Range The *range* is a measure of the width of the data that are plotted. This is the numeric range. The range is found by subtracting the smallest point of data from the largest point of data (high value − low value = range).[6]

For the data on the miles driven, range is figured by subtracting the low value from the high value: 93 − 78 = 15.

85	91
78	85
81	82
89	93
84	

The range is a measure of the extremes of the data. It shows the extent of the spread of data.

The range has one major disadvantage relative to other measures of dispersion. Range measures only two points of data and does not take into account the amount of data that are between the high and low values. If high and low values in two different distributions are the same, the first group could have five points between the high and low values, whereas the second group could have 500 points of data. If the high and low values for both groups of data are the same, the range for both would be the same.

Exercise No. 1 Find the range for the following data.

19, 23, 25, 15, 33, 30, 29, 22, 31

Exercise No. 2 Find the range for the following data.

19, 23, 25, 15, 33, 30, 29, 22, 31, 97, 9, 23, 54

Standard Deviation and Variance Standard deviation and variance are two measures of dispersion that have much in common. They both show the spread of the data, taking into account all points of data. However, range takes into account only the high and low points, ignoring the remaining data.

At the beginning of this chapter, the concept of normal distribution (bell curve) was introduced—data were distributed into six groups, and the percentage of data in each group is predictable (empirical rule). This principle is the basis of standard deviation.[7]

Standard deviation is defined as the average distance from the mean that any point of data is located. This average helps find the spread of all data. If the average distance that all points of data fall from the mean is known, then the data spread can be estimated.[8]

To find standard deviation, data are selected at random from some population (group of data). These data are then arranged into an ungrouped distribution. For example, we will use the same data used in the graphics example (the number of miles driven by 50 people in one week).

27	68	79	91	107
43	71	80	91	108
43	71	81	93	108
44	71	82	94	116
47	73	82	94	120
49	73	84	95	120
50	74	84	96	122
54	75	86	97	123
58	76	88	103	127
65	77	88	106	128

To find the average numeric distance that all the points fall from the mean, we must locate the mean. The formula for the mean, as discussed previously, is $\Sigma x/n$. In this formula, x = each value or each

Table 16–1 Average Distance from the Mean

X	$\bar{x}$	$X - \bar{x}$	X	$\bar{x}$	$X - \bar{x}$
27	83.62	-56.62	84	83.62	0.38
43	83.62	-40.62	84	83.62	0.38
43	83.62	-40.62	86	83.62	2.38
44	83.62	-39.62	88	83.62	4.38
47	83.62	-36.62	88	83.62	4.38
49	83.62	-34.62	91	83.62	7.38
50	83.82	-33.62	91	83.62	7.38
54	83.62	-29.62	93	83.62	9.38
58	83.62	-25.62	94	83.62	10.38
65	83.62	-18.62	94	83.62	10.38
68	83.62	-15.62	94	83.62	10.38
71	83.62	-12.62	96	83.62	12.38
71	83.62	-12.62	97	83.62	13.38
71	83.62	-12.62	103	83.62	19.38
73	83.62	-10.62	106	83.62	22.38
73	83.62	-10.62	107	83.62	23.38
74	83.62	-9.62	108	83.62	24.38
75	83.62	-8.62	116	83.62	32.38
76	83.62	-7.62	120	83.62	36.38
77	83.62	-6.62	120	83.62	36.38
79	83.62	-4.62	122	83.62	38.38
80	83.62	-3.62	123	83.62	39.38
81	83.62	-2.62	127	83.62	43.38
82	83.62	-1.62	128	83.62	44.38
82	83.62	-1.62	$\Sigma X = (4181)$	$\Sigma X - \bar{x}$	= (0)

distance and n = the number of distances. The sum of the x values = 4181. The number of values counted is (n) = 50; thus, 4181/50 = 83.62. The mean ($\bar{x}$) is 83.62. To find the average distance each point falls from this mean, the mean should be subtracted from each distance and the difference listed. The differences are then added together, and that sum is divided by the number of data points. An example is shown in Table 16-1.

To find the average distance that each point of data falls from the mean, find the distance of each point of data from the mean. The (–) sign shows that a data point is smaller than the mean; if the sign is (+), the data point is larger than the mean. After finding this distance for each point of data, the distances are added together. It now becomes apparent that a problem exists with this procedure. The sum of these numbers is 0.

This value cannot be used to find the average distance. Some method of removing the negative signs must be used without changing the value of the data. If all the negative signs were dropped, this would change the value of the negative numbers without changing the value of the positive numbers. But to be statistically correct, whatever function is performed on the negative numbers must be performed on the positive numbers. If all the differences were squared, all signs would be changed to positive and a function could be performed on all the numbers. This could be reversed by the square-root function. An example is shown in Table 16-2, p. 297.

To complete the formula and find the standard deviation, the value of 28,967.78 $(x - \bar{x})$ should be divided by the number of (x) values used—the (n) number. In this example the (n) value is 50, so (28,967.7800/50 = 579.3556). This value is the variance, or the average distance that the data points are away from the mean, squared. To find a standard deviation, this value (variance) must be found. The square root of 579.3556 is 24.0698 (standard deviation). The standard deviation for these data is 24.0698.

Finding the standard deviation is a matter of trying to find the average distance of the data from the mean of the data. This is found using the formula[9]

$$\sqrt{\frac{(x-\bar{x})^2}{n}}$$

The empirical rule, discussed earlier in this chapter, states that in a normal distribution, 68 percent of all data is within ±1 standard deviation of the mean. A total of 95 percent is within ±2 standard deviations of the mean, and 99.7 percent of the data is within ±3 standard deviations of the mean.

If this rule holds true, then the same percentage of data points that was used in the example should fall in this range. The mean $(\bar{x})$ of the distribution is 83.62. The standard deviation (σ) is 24.0698.

$$83.62 + 24.0689 = 107.6898$$
$$83.52 - 24.0689 = 59.5502$$

The empirical rule states that 68 percent of all data should fall within this range. Looking at the raw data reveals that 32 of the 50 data points fall within this range: 32/50 = 0.64 × 100 = 64 percent. A total of 64 percent of all data actually falls within ±1 standard deviation.

Table 16–2 Standard Deviation

x	− $\bar{x}$	= x − $\bar{x}$	$(x - \bar{x})^2$	x	− $\bar{x}$	= x − $\bar{x}$	$(x - \bar{x})^2$
27	83.62	–56.62	3,205.8244	84	83.62	0.38	0.1444
43	83.62	–40.62	1,649.9844	86	83.62	2.38	5.6644
43	83.62	–40.62	1,649.9844	88	83.62	4.38	19.1844
44	83.62	–39.62	1,569.7444	88	83.62	4.38	19.1844
47	83.62	–36.62	1,341.0244	91	83.62	7.38	54.4644
49	83.62	–34.62	1,198.5444	91	83.62	7.38	54.4644
50	83.62	–33.62	1,130.3044	93	83.62	9.38	87.9844
54	83.62	–29.62	877.3444	94	83.62	10.38	107.7444
58	83.62	–25.62	656.3844	94	83.62	10.38	107.7444
65	83.62	–18.62	346.7044	94	83.62	10.38	107.7444
68	83.62	–15.62	243.9844	96	83.62	12.38	153.2644
71	83.62	–12.62	159.2644	97	83.62	13.38	179.0244
71	83.62	–12.62	159.2644	103	83.62	19.38	375.5844
71	83.62	–12.62	159.2644	106	83.62	22.38	500.8644
73	83.62	–10.62	112.7844	107	83.62	23.38	546.6244
73	83.62	–10.62	112.7844	108	83.62	24.38	594.3844
74	83.62	–9.62	92.5444	108	83.62	24.38	594.3844
75	83.62	–8.62	74.3044	116	83.62	32.38	1048.4644
76	83.62	–7.62	58.0644	120	83.62	36.38	1323.5044
77	83.62	–6.62	43.8244	120	83.62	36.38	1323.5044
79	83.62	–4.62	21.3444	122	83.62	38.38	1473.0244
80	83.62	–3.62	13.1044	123	83.62	39.38	1550.7844
81	83.62	–2.62	6.8644	127	83.62	43.38	1881.8244
82	83.62	–1.62	2.6244	128	83.62	44.38	1969.5844
82	83.62	–1.62	2.62444	**(4181)**		**(0)**	**(28782.674)**
8	83.62	0.38	0.1444				

For ±2 standard deviations

$$83.62 + 24.0689 + 24.0689 = 131.7596$$
$$83.62 - 24.0689 - 24.0689 = 35.4804$$

The empirical rule states that 95 percent of all data should fall within this range. Looking at the raw data reveals that 49 of the 50 data points fall within this range: 49/50 = 0.98 × 100 = 98 percent. A total of 98 percent of all data actually falls within ±2 standard deviations.

For ±3 standard deviations

$$83.62 + 24.0689 + 24.0689 + 24.0689 = 151.8294$$
$$83.62 - 24.0689 - 24.0689 - 24.0689 = 11.4106$$

The empirical rule states that 99.7 percent of all data should fall within this range. Looking at the raw data reveals that all 50 data points fall within ±3 standard deviations.

The empirical rule is used to estimate the percentage of data in each deviation. The larger the number of points used to find a standard deviation, the closer to the exact percentages in the rule will be in each deviation.

If data that are used are samples from a larger population, the formula changes slightly. The lower portion (divisor) contains the value of (n − 1) instead of the (n) value. This minimizes any error that was the result of using a sample of data instead of all data. For this problem, the standard deviation for the sample is 28,967.78/49 = 591.1791837. The square root of this number is 24.3141766. This is the standard deviation if the data are samples of a larger population.

This method of finding standard deviation can be tedious when a large amount of data is used. It becomes an overwhelming task when the units of data number 500 or more. This method does provide the exact standard deviation for data, however.

Data that amount to 50 units or more can be grouped into a frequency distribution. The standard deviation can then be found using the following formula.[10]

$$\sqrt{\frac{(x^2 f)}{n-1} - \frac{(xf)^2}{n}}$$

where

x = class midmark
f = frequency of the class
n = number of data points in the distribution

Although the formula looks quite different from the earlier formula, the concept is the same. The different look results because, with the data grouped, the standard deviation must be approximated.

Variables that must be found for this formula are combinations of the three variables (x), (f), and (n). The first variable is the (xf) combination. To find (xf), multiply the midmark (x) for each class by the frequency (f) of that class, and list this figure for each class. The total (xf) is found by adding all (xf) values.

The second value combination to be found is (x^2f). This value can be found in two ways. The midmark squared is the (x^2) value. This value multiplied by frequency (f) gives the (x^2f) value.

The second way to find the (x^2f) value can also be written x × x × f. The (xf) value, if multiplied by the (x) value a second time, provides the (x^2f) value.

Table 16–3

CLASS	LIMITS	MIDMARK	FREQUENCY		
		x	f	xf(x × f)	x^2f (x × x × f)
1	27 to 43	35	3	105	3675
2	44 to 60	52	6	312	16,244
3	61 to 77	69	11	759	52,371
4	78 to 94	86	16	1376	118,336
5	95 to 111	103	7	721	74,263
6	112 to 128	120	7	840	100,800
Σf = (50)	Σxf = (4113)	Σx^2f = (365,689)	Σf = (50)	Σxf = (4113)	Σx^2f = (365,589)

The data that will be used for this example are the same data that were used in the previous example. The (n) value needed to complete the formula is equal to the sum of the frequencies (n = f).

The first approximation that can be found is the ($\overline{x}$). The formula for this is xf/n. From the distribution in Table 16–3, values can be found. The xf = 4113 and the n = 50. The $\overline{x}$ = 4113/50, so $\overline{x}$ = 82.26. This is the approximate value of ($\overline{x}$). If we compare the true ($\overline{x}$) (83.62) and the approximate ($\overline{x}$) (82.26) values, we find a difference of 1.36. This difference is the result of the method of finding the (xf) value. The true ($\overline{x}$) is found by adding all single points of data and dividing by the number of data points. To find the approximate ($\overline{x}$), multiply the class midmark by the frequency of that class. This gives an approximate value of the data sum in that class. Data in Class 1 (from the raw data) are 27, 43, 43. Added together, they equal 113. For the approximate value that was used on the grouped frequency distribution, xf = 105. There is a difference of eight between the true sum and the approximate sum. Since the frequency of the class is greater, the approximation will be closer. In Class 4, the single data points are (79, 80, 81, 82, 82, 84, 84, 86, 88, 88, 91, 91, 93, 94, 94, 94). The total of the data is 1391. The approximate value (xf) for Class 4 is 1376. Although the difference is 15, the percentage that each number's true value was off was less. In Class 1, with three numbers in the class and the difference between the true total and approximate total being eight, each number was missed by 2.66 percent (8/3 = 2.66). In Class 4, with 16 numbers and the difference between the true total and approximate total being 15, each number was missed by 0.9375 percent (15/16 = 0.9375). The greater the number of data points in a class, the closer the approximate value will be to the true value.

Table 16–4

CLASS	LIMITS	MIDMARK	FREQUENCY		
		x	f	xf(x × f)	x^2f(x × x × f)
1	27 to 43	35	3	105	3,675
2	44 to 60	52	6	312	16,244
3	61 to 77	69	11	759	52,371
4	78 to 94	86	16	1,376	118,336
5	95 to 111	103	7	721	74,263
6	112 to 128	120	7	840	100,800

There is one additional item that will help the approximation of the total data sum and the true data sum be very close. Although each approximation may be off from the true value, some classes will be estimated high and some classes will be estimated low. When class totals are added, each high class estimate will help offset a low estimate. The true data sum is 4181, and the approximate sum is 4113. Using the approximate data sum gives a slightly different estimate ($\bar{x}$) when compared to the true ($\bar{x}$) value. This difference does not distort the outcome of the standard deviation to the point that it is unusable for decision making.

In the formula for the grouped method of finding standard deviation, the (x × f/n − 1) portion is used to find the approximate mean or (x). The remainder of the formula (xf/n) is used to find an approximation of the (x) values squared, divided by the number of data points. This formula now becomes the same as the formula used to find the standard deviation longhand.[11]

$$\sqrt{\frac{(x-\bar{x})^2}{n-1}} \qquad \sqrt{\frac{(x^2f)}{(n-1)} - \frac{(xf)^2}{(n)}}$$

To find the standard deviation of the grouped data, complete the operations as shown (see Table 16-4).

$$f = 50 \qquad xf = 4113 \qquad x^2f = 365{,}689$$

$$\sqrt{\frac{50(365{,}689) - (4113)}{(50)(50-1)}} = \sqrt{\frac{(18{,}284{,}450) - (16{,}916{,}769)}{(50)(49)}}$$

$$= \sqrt{\frac{1{,}367{,}681}{2450}} = \sqrt{585.2371429} = 23.62704262$$

The approximate standard deviation, then, is determined to be 23.627. Comparing the true standard deviation and approximate standard deviation reveals that the difference is slight, namely, 24.0689 − 23.6270 = 0.4419. The difference is 1.8 percent of true standard deviation.

The advantage of this method is that there are fewer math calculations and therefore less chance of a mistake in calculating. If the difference is only a 1.8 percent error, the approximate standard deviation will be as useful in making decisions as the true value.

The grouped method of finding the standard deviation is less tedious than the longhand method. The grouped method still requires much math. To reduce the amount of calculation, and possible math error, a second method of finding the standard deviation is used. The *coded method* finds the standard deviation as a percentage of class width. The formula used is much like the formula for the grouped method[12]:

$$C \cdot \sqrt{\frac{n(\Sigma\mu^2 f) - (\Sigma\mu f)^2}{n(n-1)}}$$

The difference in the formula is in the (c) and the (μ) factors, where

c = class width
μ = class code
n = number of frequencies in all classes
f = number of frequencies in each class

The first step in finding the standard deviation is to make a frequency distribution from the data. For this example, data from the grouped method will be used for the coded method. The next step is to code the classes. To code a class, assign one of the middle classes the value (code number) 0. All classes above 0 class are numbered −1, −2, −3, etc. All classes below 0 class are coded 1, 2, 3, etc. Codes are assigned the variable (μ). It does not matter which class has the code 0, but a class near the center is the best location for the 0 code. This keeps (μf) and ($\mu^2 f$) values as small as possible. If the values are small, there is less chance of a simple math error.

The math calculations include the following (see Table 16-5).

$$(17)\sqrt{\frac{50(125) - (39)}{50(50-1)}} = (17)\sqrt{\frac{6250-1521}{50(49)}}$$

$$= (17)\sqrt{\frac{4729}{2450}} = \sqrt{1.930} = (17)1.3893 = 23.6181$$

Table 16–5

CLASS	LIMITS	MIDMARK	FREQUENCY		CODE	
		x	f	μ	μf	μ^2f
1	27 to 43	35	3	−2	−6	12
2	44 to 60	52	6	−1	−6	6
3	61 to 77	69	11	0	0	0
4	78 to 94	86	16	1	16	16
5	95 to 111	103	7	2	14	28
6	112 to 128	120	7	3	21	63
			$\Sigma f = (50)$		$\Sigma \mu f = (39)$	$\Sigma \mu^2 f = (125)$

The approximate deviation for these data by the coding method is 23.6181.

When you compare the three methods, the approximations are very close to the true value. The difference in each approximation is less than 1 percent of the true deviation.

True standard deviation = 24.0689
Approximate grouped deviation = 23.627
Approximate coded deviation = 23.6181

The 0 can be placed anywhere in the distribution, and the results will be the same. The following problem is solved with no 0 or negative codes.

$$f = (50) \qquad \mu f = (189)\ \mu^2 f = (809)$$

$$(17)\sqrt{\frac{50(809) - (189)}{50(50-1)}} = (17)\sqrt{\frac{40450 - 35721}{50(49)}}$$

$$= (17)\sqrt{\frac{4729}{2450}} = \sqrt{1.93020} = (17)1.3893 = 23.6181$$

Table 16–6

CLASS	LIMITS	MIDMARK	FREQUENCY		CODE	
		x	f	μ	μf	μ^2f
1	27 to 43	35	3	1	3	3
2	44 to 60	52	6	2	12	24
3	61 to 77	69	11	3	33	99
4	78 to 94	86	16	4	64	256
5	95 to 111	103	7	5	35	175
6	112 to 128	120	7	6	42	252

The approximate mean can be located by using the following formula:

$$\overline{x} = X + \left[\frac{\mu f}{n}\right] c$$

where

x = midmark of the class that has been coded 0
c = class width
μf = code times the frequency of all classes added together
n = sum of the frequencies of all classes

From these data, ($\overline{x}$) = 69 + (17) (39/50) = 82.26. This mean is very close to the true mean and to the approximate mean found in the grouped method.

The result of assigning all positive codes to the distribution changed only the sizes of the numbers that were used in the formula. The size of the numbers allows for more math errors. The outcome did not change, however. The way to keep numbers in the formula small is to put a code of 0 at the middle class, negative numbers above the 0 class, and positive numbers below the 0 class. If the formula is used this way, math errors are reduced and the standard deviation can be found with much less work.

Variance is a third measure of dispersion—a measure of the way that data are spread around the mean.[13] Variance is defined as the standard deviation squared. Variance is used less often than the standard deviation for a population or sample. In the three methods of finding standard deviation listed, variance can be found by finding the square of the standard deviation. Variance is used in more advanced statistical analysis. The common use is in the analysis of variance (anova) to determine the equality of means of two distributions.

Overview Once you understand the concept of the normal distribution, the remaining question is how it is used. Two common uses of this distribution are 1) to evaluate whether a process is running with outside interference and 2) to determine the probability of a given occurrence from a normally distributed population. The normal probability distribution is discussed in Chapter 18.

A process that is running without disturbance from an outside source produces an output that is normally distributed. If an outside disturbance is acting on the process, the distribution changes shape because the outside disturbance affects the central tendency, dispersion, or both. If either or both are affected, the shape of the distribution is also affected. There are two common changes that indicate this

Figure 16–10 Skewed Distribution

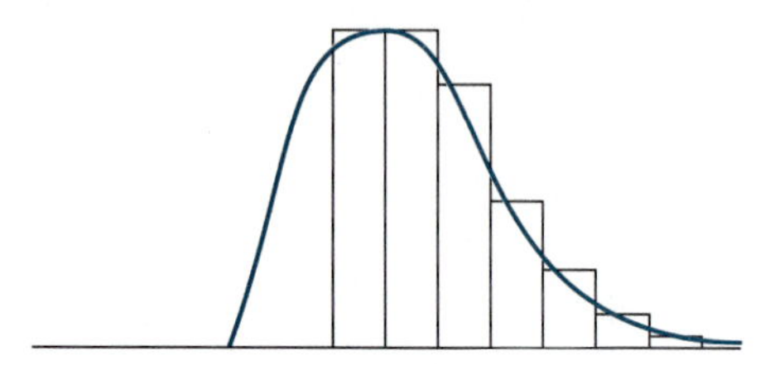

situation. The data may group around a median value that is larger or smaller than the mean, or the data may spread further away from the mean at the high or low end. This situation produces a skewed distribution (Figure 16-10).

Some outside disturbances on a process affecting the distribution may be easily identified, such as overadjustment of equipment or inconsistent raw material. Also, sorting all parts and discarding those that are out of tolerance (too small) or reworking parts (too large) produces the same effect and is easily identifiable. The outside disturbance, however, may be well hidden, such as fluctuations in machinery voltage, different temperatures in a heat treatment oven, or a humidity change that affects an operation. Any outside disturbance that does not affect all items of the product the same will change the distribution.

When a group of product is measured and the resultant distribution is other than normal, the population has an outside disturbance acting on the process used to produce the population. The outside disturbance is called an "assignable cause." One of the many tasks of a quality control professional is to identify the assignable cause and prevent it from affecting the process. One of the many ways to evaluate the distribution is to construct a histogram from a sample of the product under study. This helps the quality professional to determine whether the distribution is normal. The Supplementary Information Sheets in the back of this book provide examples of normal and abnormal distributions.

SUMMARY

As most data are grouped, they show a normal distribution (bell curve), since the data points tend to group in a pattern around the center value (central tendency). The measures of central tendency are the mean (simple average of all data points), the median (the value located in the middle of the data), and the mode (the value of data that has the highest number of occurrences). Each section of a distribution is called a deviation, with measurement identified in standard deviations (the average distance from the mean that any point of data is located).

The manner in which the data are spread out is called dispersion. The measures of dispersion are range (measure of the width of the data that are plotted), standard deviation, and variance (the measure of the way that data are spread around the mean).

A process that is running without disturbance from an outside source produces an output that is normally distributed. If an outside disturbance is acting on the process, the distribution changes shape and becomes other than normally distributed, thus indicating the need to identify and correct an assignable cause to return the process to a state of control.

KEY TERMS

Anova A method of determining the equality of means of two distributions, called analysis of variance.

Bell curve The plotted shape of a normal distribution, resembling that of a bell.

Central tendency The characteristic of data tending to group in a pattern around the center value.

Dispersion The way in which data are spread out or distributed.

Mean The simple average of data.

Median Middle value of a sample when data are ranked in ascending or descending order, according to size.

Mode Value that occurs most frequently.

Normal distribution The distribution of data existing in a normally distributed population that, when plotted on a graph, exhibits a standard bell curve form.

Quantitative Describing the quantity in numerical form.

Standard deviation The average numeric distance a point of data falls from the mean of all the data.

Variance The standard deviation squared.

CENTRAL TENDENCY AND DISPERSION TEST

1. Match the terms on the right to the correct definition.

   a. The average value of the data.
   b. The middle value of the data.
   c. The value that occurs the most often.
   d. The symbol for "the sum of."
   e. The symbol for "sample size."

   1. Mode
   2. Median
   3. n
   4. Mean
   5. $\sum$

2. Match the terms on the right to the correct definition.

   a. The difference between the high and low value.
   b. The standard deviation squared.
   c. The average distance any point lies from the mean.
   d. The symbol for standard deviation.
   e. The symbol for average.

   1. Variance
   2. $\bar{X}$
   3. Range
   4. Standard deviation
   5. σ

3. Distinguish between central tendency and dispersion by placing an X to the left of the statement that identifies dispersion.

   a. The way data are grouped around the middle value.
   b. The way data are spread over the distribution.

4. Name the three measures of the central tendency.

   a.
   b.
   c.

5. Name the three measures of dispersion.

   a.
   b.
   c.

6. Write a short paragraph defining the procedure for finding the mean, median, and mode.
7. Write a short paragraph defining the procedure for finding the range, standard deviation, and variance.
8. From the data below, find the range, standard deviation, and variance.

   1, 9, 3, 4, 7, 3, 6, 8, 5, 6, 5, 2, 5, 3, 5

9. From the data below, find the mean, median, and mode.

   1, 9, 3, 4, 7, 3, 6, 8, 5, 6, 5, 2, 5, 3, 5

10. Discuss in a short paragraph the use of the range, standard deviation, and variance.
11. Discuss in a short paragraph the use of the mean, median, and mode.

NOTES

1. Irving Burr, *Elementary Statistical Control* (New York: Marcel-Dekker, 1979), 142.
2. Bonnie Small, *Statistical Quality Control Handbook,* 2d ed. (Indianapolis, IN: AT&T Technologies, 1985), 132.
3. Dale Besterfield, *Quality Control,* 2d ed. (Englewood Cliffs, NJ: Prentice-Hall, 1986), 30.
4. Besterfield, 31.
5. Besterfield, 26.
6. Besterfield, 33.
7. Robert Johnson, *Elementary Statistics,* 4th ed. (Boston, MA: Duxbury Press, 1980), 64.
8. R. Johnson, 34.
9. R. Johnson, 52.
10. Jerome Braverman, *Fundamentals of Statistical Quality Control* (Reston, VA: Reston Publishing Co., 1981), 28.
11. Braverman, 30.
12. Braverman, 29.
13. Braverman, 28.

CENTRAL TENDENCY AND DISPERSION

Exercise No. 1

Plot the following data and find $\overline{X}$, R, mode, median, variance, and standard deviation.

1.

0.304	0.875	0.886	0.833	0.775
0.400	0.251	0.117	0.579	0.839
0.109	0.228	0.525	0.730	0.083
0.139	0.761	0.028	0.883	0.085
0.262	0.427	0.963	0.075	0.327
0.947	0.859	0.389	0.506	0.477
0.743	0.382	0.946	0.196	0.569
0.928	0.807	0.564	0.871	0.607
0.934	0.039	0.554	0.090	0.745
0.033	0.726	0.304	0.362	0.783

2.

8	16	18	17	10
12	20	17	10	7
9	6	7	10	5
19	9	14	14	20
10	13	10	7	9
9	8	11	18	5
8	17	5	4	10
9	5	5	17	9
9	4	8	9	10
4	11	15	11	8

3. Plot the following data and find $\bar{X}$, R, mode, and median.

85	60	12	59	34
82	66	14	09	87
28	29	29	47	66
83	70	15	58	60
52	63	51	11	41
46	59	76	38	27
93	55	91	25	52
64	20	66	73	53
49	69	23	50	08
54	41	53	69	81

4. Compare the following lots of shaft diameters and find the $\bar{x}$, R, and standard deviation for each lot (diameter 0.5000, tolerance ±0.0020).

Lot A

0.5000	0.5000	0.5003	0.5002
0.5000	0.4999	0.5004	0.5005
0.5001	0.5001	0.5002	0.4998
0.5001	0.5002	0.4997	0.5008
0.4999	0.5002	0.5007	0.4996
0.5003	0.5006	0.5000	0.4995
0.5002	0.4999	0.4998	0.5000

Lot B

0.4999	0.5000	0.5008	0.4997
0.5006	0.4998	0.5002	0.4997
0.4999	0.5004	0.5000	0.4995
0.5000	0.4999	0.5003	0.4996
0.5000	0.5000	0.5007	0.5000
0.5001	0.5001	0.5005	0.4999
0.4999	0.4998	0.5000	0.5000

Exercise No. 2

Plot, and find $\overline{X}$, R, median, mode, variance, and standard deviation after grouping the data.

1.

9	6	2	9	3
6	7	3	6	3
1	3	1	8	8
4	2	0	5	5
1	6	8	3	6
8	4	4	4	5
1	2	8	7	2
6	4	1	5	9
6	7	6	8	5
4	1	3	4	6

2.

0.4	0.3	0.5	0.6
0.5	0.5	0.7	0.2
0.2	0.7	0.4	0.5
0.6	0.0	0.6	0.5
0.4	0.5	0.5	0.3
0.6	0.6	0.1	0.5
0.3	0.4	0.5	0.4
0.5	0.6	0.8	1.0
0.7	0.5	0.4	0.5
0.9	0.7	0.6	0.8

Exercise No. 3

Find the standard deviation for the following data using the coded method.

Data: Curve 1

7	10	9	12	10	8
10	13	6	10	11	11
8	11	10	9	10	14
13	9	11	7	12	10
9	10	8	11	9	12

Data: Curve 2

9	6	10	11	11	12
7	12	9	8	9	11
10	11	8	10	8	11
13	9	11	10	7	13
8	10	12	9	10	12

Data: Curve 3

7	9	13	6	9	8
9	8	10	10	10	11
10	7	8	9	8	9
12	11	11	11	9	7
8	9	12	8	14	11

Data: Curve 4

6	8	6	14	6	12
13	7	13	7	14	13
7	6	11	14	12	9
11	14	14	8	14	6
8	9	12	13	7	10

Exercise No. 4

Find the standard deviation for the following data using the coded method. Compare the data and determine whether these samples came from the same lot of product.

Data: Curve 1

7	10	9	12	10	8
10	13	6	10	11	11
8	11	10	9	10	14
13	9	11	7	12	10
9	10	8	11	9	12

Data: Curve 2

9	6	10	11	11	12
7	12	9	8	9	11
10	11	8	10	8	11
13	9	11	10	7	13
8	10	12	9	10	12

Data: Curve 3

7	9	13	6	9	8
9	8	10	10	10	11
10	7	8	9	8	9
12	11	11	11	9	7
8	9	12	8	14	11

Data: Curve 4

6	8	6	14	6	12
13	7	13	7	14	13
7	6	11	14	12	9
11	14	14	8	14	6
8	9	12	13	7	10

Exercise No. 5

The following percentages of carbon were taken from heats during weeks 1 and 2. Compare each week. Plot, and find $\overline{X}$, R, mode, median, standard deviation, and variance.

Week No. 1			% Carbon	
0.02	0.04	0.01	0.04	0.06
0.05	0.05	0.04	0.05	0.06
0.07	0.01	0.06	0.05	0.06
0.03	0.04	0.07	0.06	0.03
0.08	0.07	0.02	0.08	0.05
0.06	0.06	0.04	0.03	0.07
0.09	0.08	0.06	0.07	0.06

Week No. 2			% Carbon	
0.03	0.05	0.01	0.04	0.06
0.02	0.04	0.08	0.03	0.02
0.05	0.03	0.04	0.07	0.04
0.04	0.08	0.06	0.02	0.06
0.06	0.05	0.03	0.04	0.05
0.07	0.04	0.05	0.05	0.09
0.05	0.06	0.07	0.08	0.07

Exercise No. 6

The following temperature measurements (°F) of the solder bath at an electronics manufacturer were taken at random intervals. Plot a histogram. Determine: $\overline{X}$, R, median, mode, variance, and standard deviation. Plot the data using an ogive.

799	801	806	807	808
803	807	802	817	816
807	799	808	800	807
804	811	801	809	809
800	802	800	801	802
808	808	803	804	807
802	806	811	814	801
806	814	807	802	804
810	803	813	807	806
803	806	805	806	815
809	805	810	805	805
805	804	806	811	808
811	809	808	807	811
808	807	804	808	805
806	809	807	805	807
807	808	809	812	810
809	811	812	810	809
813	810	809	811	812
818	797	810	815	799

PROBABILITY

17

The term *probability* is defined as the percent chance that an event will occur. In simple terms, it is the numeric value (e.g., 1 in 5) or percentage (e.g., 20 percent) of the chance that something will happen. Probabilities are assigned a value between 0 and 1, where 0 means that the event will never occur and 1 means that the event will always occur. A 0.5 means the event has a 50 percent probability of occurrence.

Before a comprehensive discussion of probabilities can begin, let us explain some basic terms and rules of probabilities. We will discuss the approaches used to determine probabilities and the rules of addition and multiplication of probabilities.

Historically, three approaches are used to define probability values: classical, relative frequency, and objective. Each approach is discussed in this section.

OBJECTIVES

After completing Chapter 17, the reader should be able to:

- Define the term *probability.*
- Define terms associated with probabilities.
- Distinguish among the classical, relative frequency, and objective approaches to probabilities.
- Discuss the rules for addition of probabilities.
- Discuss the rules for multiplication of probabilities.
- Discuss the use of a tree diagram.
- Discuss the use of a contingency table.
- Discuss the difference between a combination and a permutation.

17–1 CLASSICAL APPROACH

The *classical approach* is used when there is a finite number of possible outcomes, all outcomes are mutually exclusive, and all outcomes are equally likely to occur. In other words, this approach is used when a known number of outcomes are possible, only one outcome can happen at a time, and all outcomes have the same chance of happening. If a fair coin is tossed, two outcomes are possible: heads or tails. In a well-shuffled deck of cards, 52 outcomes are possible when selecting a single card. In either case, a known number of outcomes is possible.

A probability that would not fit this model would be to determine the probability of an asteroid hitting the torch on the Statue of Liberty at 12:05 a.m. on January 13, 2050. Another example would be the probability of the price of a given portion of land going up by 85 percent in three years. These types of outcomes can be determined, but they are a "best guess" type of probability.

The requirement of being *mutually exclusive* means that only one outcome can occur at a time. In the coin toss, only one side of a coin can end up on top at a time. A fair coin toss, then, results in a mutually exclusive outcome.

A nonmutually exclusive event would be the probability of selecting an ace or a spade from a well-shuffled deck of cards. A probability exists for a selection to result in:

- no ace/no spade
- one ace/no spade
- no ace/one spade
- one ace/one spade

The fourth result is that the ace of spades is drawn. This possibility of both ace and spade occurring at the same time keeps this event from being mutually exclusive.

The *equally likely requirement* simply means that each possible outcome has the same chance of happening. In the fair coin-toss example, each outcome (heads or tails) will occur approximately 50 percent of the time. If one die were rolled, each number would have the same chance (1 in 6) of occurring. If two dice were rolled, each combinational result would not be equally likely to occur. The number *seven* has a greater chance of coming up than the number two, because there are more combinations that result in seven than result in two.[1]

Probabilities are easy to figure for an experiment that can have only one of two possible outcomes. A more exotic probability comes from rolling dice. When two dice are rolled, the possible outcome is 1 of 36 combinations. The way that the probability can be figured for any specific number is to find the total combinations that will produce that number and divide by 36. If two dice are used (one red and one white), all the combinations could be found. The possible combinations are shown in Figure 17-1.

To find the percentage of times that each number is expected to be thrown, divide the number of ways that the number can be thrown by 36, then multiply by 100. The percentage that any number is expected to be thrown can be figured in the following manner.

Figure 17–1 Dice Combinations

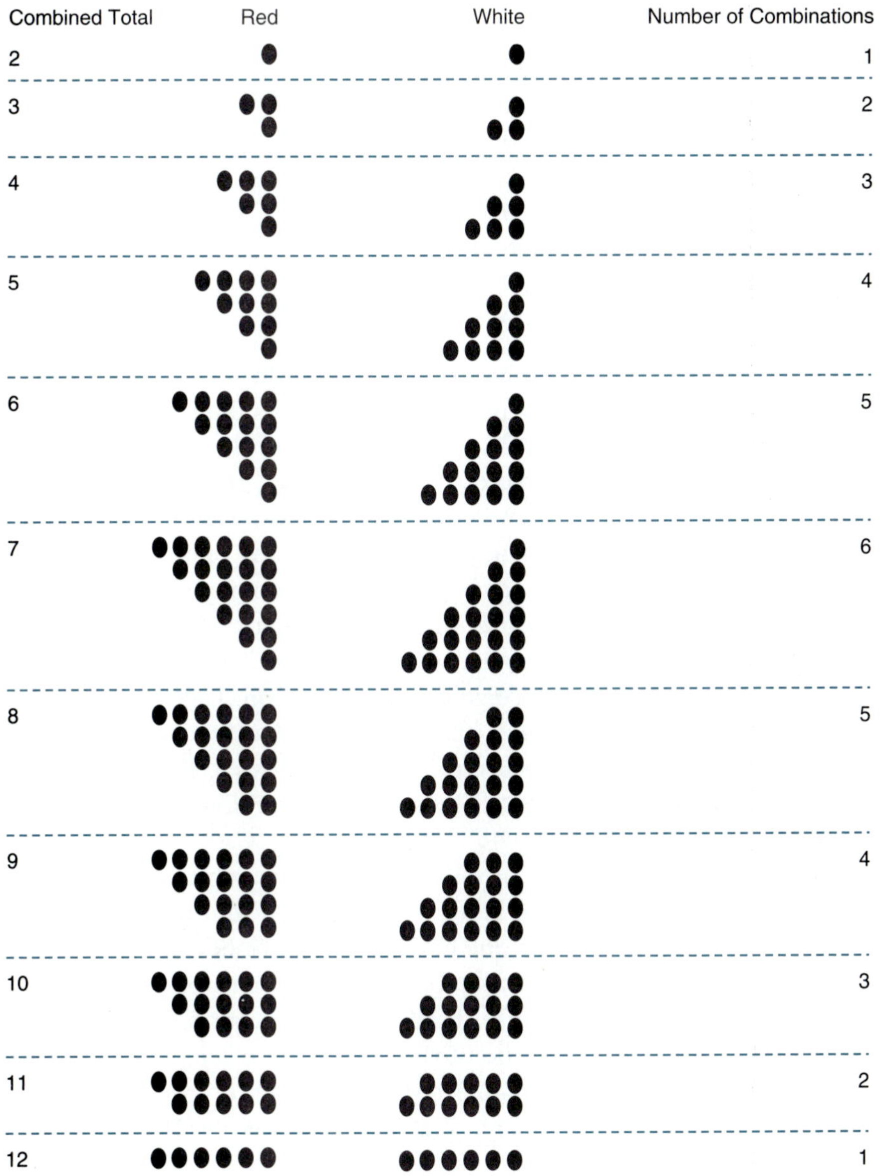

Combined Total	Red	White	Number of Combinations
2			1
3			2
4			3
5			4
6			5
7			6
8			5
9			4
10			3
11			2
12			1

2	$1/36 = 0.0277 \cdot 100 = 2.77\%$
3	$2/36 = 0.0555 \cdot 100 = 5.55\%$
4	$3/36 = 0.0833 \cdot 100 = 8.33\%$
5	$4/36 = 0.1111 \cdot 100 = 11.11\%$
6	$5/36 = 0.1388 \cdot 100 = 13.88\%$
7	$6/36 = 0.1666 \cdot 100 = 16.66\%$
8	$5/36 = 0.1388 \cdot 100 = 13.88\%$
9	$4/36 = 0.1111 \cdot 100 = 11.11\%$
10	$3/36 = 0.0833 \cdot 100 = 8.33\%$
11	$2/36 = 0.0555 \cdot 100 = 5.55\%$
12	$1/36 = 0.0277 \cdot 100 = 2.77\%$

This information can be interpreted in several ways. If the dice are rolled once, there is a 16.6 percent chance of rolling a 7 and a 2.77 percent chance of rolling a 2 or 12. The second way to interpret the information is, if the dice were rolled 100 times, the result should be that the number 7 would be rolled about 17 times, whereas the number 2 would be rolled 3 times. The probability that any one number would result from a roll does not change as the number of rolls increases, because when a combination is rolled once, that combination can be rolled again. A combination that would result in the total *five* has an 11.11 percent chance of occurring on any one roll. Each time the dice are thrown, the number *five* has the same chance of occurring no matter how often it has occurred in previous tosses. Each toss of the dice is an independent event. The results of previous tosses have no effect on future tosses.

In a different type of probability, the odds change with every selection. If one red ball and three white balls are put into a bag, what is the probability that when one ball is selected from the bag it will be white? For this example, the probability that any one ball will be selected is 25 percent, because the bag contains four balls and there is a 100 percent chance that one of the balls will be selected. If the 100 percent is divided by the number of balls, the percentage that any one ball has of being selected is found: 100/4 = 25 percent. However, three of the balls are white and one is red. The white balls make up three-quarters, or 75 percent, of the balls, whereas the one red ball makes up one-quarter, or 25 percent. The probability is 75 percent that the first ball selected will be white and 25 percent that it will be red. If the first ball selected is white, then the probability that the other balls will be selected next changes, provided that the first ball selected is not returned to the bag. If the ball is returned to the bag, the probabilities stay the same, and each selection becomes an independent event. If the first ball selected is not returned to the bag, the chance that any one of the other balls will be selected (on the next selection) goes up. The bag then contains three balls—two white and one red. The probability that the next ball selected from the bag will be white is 66.66 percent, whereas the probability that the next ball selected will be red is 33.33 percent. If the next ball selected is also white, the probability for the two remaining balls goes up to 50 percent for each ball. This type of experiment shows that the probabilities of an experiment can change as the trials are run. This is known as a *dependent* operation, because the probability of a future selection is dependent on all past selections.

When working with dependent events, the concept of "conditional probability" is employed. Conditional probability is used in the "balls selected from a bag"-type probability. If a bag contained 10 white balls and 5 red balls, the probability of selecting a white ball on the first selection could be figured. The probability of selecting a white ball on later selections depends on what color balls were selected on previous selections. Conditional probabilities occur only with dependent events.[2]

What sets this type of probability off from the first experiment discussed is the fact that in the first experiment, the same result can repeat. An example of this is in flipping the coin. It is possible to flip a coin five times with the result each time being heads. In rolling dice, the red die could show *five* and the white die *two* numerous times when the dice are rolled 100 times. In the second type of experiment, when a ball was removed from the bag and not returned, the number of balls in the bag was lower for each selection. This caused the probability that any one of the remaining balls would be selected to go up.

17–2 RELATIVE FREQUENCY APPROACH

The *relative frequency approach* is determined by collecting sample data from a population and evaluating how many times the event being studied occurs in the sample. Provided that no change occurs in the population, the percentage of occurrence of the event under study in the sample data is assumed to be the probability of occurrence in the population. Insurance companies use this type of probability in setting insurance rates for each age group. For example, to set the automobile insurance rates for a given age group, the accident rate of 100,000 people from that age group is studied and a probability is determined.

If 1500 people have had accidents in a one-year period, the probability for the accident rate of the population is figured by dividing 1500 by 100,000. The result is multiplied by 100 to convert the decimal to a percentage ($1500/100,000 = 0.015 \cdot 100 = 1.5\%$). The insurance rates for that age group would be determined based on a 1.5 percent probability of an accident in a one-year period. This is the type of probability employed when using sample inspection to accept or reject a lot of product.

17–3 OBJECTIVE APPROACH

The *objective approach* is basically the "best guess" method of predicting the future. This approach is used when an investor purchases a stock on the stock exchange with the hope of making a profit. For this probability approach to be effective, the investor evaluates many factors, such as past performance in the known environment and how other factors influence the results. The future trends are then predicted, based on the investor's best guess of how the influencing factors are expected to affect the outcome in the future and whether the past trend can be expected to continue.

In the case of the stock purchase, the factors to be evaluated would include the long-term growth rate and dividend, how the stock has performed in a growing economy, how it has performed in a recession, how the company has responded to global competition, how diversified the company is, and the company's market share trend (is the market share increasing or decreasing?). Other factors would have to be evaluated as well. These include the expected economic changes, the expected competition, and future market trends. Although this is truly a best guess, many market analysts can predict the probability of a stock increasing or decreasing with a high rate of success.

17–4 RULES OF ADDITION

The rules of addition are used when the probability of one or more events occurring on a single trial is to be determined. In other words, if two mutually exclusive events (Event A and Event B) are being observed (Figure 17-2), the probability (P) of Event A or Event B occurring is designated by P(A or B).

The solution to this question is to add the probability of Event A occurring to the probability of Event B occurring. This can be stated by $P(A \text{ or } B) = P(A) + P(B)$.

Figure 17–2

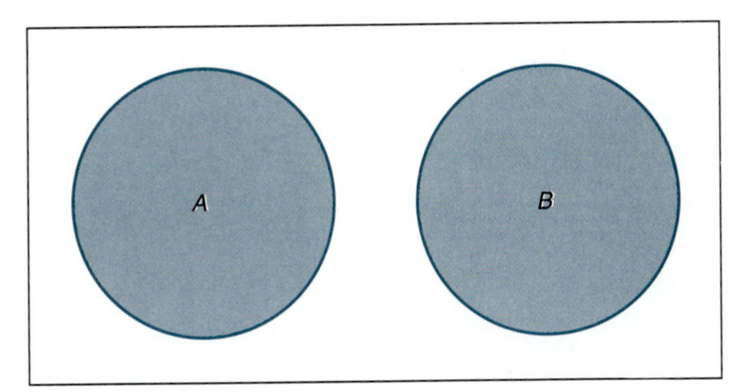

If a card is drawn from a well-shuffled deck of cards and the probability (P) of selecting a king (K) or a queen (Q) on a single selection is to be determined, this question can be stated by the mathematical statement P(K or Q). To find this probability, the probability of selecting a king must be added to the probability of selecting a queen. In mathematical terms,[3]

$$P(K \text{ or } Q) = P(K) + P(Q)$$

By simple deduction, a deck of cards contains 52 cards, of which four are kings and four are queens. From this information the probability of selecting a king on a single trial is 4/52, and the probability of selecting a queen is also 4/52. The solution is to add the two probabilities together as follows:

$$P(K \text{ or } Q) = P(K) + P(Q) = 4/52 + 4/52 = 8/52 = 2/13$$

This fractional probability can be converted into a percentage by dividing 2 by 13 and multiplying the result by 100.

$$2/13 = 0.1538 \cdot 100 = 15.38\%$$

A 15.38 percent chance exists that a king or a queen will be selected on one draw from a deck of cards.

For a nonmutually exclusive event, the probability formula has to be slightly changed to take into account the overlap of Event A and Event B (Figure 17-3). This overlap can be illustrated by determining the probability of selecting a king or a spade when selecting a card from a well-shuffled deck. A deck of cards contains 4 kings and 13 spades. If all the kings and all the spades were removed from the deck and counted, only 16 cards would be present. The reason 16 cards would have been removed instead of 17 is that one of the cards is both a king

Figure 17-3

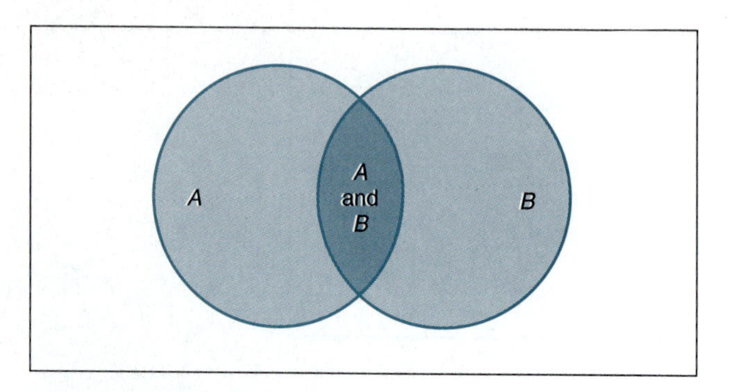

and a spade (the king of spades). This card was counted twice, as a spade and as a king. To find the probability of a king (K) or spade (S) being selected, the following formula is used:

$$P(K \text{ or } S) = P(K) + P(S) - P(K \text{ and } S)$$
$$P(K \text{ or } S) = 4/52 + 13/52 - 1/52 = 16/52 = 4/13$$

The reason that 1/52 must be subtracted (the probability of selecting the king of spades) is that the king of spades is counted twice—once in the probability of selecting one of four kings and once in the probability of selecting one of 13 spades. Since only one king of spades is present in a fair deck of cards, the formula has to compensate for counting this card twice. Thus, the probability of selecting a king and spade is subtracted from the formula for the proper conclusion to the probability.

A further step to this probability is to determine the probability of Event B occurring given the fact that Event A has already occurred. This statement is written as

$$P(B) = P(B \mid A)$$

This is not a formula. The right side is a mathematical statement that says, "The probability of Event B occurring given the fact that Event A has already occurred." As discussed earlier in this chapter, this is a conditional probability. Figure 17-4 shows the Venn diagram for a nonmutually exclusive set of events.

Figure 17-3 shows Event A and the overlap area of Event B. It is the area within Event A that the probability is calculated to find. The formula to find this probability of occurrence looks at Event A as the

Figure 17–4

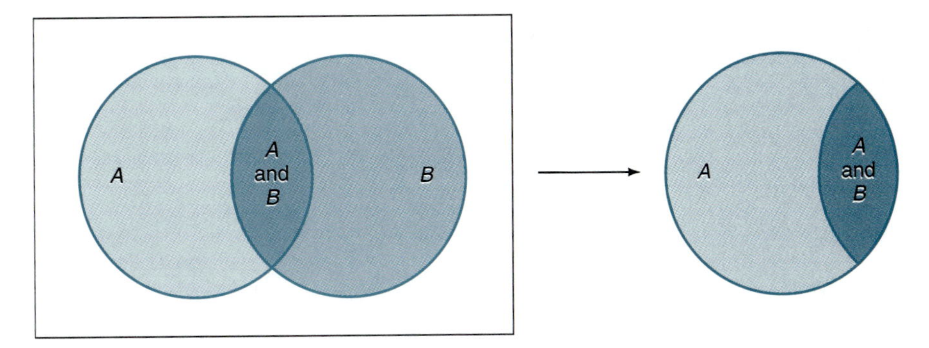

total population of possible events, and determines the percentage of Event A that is overlapped by Event B as follows[4]:

$$P(B|A) = \frac{P(A \text{ and } B)}{P(A)}$$

In the example of the probability of a king (K) or a spade (S) being drawn, the mathematical statement would be

$$P(S|K) = \frac{P(S \text{ and } K)}{P(K)}$$

Common sense dictates that if Event K has occurred (a king was selected from the deck), that there is a 25 percent chance that it is the king of spades—four kings are in a deck of cards, of which one is a spade and three are of other suits. This formula provides the same result:

$$P(S \text{ and } K) = 1/52 = 0.01923$$
$$P(K) = 4/52 = 0.07692$$

$$P(S|K) = \frac{P(S \text{ and } K)}{P(K)} = \frac{0.01923}{0.07692} = 0.24999 \cdot 100 = 24.9999\%$$

If all the decimal places are used, the result is exactly 25 percent.

17–5 RULES OF MULTIPLICATION

The rules of multiplication are used when determining the probability of two events occurring in a given order, such as the probability of Event A occurring and being followed by Event B. The method of

Figure 17–5

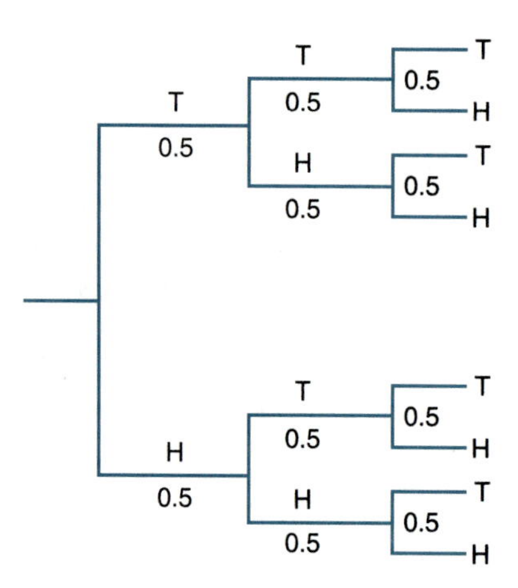

finding this type of probability is different for independent and dependent events.

The formula used for dependent events is simply to multiply the probability of Event A occurring by the probability that Event B will occur. In mathematical terms, the formula is

$$P(A \text{ and } B) = P(A)P(B)$$

In the fair coin toss, a 0.5 (50 percent) probability exists that a head will be tossed, and the same for a tail. Using the rule of multiplication of independent events, the probability of tossing a head (H) followed by a tail (T) can be calculated.

$$P(H \text{ and } T) = P(H)P(T) = (0.5)(0.5) = 0.25$$

A 25 percent chance exists for a head to come up on the first toss followed by a tail on the second toss. This can be taken one step further to determine the probability of tossing three heads in a row. The formula used to solve this probability is

$$P(H \text{ and } H \text{ and } H) = P(H)P(H)P(H) = (0.5)(0.5)(0.5) = 0.125$$

A graphical method can be used to determine this type of probability. The tree diagram (Figure 17–5) shows the results of a given set of events.

This chart shows the possible results of tossing a coin three times. This tree chart could be further developed to show even more tosses. To find the probability of a given set of outcomes, multiply the probability of each branch following the desired set of results.

The tree diagram can also be used for dependent (or conditional) probabilities, with the same result. As previously discussed, a dependent event is one in which the probability of occurrence depends on what events have already occurred. For example, if a bag contains 10 balls, of which two are white and eight are black, what is the probability that of two balls selected at random, the first would be white and the second one black? (Note: This probability is a dependent set of events, provided the first selection is not returned to the bag.) A tree diagram has been constructed to show the possible outcomes graphically (Figure 17-6).

$$(W \text{ and } B) = P(W)P(B) = (0.2)(0.8888) = 0.1776 \text{ or } 17.76\%$$

There is a 17.76 percent chance that two selections would result in a white ball on the first selection and a black ball on the second.

A second method to graphically depict the same information is by constructing a contingency table. This table displays information in a way that is easier to understand. If the information on the probability of

Table 17–1 Contingency Table

	SUIT		
CARD	**SPADE**	**NONSPADE**	**TOTAL**
KING	1	3	4
NONKING	12	36	48
TOTAL	13	39	52

Figure 17–6

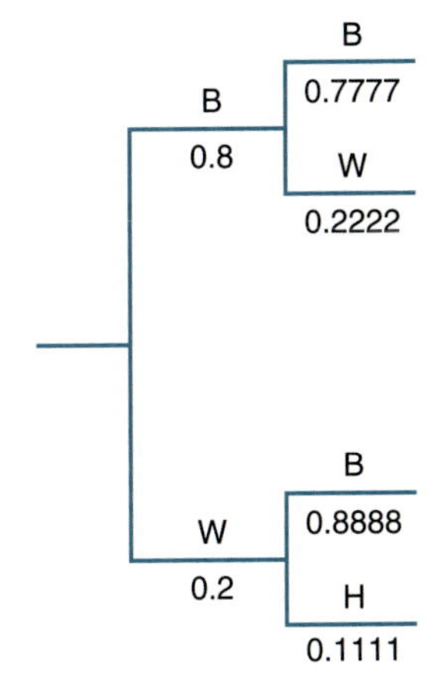

Table 17–2 Joint Probability Table

CARD	SUIT: SPADE	NONSPADE	TOTAL
KING	0.0192	0.0577	0.0769
NONKING	0.2308	0.6923	0.9231
TOTAL	0.25	0.75	1.00

selecting a king or spade were put into graphical form, a contingency table could be constructed as in Table 17-1 (previous page).

From the contingency table a joint probability table could be constructed. This is accomplished by finding what percentage each value is of the total for that column. A joint probability table is shown in Table 17-2.

From this table, it can be determined, if a card selected at random were a king (K), what the chance would be that it would be a spade (S). To find this, the formula for conditional probability would be used. This probability would be solved as follows:

$$P(S \mid K) = \frac{P(K \text{ and } S)}{P(K)} = \frac{0.0192}{0.0769} = 0.2496 \cdot 100 = 24.96\%$$

There is a 25 percent chance that the card will be a spade.

The rules of addition and multiplication are used throughout many of the methods used in quality control. They are used when making decisions about the validity of data and test results. The question that often arises when the results of an experiment are different from those expected is the following: "Is this result an expected value or is there a problem?" The proper application of these rules will help the quality professional make those judgments based on facts rather than a "gut feeling."

17–6 COMBINATIONS AND PERMUTATIONS

Combinations are the number of ways objects can be grouped, not considering the order of the objects. If two letters (A and B) were to be grouped, they would make one group, or combination. The order of the group would not matter. Group (A, B) and Group (B, A) are considered the same combinational group. The following is the combinational formula[5]:

$$\frac{n!}{x!(n - x)!}$$

where

n = number of total objects from which the groups are to be made (population)
x = number of the objects grouped together at a time (group), and
! = factorial

This formula is used to determine how many combinations can be made up from "n" things grouped together "x" at a time.

The factorial is a method of writing a long multiplication problem. For example, $3! = 3 \cdot 2 \cdot 1$ and $9! = 9 \cdot 8 \cdot 7 \cdot 6 \cdot 5 \cdot 4 \cdot 3 \cdot 2 \cdot 1$. Factorials can be worked by hand. Most calculators will perform this function speedily. The question that now arises is, what about zero? By definition, *zero factorial equals one.* Memorize this rule for future use.

What is meant by a combinational problem? For example, how many combinations can be made from the letters A, B, and C if they are grouped two at a time (AB, AC, BC)? Combinational problems do not take order within each group into account. In other words, the AB group and the BA group are the same.

In this example, n = 3. The three items in the population are (A, B, C). The (x) = 2, the number of items grouped together. The question is, how many combinations are there for three things (n) taken two (x) at a time? That problem was worked out by trial and was proved to have three combinations. If the same problem is worked by the formula, it will be figured as follows:

$$\frac{3!}{2!(3 - 2)!} = \frac{6}{2(1)} = \frac{6}{2} = 3$$

This may seem a long way around to solve a simple problem (which it is), but it shows how the formula is used. Let us take a more elaborate problem: How many pictures would be required to photograph 25 people, 2 at a time? This problem would be tougher to work out by hand; the formula makes it simpler:

$$\frac{25!}{2!(25-2)!} = \frac{15{,}511{,}210{,}040{,}000{,}000{,}000{,}000{,}000}{2(25{,}852{,}016{,}740{,}000{,}000{,}000{,}000)}$$

$$= \frac{15{,}511{,}210{,}040{,}000{,}000{,}000{,}000{,}000}{51{,}704{,}033{,}470{,}000{,}000{,}000{,}000} = 300$$

Therefore, it would take 300 pictures to photograph all possible combinations of 25 people taken 2 at a time. As can be seen, this formula would be best worked on a calculator.

Permutations are similar to combinations. The main difference in permutations is that the order of the group is taken into account. In other words, two permutations can be made from the population (A and B): Group 1 (A, B) and Group 2 (B, A). An example where the order needs to be taken into account would be in an area code for a telephone number. If area codes were limited based on combinations, 120 area codes would be the minimum available. Under the combinational theory, if an area code were 234, then those numbers could not be used in any other combination, such as 432. However, 234 and 432 *can* be used as different area codes. This difference is a permutation rather than a combination. The formula for a permutation is:

$$\frac{n!}{(n - x)!}$$

where

n = number of total objects from which the groups are to be made (population),
x = number of the objects grouped together at a time (group), and
! = factorial

Looking at the permutational theory, the total number of area codes available using three-digit combinations would be figured as follows:

$$n = 10$$
$$x = 3$$
$$\frac{10!}{(10 - 3)!} = \frac{3{,}628{,}800}{5040} = 720$$

Knowing that area codes are permutations rather than combinations allows us to determine that 720 area codes are possible with a three-digit code.

Combinational and permutational theories are used in many areas of probabilities. The binomial distribution (discussed in Chapter 18) is based on the combinational formula. With an understanding of these concepts, a quality professional can make effective decisions based on probability distributions.

SUMMARY

Probability is the percentage chance that an event will occur. It is represented by a numerical value or percentage of the chance that something will happen.

The classical approach in probability is used when there is a finite number of possible outcomes, all outcomes are mutually exclusive, and all outcomes are equally likely to occur. "Mutually exclusive" means that only one outcome can occur at a time. The "equally likely" requirement means that each possible outcome has the same chance of happening.

The relative frequency approach is determined by collecting sample data from a population and evaluating how many times the event being studied occurs in the sample. Provided no change occurs in the population, the percentage of occurrence of the event under study in the sample data can be assumed to be the probability of occurrence in the population.

The objective approach is basically a "best guess" method of predicting probability. This approach requires the evaluation of many factors if it is to be successful.

Combinations and permutations describe the relationship of data within an analysis. Combinations are the number of ways objects can be grouped, not considering the order of the objects. Permutations, although similar to combinations, take into account the order in which the objects are grouped.

KEY TERMS

Classical approach Method used to find the probability when there is a finite number of possible outcomes, all outcomes are mutually exclusive, and all outcomes are equally likely to occur.

Combination The grouping of data in which the order of the objects is not a consideration.

Dependent event An event in which the probability of occurrence depends on what events have previously occurred.

Equally likely requirement The condition in which each possible outcome has the same chance of occurring.

Independent event An event in which the probability of occurrence does not depend on previously occurring events.

Mutually exclusive The condition in which only one outcome is possible for a single trial.

Objective approach The method of determining probability through a "best guess" approach.

Permutation The grouping of data in which the order of the data is taken into account.

Probability The numerical likelihood of an uncertain occurrence of an event.

Relative frequency approach The method of determining probability through the collection of sample data from a population and evaluating the number of times that the event being studied occurs in the sample.

PROBABILITY TEST

1. Define in a short paragraph the meaning of the term *probability.*
2. Match the terms on the right to the correct definition.
   - a. An event that has the same chance of occurring each time an experiment is conducted.
   - b. An event that can occur after a unique event has already occurred.
   - c. The "best guess" approach to determining a probability.
   - d. Events for which only one outcome at a time is possible are known as _____ _____ events.
   - e. Events having the exact same chance of occurring are said to be _____ _____.

   1. Equally likely
   2. Mutually exclusive
   3. Independent event
   4. Dependent event
   5. Objective
3. Pick out the objective approach from among the classical, relative frequency, and objective approaches by placing an X to the left of the description of the objective approach to probabilities.
   - a. This approach is used when a known number of outcomes are possible, only one outcome can happen at a time, and all outcomes have the same chance of happening.
   - b. Collecting sample data from a population and evaluating how many times the event being studied occurs in the sample.
   - c. Many factors are evaluated in the known environment, along with how other factors influence the results.
4. Discuss in a short paragraph the method to determine the probability of one or more events occurring on a single trial.
5. Discuss in a short paragraph the method to determine the chance of two or more events happening in a given order.
6. Discuss in a short paragraph the use of a tree diagram.

7. Discuss in a short paragraph the use of a contingency table.
8. Discuss in a short paragraph the difference between a combination and a permutation.

NOTES

1. Robert Johnson, *Elementary Statistics,* 4th ed. (Boston, MA: Duxbury Press, 1980).
2. M. Spiegel, *Probability and Statistics* (New York: McGraw-Hill, 1975).
3. A. Naiman, R. Rosenfield, and G. Zirkel, *Understanding Statistics* (New York: McGraw-Hill, 1983).
4. Naiman et al.
5. Naiman et al.

PROBABILITY

Exercise No. 1

1. Using the classical approach, what is the probability that a sample from the following data will be 0.389 inches?

0.304	0.875	0.886	0.833	0.775
0.400	0.251	0.117	0.579	0.839
0.109	0.228	0.525	0.730	0.083
0.139	0.761	0.028	0.883	0.085
0.262	0.427	0.963	0.075	0.327
0.947	0.859	0.389	0.506	0.477
0.743	0.382	0.946	0.196	0.569
0.928	0.807	0.564	0.871	0.607
0.934	0.039	0.554	0.090	0.745
0.033	0.726	0.304	0.362	0.783

2. Using the classical approach, what is the probability that a sample from the following data will be 11 inches?

8	16	18	17	10
12	20	17	10	7
9	6	7	10	5
19	9	14	14	20
10	13	10	7	9
9	8	11	18	5
8	17	5	4	10
9	5	5	17	9
9	4	8	9	10
4	11	15	11	8

3. Using the classical approach, what is the probability that a sample from the following data will be greater than 0.5004 inches?

0.5000	0.5000	0.5003	0.5002
0.5000	0.4999	0.5004	0.5005
0.5001	0.5001	0.5002	0.4998
0.5001	0.5002	0.4997	0.5008
0.4999	0.5002	0.5007	0.4996
0.5003	0.5006	0.5000	0.4995
0.5002	0.4999	0.4998	0.5000

4. Using the classical approach, what is the probability that a sample from the following data will be between 0.5000 and 0.5004 inches?

0.4999	0.5000	0.5008	0.4997
0.5006	0.4998	0.5002	0.4997
0.4999	0.5004	0.5000	0.4995
0.5000	0.4999	0.5003	0.4996
0.5000	0.5000	0.5007	0.5000
0.5001	0.5001	0.5005	0.4999
0.4999	0.4998	0.5000	0.5000

Exercise No. 2

1. Using the relative frequency approach, what is the probability of selecting a 5 from the following data?

9	6	2	9	3
6	7	3	6	3
1	3	1	8	8
4	2	0	5	5
1	6	8	3	6
8	4	4	4	5
1	2	8	7	2
6	4	1	5	9
6	7	6	8	5
4	1	3	4	6

2. Using the relative frequency approach, what is the probability of selecting a 0.5 from the following data?

0.4	0.3	0.5	0.6
0.5	0.5	0.7	0.2
0.2	0.7	0.4	0.5
0.6	0.0	0.6	0.5
0.4	0.5	0.5	0.3
0.6	0.6	0.1	0.5
0.3	0.4	0.5	0.4
0.5	0.6	0.8	1.0
0.7	0.5	0.4	0.5
0.9	0.7	0.6	0.8

Exercise No. 3

From which of the following groups of data would the probability of selecting a 7 be greatest?

Curve 1

7	10	9	12	10	8
10	13	6	10	11	11
8	11	10	9	10	14
13	9	11	7	12	10
9	10	8	11	9	12

Curve 2

9	6	10	11	11	12
7	12	9	8	9	11
10	11	8	10	8	11
13	9	11	10	7	13
8	10	12	9	10	12

Curve 3

7	9	13	6	9	8
9	8	10	10	10	11
10	7	8	9	8	9
12	11	11	11	9	7
8	9	12	8	14	11

Curve 4

6	8	6	14	6	12
13	7	13	7	14	13
7	6	11	14	12	9
11	14	14	8	14	6
8	9	12	13	7	10

Exercise No. 4

What is the probability of selecting a 7, 8, or 9 from the following data?

Curve 1

7	10	9	12	10	8
10	13	6	10	11	11
8	11	10	9	10	14
13	9	11	7	12	10
9	10	8	11	9	12

Curve 2

9	6	10	11	11	12
7	12	9	8	9	11
10	11	8	10	8	11
13	9	11	10	7	13
8	10	12	9	10	12

Curve 3

7	9	13	6	9	8
9	8	10	10	10	11
10	7	8	9	8	9
12	11	11	11	9	7
8	9	12	8	14	11

Curve 4

6	8	6	14	6	12
13	7	13	7	14	13
7	6	11	14	12	9
11	14	14	8	14	6
8	9	12	13	7	10

Exercise No. 5

The following percentages of carbon were taken from heats during weeks no. 1 and 2 and compared each week. If 0.05 ± 0.01 is accepted, what is the probability that a sample taken at random will be outside that range?

Week No. 1				% Carbon
0.02	0.04	0.01	0.04	0.06
0.05	0.05	0.04	0.05	0.06
0.07	0.01	0.06	0.05	0.06
0.03	0.04	0.07	0.06	0.03
0.08	0.07	0.02	0.08	0.05
0.06	0.06	0.04	0.03	0.07
0.09	0.08	0.06	0.07	0.06

Week No. 2				% Carbon
0.03	0.05	0.01	0.04	0.06
0.02	0.04	0.08	0.03	0.02
0.05	0.03	0.04	0.07	0.04
0.04	0.08	0.06	0.02	0.06
0.06	0.05	0.03	0.04	0.05
0.07	0.04	0.05	0.05	0.09
0.05	0.06	0.07	0.08	0.07

Exercise No. 6

Develop a contingency table for the following electronic boards.

500 boards were bought and placed into inventory
312 of the boards were tested
93 of the boards were purchased from supplier A
17 boards supplied by supplier A were tested

Show actual numbers

	Tested	Not tested	
Supplier A			
Other supplier			

Show percentages.

	Tested	Not tested	
Supplier A			
Other supplier			

What is the probability that a board selected at random will be a tested board supplied by supplier A?

Exercise No. 7

1. The Texas lotto pays large sums of money for selecting 6 correct numbers from a pool of 50 numbers. How many possible combinations are available for each drawing?
2. How many different phone numbers are available for the 437-XXXX exchange?
3. If a phone number with area code has 10 digits total, how many different phone numbers can be assigned without duplicating one?

PROBABILITY DISTRIBUTIONS

18

OBJECTIVES

After completing Chapter 18, the reader should be able to:

- Distinguish among normal, binomial, and Poisson distributions.
- Name the distribution to be used with given data.
- Discuss the procedure for finding the probability using a normal, binomial, or Poisson distribution.
- Demonstrate the ability to find the probability using the normal, binomial, or Poisson distribution.
- Discuss the use of the normal, binomial, and Poisson distribution.

Many of the charts used in statistical quality control center around three probability distributions, each of which is discussed in this chapter. They relate to attributes and variables data. This by no means covers all the types of probability distributions, but for the purposes of this text the types to be discussed are:

- Normal
- Binomial
- Poisson

These three distributions were selected because they cover the most common types of problems that will be encountered.

The *normal* distribution is used to find the probability that a selection of a given quantity will be made from a group of numbers that is normally distributed. The *binomial* distribution is used to find the probability that a sample selected from a population with a known percentage of defective product will contain a given number of defects. The *Poisson* distribution is used to identify the probability of finding a product with a given number of defects if the average number of defects per product is known.

If this explanation is bothersome, ignore it. As stated in the introduction to Part III, each of the distributions contains math functions that are commonly used every day. All calculations can be performed on a normal calculator.

18–1 NORMAL PROBABILITY DISTRIBUTION

A normal distribution is used with variables data (measurements). This type of probability distribution uses a formula and a chart to find the percentage.

We will use the same data used in Chapter 16 (the number of miles that 50 students drove in one week). In the section covering standard deviation, the following information was figured for the data.

$$(\sigma) = 24.06$$
$$(\overline{x}) = 83.62$$

Both variables (σ and $\overline{x}$) are used in the formula.

What is the probability that a student selected at random from the group of 50 students will have driven between 107 miles and the average ($\overline{x}$)? To solve this probability, the variables for the distribution must be identified:

(σ) = standard deviation of the data
($\overline{x}$) = mean or average of the data
(x) = value that the probability is being worked to find
(z) = a value that relates to a percentage of a standard deviation that the probability represents.[1]

In this example, the problem is solved as follows:

$$(x) = 107$$
$$(\overline{x}) = 83.62$$
$$(\sigma) = 24.06$$

$$z = \frac{(x - \overline{x})}{\sigma} = z = \frac{(107) - (83.62)}{24.06} = \frac{23.38}{24.06} = 0.9717$$

This value is called the (z) score of the probability (Table 18-1).

- The first place of the decimal is located down the left side of the chart.
- The second place of the decimal is located along the top of the chart.
- Moving to the number where the two rows intersect, the probability (in decimal form) can be located.

In our example, (z) = 0.9717. Working with the first two decimal places, the probability can be located. On the left-hand side of the chart is the first decimal place of the (z) score. In the example, 0.9 is the first decimal. Along the top is the second decimal place of the (z) score (0.07). Locate the number where the two rows intersect (0.3340). Multiply this number by 100 to find the percentage of 33.40.

What is the probability that a student selected at random from the group of 50 students will have driven between 107 miles and the average ($\overline{x}$)? From the preceding work there is a 33.4 percent chance

Table 18–1 Normal Probability Distribution

z	.00	.01	.02	.03	.04	.05	.06	.07	.08	.09
0.0	.0000	.0040	.0080	.0120	.0160	.0199	.0239	.0279	.0319	.0359
0.1	.0398	.0438	.0478	.0517	.0557	.0596	.0636	.0675	.0714	.0753
0.2	.0793	.0832	.0871	.0910	.0948	.0987	.1026	.1064	.1103	.1141
0.3	.1179	.1217	.1255	.1293	.1331	.1368	.1406	.1443	.1480	.1517
0.4	.1554	.1591	.1628	.1664	.1700	.1736	.1772	.1808	.1844	.1879
0.5	.1915	.1950	.1985	.2019	.2054	.2088	.2123	.2157	.2190	.2224
0.6	.2257	.2291	.2324	.2357	.2389	.2422	.2454	.2486	.2518	.2549
0.7	.2580	.2612	.2642	.2673	.2704	.2734	.2764	.2794	.2823	.2852
0.8	.2881	.2910	.2939	.2967	.2995	.3023	.3051	.3078	.3106	.3133
0.9	.3159	.3186	.3212	.3238	.3264	.3289	.3315	.3340	.3365	.3389
1.0	.3413	.3438	.3461	.3485	.3508	.3531	.3554	.3577	.3599	.3621
1.1	.3643	.3665	.3686	.3708	.3729	.3749	.3770	.3790	.3810	.3830
1.2	.3849	.3869	.3888	.3907	.3925	.3944	.3962	.3980	.3997	.4015
1.3	.4032	.4049	.4066	.4082	.4099	.4115	.4131	.4147	.4162	.4177
1.4	.4192	.4207	.4222	.4236	.4251	.4265	.4279	.4292	.4306	.4319
1.5	.4332	.4345	.4357	.4370	.4382	.4394	.4406	.4418	.4429	.4441
1.6	.4452	.4463	.4474	.4484	.4495	.4505	.4515	.4525	.4535	.4545
1.7	.4554	.4564	.4573	.4582	.4591	.4599	.4608	.4616	.4625	.4633
1.8	.4641	.4649	.4656	.4664	.4671	.4678	.4686	.4693	.4699	.4706
1.9	.4713	.4719	.4726	.4732	.4738	.4744	.4750	.4756	.4761	.4767
2.0	.4772	.4778	.4783	.4788	.4793	.4798	.4803	.4808	.4812	.4817
2.1	.4821	.4826	.4830	.4834	.4838	.4842	.4846	.4850	.4854	.4857
2.2	.4861	.4864	.4868	.4871	.4875	.4878	.4881	.4884	.4887	.4890
2.3	.4893	.4896	.4898	.4901	.4904	.4906	.4909	.4911	.4913	.4916
2.4	.4918	.4920	.4922	.4925	.4927	.4929	.4931	.4932	.4934	.4936
2.5	.4938	.4940	.4941	.4943	.4945	.4946	.4948	.4949	.4951	.4952
2.6	.4953	.4955	.4956	.4957	.4959	.4960	.4961	.4962	.4963	.4964
2.7	.4965	.4966	.4967	.4968	.4969	.4970	.4971	.4972	.4973	.4974
2.8	.4974	.4975	.4976	.4977	.4977	.4978	.4979	.4979	.4980	.4981
2.9	.4981	.4982	.4982	.4983	.4984	.4984	.4985	.4985	.4986	.4986
3.0	.49865	.4987	.4987	.4988	.4988	.4989	.4989	.4989	.4990	.4990
4.0	.4999683									

that a student selected at random from the 50 students will have driven between 83.62 miles and 107 miles.

This probability is much like the example of the bag of colored balls previously discussed. The probability of selecting any one ball was

Figure 18–1 Normal Distribution

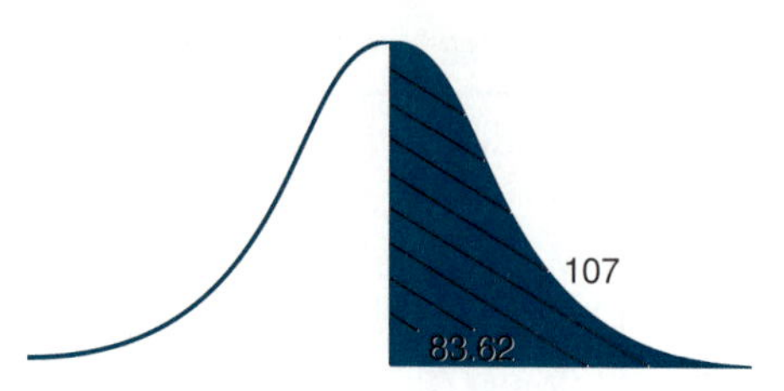

dependent on the number of balls in the bag. As the number of balls decreased, the probability of selecting any one ball increased.

In the normal distribution, the probability of selecting at random a student who drove between 83.62 miles and 107 miles was dependent on the number of students in the group who had driven between the two mileages. We look the (z) score up on the chart and find the percentage of the 50 students that were in the mileage category. In the example, 33.4 percent of the 50 students were in that category. This means that approximately 17 of the 50 students had driven between 83.62 miles and 107 miles. If the distribution is drawn and the two points are plotted, the percentage becomes apparent (Figure 18-1).

What is the probability of selecting, at random, a student who had driven between 27 miles and 83.62 miles? The variables put into the formula, as previously shown, are solved as follows:

$$z = \frac{27 - 83.62}{24.06} = \frac{-56.62}{24.06} = -2.35$$

In this case, the (z) equals a negative number (−2.350). To find the probability, use the positive number (2.35). On the left-hand side of the chart in Table 18-1, find 2.3. On the top of the chart, find 0.05. The number where the two rows intersect is 0.4906, or 49.06 percent. To answer the probability question, there is a 49.06 percent chance that a student selected at random had driven between 27 miles and 83.62 miles.

With this problem plotted in a normal distribution, it becomes apparent why the percentage is so high (Figure 18-2).

The normal probability always shows the probability of finding a value between the given value and the mean. If the probability is asked in such a way that one limit is not the mean, then some addition or subtraction is needed. What is the probability that a student selected at random will have driven between 27 miles and 107 miles? This

Figure 18–2 Normal Distribution

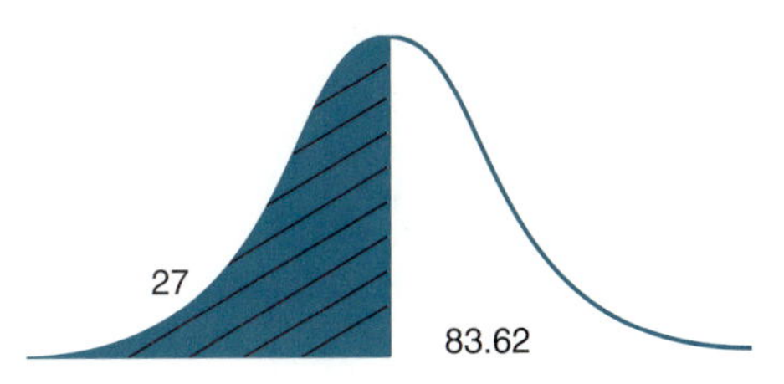

probability problem is best understood if it is plotted on a normal curve (Figure 18-3).

The method of finding this probability is a combination of two different probability distributions. Normal probability distributions show only from the $\overline{x}$ value to the mean of the distribution.

In this example, the first step is to find the probability for the (z) value of 27. Then, find the (z) value of 107. From the graph, it is apparent that the percentage of both should be added together. The probability for each (z) value has already been figured earlier in this chapter. The probability of selecting a student between the $\overline{x}$ and 27 miles was 49.06 percent. The probability of selecting a student between the $\overline{x}$ and 107 was 33.4 percent. The probability of selecting a student between 27 and 107 miles is 82.46 percent, because this percentage of the 50 students is included in this group.

What is the probability that a student selected at random will have driven between 27 miles and 60 miles? This probability, graphed on a distribution, shows that the percentage between 27 and $\overline{x}$ must be found, and the percentage between 60 and $\overline{x}$ subtracted from it (Figure 18-4).

The percentage between the x and 27 has already been figured—49.06 percent. The percentage of students between x and 60, after figuring, is $z = 0.9817$. From the chart of (z) scores, the percentage is 33.65. To find the percentage between 27 and 60, subtract 33.65 from

Figure 18–3 Normal Distribution

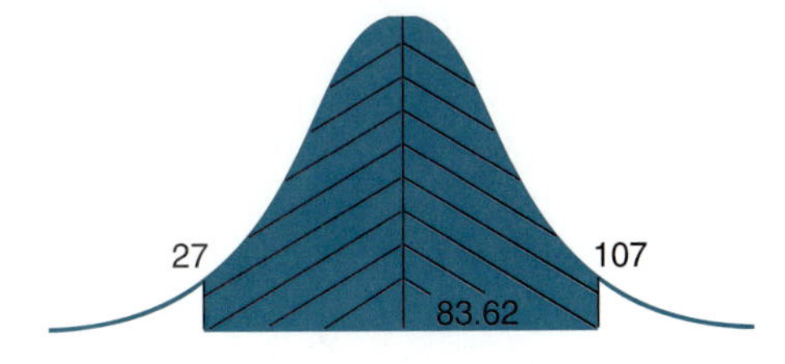

Figure 18–4 Normal Distribution

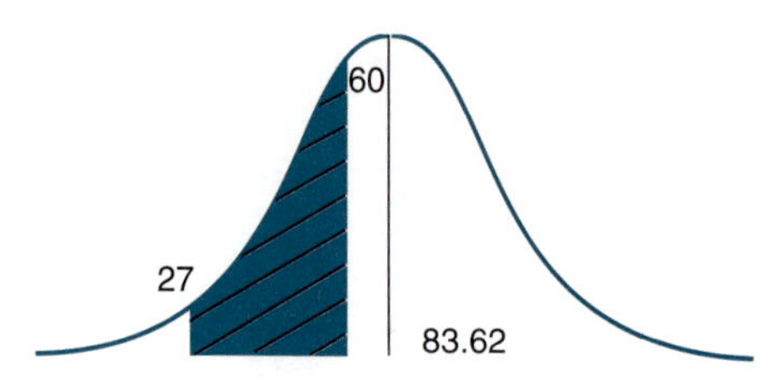

49.06. The remainder (15.41 percent) is the probability. In this probability, the percentage of the population between 60 and $\bar{x}$ must be subtracted from the population between 27 and $\bar{x}$. This leaves the population between 27 and 60.

Another way a probability can be asked is, what is the probability that a student selected at random will have driven more than 60 miles? This probability, drawn on a normal curve, includes all students between 60 and $\bar{x}$, plus all students above $\bar{x}$ (Figure 18-5).

The amount of data that falls above and below $\bar{x}$ is 50 percent. To find the percentage that falls above 60 miles, the percentage for 50 to $\bar{x}$ must be figured. As in the previous example, this is 33.65 percent. To this number, the percentage above x (50 percent) must be added. The probability that a student selected at random will have driven more than 60 miles is 83.65 percent.

The same method is used to find the probability of selecting at random a student who had driven less than 107 miles. Find the probability that the student drove between $\bar{x}$ and 107 miles. Add 50 percent to include all students below the $\bar{x}$ (Figure 18-6).

The last way that the probability could be asked is, what is the probability that a student selected at random will have driven more than 107 miles? When the probability is drawn on a normal curve, the method of finding this should be apparent. The total amount of data

Figure 18–5 Normal Distribution

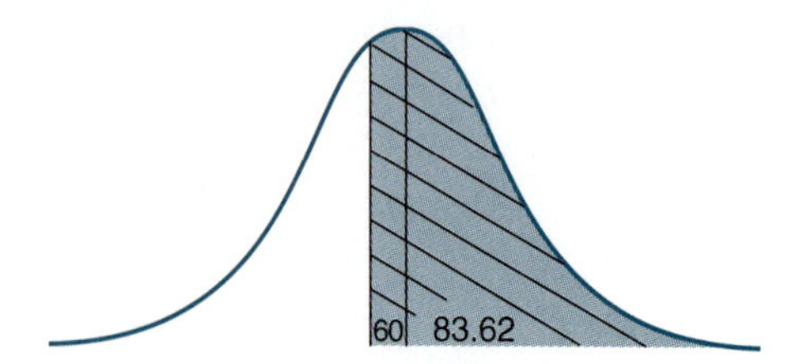

Figure 18–6 Normal Distribution

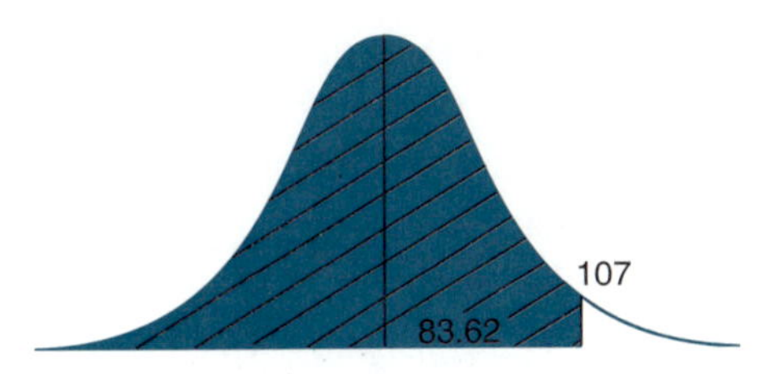

above $\overline{x}$ is 50 percent. The amount of data between 107 and $\overline{x}$ is 34.4 percent. This means that 34.4 percent of the data above $\overline{x}$ is below 107. To find the amount of data above 107, subtract the amount below 107, but above $\overline{x}$, from all data above $\overline{x}$; then find the amount above 107. In this problem, 33.4 percent of the data is above the $\overline{x}$ and below 107. This percentage subtracted from 50 percent leaves 16.6 percent, the amount above 107.

This method holds true if the problem asks for the probability of selecting a student who drove less than 60 miles. The percentage between 60 miles and $\overline{x}$ (33.65 percent) is subtracted from 50 percent, the amount of data below $\overline{x}$; the remainder (16.35 percent) is the probability.

In simple terms, if a bag contains 10 balls, 3 red and 7 white, the probability that a ball selected at random will be red is dependent on the percentage of red balls in the bag. The percentage of red balls being 30 percent, there is a 30 percent probability of selecting a red one. If the percentage of students who have driven between 27 and 60 miles is known, the probability of a student, selected at random, being in that group is the same as the percentage of students in that group. Normal probability distributions are based on an assumption: the probability of making a random selection from a given group within a population is based on the size of the group compared to the population. The percentage chance that a random selection will be from a group is the same percentage that the group constitutes of the population.

Conclusions The normal probability distribution is the basis for many of the procedures used in statistical process control. Charting methods such as X-bar and R charts are designed to plot the performance of a process and to show that the process is having problems when the results fall outside the expected probability value. The advantage of using X-bar and R charts is that the results are shown in graphic form and require only limited calculations.

18–2 BINOMIAL DISTRIBUTION

The binomial distribution is used with *attributes data.* This type of distribution is used in quality control to find the probability that a sample selected from a population will possess a certain characteristic. For example, if a process that makes ball bearings has been producing 10 percent oversize balls, what is the probability that in a sample of 15 balls, 2 will be oversize?

Information from this distribution allows decisions to be made about a process while it is running without having to measure all the balls or without knowing the number of balls made. This distribution works for an unknown population size or a theoretically infinite population.

The binomial distribution is based on the number of combinations that could be selected from a population. It is used to make decisions about a production process without measuring all of the product. The information that is found most often is (from the number of defective product in a sample) the expected rate of defectives that the production process is producing.

Referring back to the example of the ball bearings, the number of defectives in the sample taken from the process will determine whether the process making the balls is running at the same efficiency or has changed. If the number of defectives in the sample is greater or less than expected, then the process has changed and a corrective action can be made.

The manufacturer needs to find this probability because, when sampling from a large group of items, there is a chance of an error. For example, suppose a jar contained 100 marbles, 80 white and 20 red. If a sample of 10 were taken from the jar, the expected percentage of each color in the sample would be the same as the percentage of each color in the total number of marbles. In theory, the sample would contain 8 white marbles and 2 red marbles. In practice, the sample may contain 7 white and 3 red, or it may contain 9 white and 1 red. This difference is known as the sampling error. This error is normal for sampling; the quality professional must understand and deal with it when using sampling to determine the level of production of a process. The error would be greater if the marbles were divided into colors such as 80 white, 15 red, and 5 black. If a sample of 10 marbles were taken from this combination, a sampling error would exist every time a sample was taken, because each of the 10 marbles would make up 10 percent of the sample. The group of 100 marbles has a 5 percent black population. The sample of 10 marbles that had one black marble would have 5 percent more than the true percentage of black marbles. For this reason, the binomial distribution will show the probability that if a sample is taken from a group where the percentage of a characteristic (color, size, shape, etc.) is known, then the probability that the sample will contain a certain number of parts with that characteristic can be found. The formula for binomial distribution is[2]:

$$\frac{n!}{x!(n - x)!} (p')^x (1 - p')^{n-x}$$

As you can see, the combinational formula is the first portion of the binomial formula.

All distributions have a mean ($\overline{x}$) and a variance. For the binomial distribution, the mean ($\overline{x}$) is np'. The variance is $np'\ (1 - p')$. The standard deviation is the square root of the variance.

The variables that are used in the formula are defined as

(n) = number in the sample size
(p′) = percentage of the group that has a certain characteristic
(x) = number of parts from the sample that are expected to have the characteristic

For example, in the case of the marbles, the jar contains 100 marbles, 20 red and 80 white. What is the probability that 10 marbles selected at random will have 2 red marbles? This question can be solved using the binomial distribution. In this problem, the (n) value is 10. This is the size of the sample. The (p′) is 0.20. This is the percentage (in decimal form) of the 100 marbles that are red. The ($\overline{x}$) value is 2. This is the number of marbles in the sample (10 marbles) for which the probability of finding the characteristic is being figured.

$$\frac{n!}{x!\ (n - x)!} (p')^x(1 - p')^{n - x}$$

$$= \frac{10!}{(0.40)(0.08)^8} (0.20)^2(1 - 0.20)^{10 - 2} = \frac{3{,}628{,}800}{2!(10 - 2)!}$$

$$= \frac{3{,}628{,}000}{2!(8)!} (0.04)(0.167) = \frac{3{,}628{,}800}{2(40{,}320)} (0.04)(0.167)$$

$$= 45(0.04)\ (0.167) = 0.3006 = 0.3006 \cdot 100 = 30.06\%$$

There is a 30 percent probability that, from a sample of 10 marbles, 2 will be red and 8 will be white. (Note: This problem has been worked step by step to show the progression of math. In the remainder of this section, the progression of math will not be shown in such detail.)

What is the probability that the sample will contain 3 red marbles? For this problem, the variables are

$$(n) = 10$$
$$(p') = 0.20$$
$$(x) = 3$$

$$\frac{10!}{3!\,(10-3)!}(0.20)^3(1-0.20)^{10-3} = \frac{3{,}638{,}800}{6(5040)}(0.008)(0.2097)$$
$$= 120.33\,(0.008)(0.2097)$$
$$= 0.2018 \cdot 100$$
$$= 20.18\%$$

The probability that three red marbles will be in the sample is 20.18 percent. This is less than the probability of finding two marbles for the reason that there is less chance of a sampling error. This shows that there is no reason to assume that the jar contains more red marbles if the sample contains three red marbles. This result can be expected about 20 percent of the time. If 100 samples were taken from the jar, and the marbles replaced after each sample, the results would be that about 30 of the samples would contain 2 red, and about 20 would contain 3 red. The other 50 samples would contain more than 3 or less than 2. To find the exact amount that each sample would be expected to contain, the probability for each number, 0 to 10, should be figured.

For example, what is the probability that, from a sample of 15 marbles selected from a process (which makes 10 percent defective product), 2 marbles will be defective?

$$(n) = 15$$
$$(p') = 0.10$$
$$(x) = 2$$

$$\frac{15!}{2!(15-2)!}(0.10)^2(1-0.10)^{15-2} = 105(0.01)(0.254)$$
$$= 0.2667 \times 100$$
$$= 26.67\%$$

There is a 26.67% probability that if a sample of 15 marbles is taken from a process running at 10 percent defective, 2 marbles in that sample will also be defective. If the amount of defective product in the sample has a low probability of being selected, then there is a high probability that the process has changed. Let us take another example from the process making ball bearings. The number of defectives in the sample is 7. What is the probability that, if the process is running at 10 percent defective, in a sample of 15, 7 will be defective?

$$(n) = 15$$
$$(p') = .10$$
$$(x) = 7$$

$$\frac{15!}{7!(15-7)!}(0.10)^7(1-0.10)^{15-7} = 6435(0.0000001)(0.430)$$
$$= 0.000276 \times 100 = 0.0276\%$$

This means that if the process making ball bearings is still producing 10 percent defective product, the sample that was taken (which had 7 defects) had an unusually large sampling error. The error would be so great that, instead of assuming that the sample was in error, a quality professional would conclude that the number of defective product had gone up in the process. This allows the professional to make a decision about the process without measuring all of the product. The decision on this process would be to stop production and investigate to find the problem. Additional statistics would be performed on this process to help locate the cause.

As in any of the distributions, the way the problem is presented determines the amount of work required. Binomial distributions find the probability of one (x) value only. A problem may be presented in which more than one value must be found. In this case, the final answer would be the sum of the different (x) values. For example, what is the probability that if a sample of 15 ball bearings is taken from a process that is producing 10 percent defective product, either 3 or 2 defective ball bearings will be found?

To find the answer for this problem, find the percentage for 3 and for 2 and then add the two percentages together.

$$x = 2\ \frac{15!}{2!(15-2)!}(0.10)^2\,(1-0.10)^{15-2}$$
$$= 105\,(0.01)\,(0.254)$$
$$= 0.2667 \cdot 100 = 26.67\%$$
$$x = 3\ \frac{15!}{3!(15-3)!}(0.10)^3\,(1-0.10)^{15-3}$$
$$= 455\,(0.001)\,(0.282) = 0.1283 \cdot 100 = 12.8\%$$

The probability that the sample will have 3 or 2 defects is (26.75 + 12.83) = 39.50 percent.

Conclusions The binomial distribution has many applications. In quality control, it provides information on which decisions about a process can be based. The binomial distribution is not limited to this use. Binomial distribution can be used in any type of random sample for attributes data. It will show whether a sampling error exists, or if there is a statistically significant deviation from the expected conclusion.

The binomial probability distribution is the basis for many of the procedures used in statistical process control. Charting methods such as P′ and NP′ charts are designed to plot the performance of a process. The charts show that the process is having problems when the results fall outside the expected probability value. The advantage of using P′ and NP′ charts is that they show the results in graphic form and require only limited calculations.

18–3 POISSON DISTRIBUTION

The Poisson distribution is used for attributes data in which the presence of a characteristic can be counted, but the absence of the characteristic cannot. Consider an automobile as a product. If a car were taken apart, it would contain thousands of parts. Any one of the components inside the car may contain hundreds of parts. For example, the radio is full of resistors, transistors, and other parts. If each part in a car were listed, the list would be quite long. For each component in a car, there is a large number of defects that could exist. If all the possible ways that a defect could exist were tallied, the list would be almost endless. To this list must be added all ways that the parts could be put together wrong and all ways that the parts might be incompatible. As you can see, it would be almost impossible to list all the opportunities for a defect to exist. Defects that do exist, however, are not that hard to locate. The Poisson distribution is ideal for this type of product.

This distribution differs from the binomial distribution only in the type of product that is being inspected. The reason that binomial distribution cannot be used is that the percentage defective cannot be found on this type of product. To find the percentage defective of a product, opportunity for a defect to exist must be known, along with the number of defects that do exist. For example, if a jar contains 100 marbles, there are 100 chances for a red marble to exist. If 10 marbles were red, the jar would contain 10 percent red. To know the percentage of the sample, the opportunity for red marbles to exist must be known, as well as the total number of red marbles that exist. In this example, the opportunity is 100 and the amount that does exist is 10. The jar contains 10 percent red.

In making a car, there is an almost infinite number of ways that a defect could exist. Opportunity for a defect is almost endless. If 50 defects exist in one car, what percentage defective is it? To find the percentage defective, the number of defects must be known, along with the number of opportunities for a defect. In the case of the car, the percentage defective would not be practical to find. This is the type of condition in which a Poisson distribution is used.

Poisson distributions are used in many industries, such as auto manufacturing, electronics, computers, textiles, and any industry that makes complex products. The type of information that a Poisson distribution provides is the same as for binomial, but for a different product. For example, if carpet is made with an average of three defects per square yard, what is the probability that a square yard selected at random will contain exactly two defects? This problem is worked out and the probability is used to decide whether the process is still producing at the expected level of quality.

The formula for the Poisson distribution is[3]

$$\frac{e^{-\mu}\,\mu^{x}}{x!}$$

In this formula, two variables must be defined; (e) is a constant.

(e) = absolute value of 2.718281 . . .
(for the purposes of this text, the value will be 2.7183)
(μ) = average of the data
(x) = number of defects for which the probability is being figured

In this problem, the variables are

$$(e) = 2.7183$$
$$(\mu) = 3$$
$$(x) = 2$$

$$\frac{2.7183^{-3}\,(3)^2}{2!} = \frac{0.04978(9)}{2} = \frac{0.44802}{2}$$
$$= 0.22401 \cdot 100 = 22.401\%$$

The probability that one square yard of carpet selected at random will contain two defects is 22.4 percent. The other way in which this distribution can be used is to determine that, if 100 samples were taken from a process that makes an average of three defects per yard, approximately 22 of the samples would contain two defects. This would be a sampling error. If no sampling error existed, all samples would contain three defects. In real life, the error rate would be an average of three. The term *average* means that some of the samples would have no defects, whereas others would have six; some samples would have one and others five, and so on.

With the information from the sample, a decision is made about the process defect level. If the number of defects in the sample has a high probability of occurrence (for the expected defect level), the process is allowed to run with no change. If the defect level has only a small probability of occurrence, the quality professional assumes that the level of defects has changed, and the process is investigated to find the cause. For example, if a process of producing carpet were making three defects per yard, on the average, and a sample was taken that had four defects, the probability that one sample would have four defects is high. The assumption would be made that it was a sampling error, and not that the process had changed. If a sample was taken that had 40 defects, however, the probability that this is a sampling error would be very small. The quality professional would assume that the process has changed and investigate the process. This would prove to be a great cost savings to a manufacturer by controlling a process without checking all product. A better decision can be made with sampling than can be made with 100 percent inspection.

An auto manufacturer produces cars with an average of six defects per car. What is the probability that a new car purchased from that manufacturer will have zero defects? For this problem the variables include the following:

$$(e) = 2.7183$$
$$(\mu) = 6$$
$$(x) = 0$$

(Note: Zero factorial, by definition, equals one.)

$$\frac{2.7183^{-6}(6)^0}{0!} = \frac{0.00247(1)}{1} = \frac{0.00247}{1}$$
$$= 0.00247 \cdot 100 = 0.247\%$$

There is 0.247 percent (247/1000 of one percent) probability that a new car purchased will contain zero defects. For this reason, new cars have a warranty. If one is found to have zero defects, there is a greater probability that the process of making cars has improved than that a sampling error has occurred.

As in binomial distribution, the problem may have more than one (x) value that is desired. For example, what is the probability that a new car purchased will contain two or fewer defects? For this type of probability, the percentage for 2, 1, and 0 must be found and added together.

What is the probability that a car purchased will have more than two defects? In this type of probability, there is a 100 percent chance that the car will contain 0 or more defects. If the probability of finding 2, 1, and 0 were figured and added together, that sum could be subtracted from 100 percent to find the answer.

This type of probability can be used in many other ways as well. It is used for attributes data in which the presence of an attribute can be counted, but the absence cannot.

Conclusions The Poisson distribution has many applications in quality control. It provides information on which decisions about a process can be based. This distribution is not limited only to this use. It can be used in any type of random sample for attributes data. It shows if a sampling error exists, or if there is a statistically significant deviation from the expected conclusion.

The Poisson probability distribution is the basis for many of the procedures used in statistical process control. Charting methods, such as C and U charts, are designed to plot the performance of a process and to show the process is having problems when the results fall outside the expected probability value. The advantage of using C and U charts is that they show the results in graphic form and require only limited calculations.

SUMMARY

One of the more common uses of statistical probabilities is in the development of control charts. Although the construction of charts is beyond the scope of this text, a brief explanation is in order. The normal, Poisson, and binomial distributions that have been discussed all are used to discover and quantify the extent and type of variation that may be present in a process. The different distributions are used for different types of measured product. No matter whether the measurement is average number of defects per lot of product, actual number of defects per product, or an actual size of every part, the results of the charts are the same.

Control charts are made and monitored to make a decision about the process. The decision is whether to assume that the process is running without change or whether the process has changed. The purpose of using three distributions is to make good decisions about allowing a process to continue to produce product, changing the process average, or stopping the process. It is important that the statistical analysis being conducted serve a productive purpose. Otherwise, the analysis is nothing more than mathematical gymnastics.

Probability distributions are used to analyze data in statistical quality control. Three of the more common types of probability distributions are the normal, binomial, and Poisson.

The normal distribution is used to find the probability that a selection of a given quantity will be made from a group of numbers that is normally distributed. The normal distribution is used with variables data (measurements). This type of probability distribution uses a formula and a chart to find the percentage. The normal probability distribution is the basis of charting methods such as X-bar and R charts designed to plot the performance of a process.

The binomial distribution is used to find the probability that a sample selected from a population with a known percentage defective product will contain a given number of defects. Binomial distributions are used with attributes data. The binomial distribution is based on the number of combinations that could be selected from a population. It is the basis of charting methods such as P′ and NP′ charts used to plot the performance of a process.

The Poisson distribution is used to identify the probability of finding a product with a given number of defects if the average number of defects per product is known. The Poisson distribution is used for attributes data in which the presence of a characteristic can be counted, but the absence of the characteristic cannot. The Poisson probability distribution is the basis of C and U charts designed to plot the performance of a process. It shows that the process is having problems when the results fall outside the expected probability value.

KEY TERMS

Attributes data Data gathered from observing an elementary unit characteristic, which is either present or absent.
Binomial distribution The method used to find the probability that a sample selected from a population with a known percentage defective product will contain a given number of defects.
Normal distribution The method used to find the probability that a selection of a given quantity will be made from a group of numbers that is normally distributed.
Poisson distribution The method used to identify the probability of finding a product with a given number of defects if the average number of defects per product is known.

PROBABILITY DISTRIBUTIONS TEST

1. Define in a short paragraph the meaning of the term *probability*.
2. Match the terms on the right to the correct definition.
    - **a.** Data that are either present or absent. They are found by counting.
    - **b.** Data that can assume any value between two whole numbers.
    - **c.** Symbol for the value found using the formula $\frac{x-\bar{x}}{\sigma}$
    - **d.** Symbol for a math function that multiplies a number by all numbers below it.
    - **e.** Symbol for the absolute value 2.7183.

    1. Variables
    2. z score
    3. Attributes
    4. !
    5. e
3. Pick out the binomial distribution from among normal, binomial, and Poisson distributions by placing an X to the left of the description of the binomial distribution.
    - **a.** This distribution will find the probability of selecting a given percentage defective from a lot of product with a known percentage defective.
    - **b.** This distribution will find the probability of selecting a product with a given number of defects if the average number of defects for all product is known.

   c. This distribution will find the probability of selecting a product between two given values.

4. Name the distribution used for each of the following types of data.
   a. Average number of defects on a computer system.
   b. Percentage defective of a sample of ball bearings.
   c. Chance of selecting a knob with a hole size between 0.355 and 0.357 in.

5. Discuss in a short paragraph the procedure for using the normal distribution.

6. Discuss in a short paragraph the procedure for using the binomial distribution.

7. Discuss in a short paragraph the procedure for using the Poisson distribution.

8. Demonstrate the ability to find a probability using a normal distribution by solving the following problem.
   What is the probability of selecting a knob with a hole size between 0.356 in. and the x, if the $\bar{x} = 0.354$ and the $\sigma = 0.032$?

9. Demonstrate the ability to find a probability using a binomial distribution by solving the following problem.
   What is the probability of selecting a sample (of 20) that has two defective pieces from a population that is 4 percent defective?

10. Demonstrate the ability to find a probability using a Poisson distribution by solving the following problem.
   What is the probability of selecting a car at random that has two defects, if the average number of defects per car is six?

11. Discuss in a short paragraph how the normal distribution is used.

12. Discuss in a short paragraph how the binomial distribution is used.

13. Discuss in a short paragraph how the Poisson distribution is used.

NOTES

1. Jerome Braverman, *Fundamentals of Statistical Quality Control* (New York: Marcel-Dekker, 1979).
2. Robert Johnson, *Elementary Statistics,* 4th ed. (Boston, MA: Duxbury Press, 1980).
3. Braverman.

PROBABILITY DISTRIBUTIONS

Exercise No. 1

A sample of 50 punched-hole diameters was inspected from a lot of 1000. A second sample was made the next day.

1. What is the probability that a sample taken from each day will be between 0.499 and the x?
2. What is the probability that a sample taken from each day will be between 0.499 and 0.503?
3. What is the probability that a sample taken from each day will be below 0.499 or above 0.503?

Day 1

0.499	0.502	0.504	0.503	0.501
0.500	0.501	0.502	0.501	0.498
0.498	0.500	0.500	0.504	0.502
0.501	0.504	0.497	0.499	0.501
0.502	0.499	0.499	0.500	0.499
0.501	0.503	0.503	0.498	0.503
0.503	0.502	0.500	0.503	0.502
0.500	0.499	0.505	0.501	0.500
0.502	0.500	0.501	0.502	0.501
0.499	0.501	0.502	0.500	0.503

Day 2

0.497	0.499	0.502	0.503	0.501
0.502	0.502	0.498	0.497	0.502
0.498	0.496	0.499	0.499	0.498
0.501	0.503	0.497	0.496	0.502
0.495	0.499	0.502	0.501	0.497
0.499	0.498	0.496	0.498	0.499
0.497	0.501	0.499	0.502	0.497
0.499	0.497	0.501	0.498	0.502
0.502	0.498	0.498	0.502	0.499
0.498	0.503	0.502	0.499	0.501

Exercise No. 2

Compare the two samples of inspection data on length of rebar taken from the same truckload. Why would you question these data?

1. What is the probability that a sample taken from each day will be between 29′3″ and the x?
2. What is the probability that a sample taken from each day will be between 29′9″ and 30′3″
3. What is the probability that a sample taken from each day will be below 29′0″ or above 30′2″?

Sample No. 1

29′10″	29′11″	30′2″	30′4″	29′9″
30′0″	30′4″	29′7″	30′0″	30′1″
29′11″	29′10″	30′0″	30′2″	29′11″
30′3″	30′0″	30′3″	29′8″	30′6″
30′1″	29′9″	30′1″	29′11″	30′2″
29′9″	30′2″	29′10″	30′1″	30′0″

Sample No. 2

29′11″	29′9″	30′1″	30′3″	30′2″
30′2″	30′6″	30′3″	30′2″	30′4″
30′4″	29′11″	29′10″	30′0″	30′6″
29′8″	30′5″	30′4″	30′5″	30′0″
30′3″	30′1″	30′5″	29′10″	30′3″
30′1″	30′3″	30′0″	30′4″	29′11″

Exercise No. 3

1. Find the probability of finding the following number of bad parts in the sample.
   Process average defects: 3%
   Sample size: 57

Number bad	Probability
0	________
1	________
2	________
3	________
4	________
5	________

2. Find the probability of finding the following number of bad parts in the sample.
   Process average defects: 12%
   Sample size: 20

Number bad	Probability
0	________
1	________
2	________
3	________
4	________
5	________

3. Find the probability of finding the following number of bad parts in the sample.
   Process average defects: 8%
   Sample size: 31

Number bad	Probability
0	________
1	________
2	________
3	________
4	________
5	________

Exercise No. 4

1. A certain manufacturer of new cars produces cars with an average of 17 defects. What is the chance that a single car bought in the showroom will have 0 defects?
2. What is the probability that five cars selected at random would have the following number of defects?

Number of defects	Probability
3	________
12	________
2	________
7	________
22	________

3. If these five cars were selected at random from a production line, could it be assumed that the process used to build these cars is still averaging 17 defects per car? Why?

BIBLIOGRAPHY

American Heritage Dictionary.

American Society for Quality Control. *Standard A3.* 1978.

American Society for Quality Control. *SNT-TC-1A.* 1978.

Aspell, P. *Buried in the Ice.* Tinsel Media Production, Ltd., in association with Telefilms, Canada and the CBC for "Nova."

Aubrey, C. *Quality Management in Financial Service.* Wheaton, IL: Hitchcock Publishing Co., 1985.

Besterfield, D. *Quality Control,* 2d ed. Englewood Cliffs, NJ: Prentice-Hall, 1986.

Bittel, L. *Essentials of Supervisory Management.* 4th ed. New York: McGraw-Hill, 1981.

Blank, L. *Statistical Procedures for Engineering, Management, and Science.* New York: McGraw-Hill, 1980.

Bralla, J. *Product Design for Manufacturing.* New York: McGraw-Hill, 1986.

Braverman, J. *Fundamentals of Statistical Quality Control.* Reston, VA: Reston Publishing Co., 1981.

Burr, I. *Elementary Statistical Control.* New York: Marcel-Dekker, 1979.

Busch, T. *Fundamentals of Dimensional Metrology,* 3d ed. New York: Delmar Publishers, Inc., 1966.

Carrubba, E., and Gordon, R. *Product Assurance Principles.* New York: McGraw-Hill, 1976.

Carter, C., Jr. *The Control and Assurance of Quality, Reliability and Safety.* Richardson, TX: C. L. Carter, Jr. & Associates, Inc., 1978.

Charbonneau, H., and Webster, G. *Industrial Quality Control.* Englewood Cliffs, NJ: Prentice-Hall, 1978.

Crosby, P. *Quality Is Free.* New York: McGraw-Hill, 1979.

______. *Quality Without Tears.* New York: McGraw-Hill, 1984.

DataMyte Corporation. *DataMyte Handbook,* 2d ed. Minneapolis, MN: DataMyte Corp., 1986.

Deming, W. E. *Out of the Crisis.* Cambridge, MA: MIT Press, 1985.

______. *Quality, Productivity and Competitive Position.* Cambridge, MA: MIT Press, 1982.

______. Speech delivered to the 40th Quality Congress, 19 May 1986, Anaheim, CA.

Dillon, A. *Shigeo Shingo Zero Quality Control: Source Inspection and the Poka-yoke System.* Cambridge, MA: Productivity Press, 1986.

Feigenbaum, A. V. *Total Quality Control,* 3d ed. New York: McGraw-Hill, 1983.

Ford Motor Company, nationwide television advertisement.

Fox, I., and Twomey, D. *Business Law,* 4th ed. Cincinnati, OH: South Western Publishing Co., 1982.

General Atomic Project 2117. *Quality Assurance System Audit.* General Atomic Company, 1982.

General Dynamics Convair Division. *Nondestructive Testing Eddy Current Testing* CT-6-5, 2d ed. 1979.

______. *Nondestructive Testing Magnetic Particle* CT-6-3, 2d ed. 1977.

______. *Nondestructive Testing Liquid Penetrant* CT-6-2, 4th ed. 1977.

______. *Nondestructive Testing Radiographic Testing* CT-6-6, 2d ed. 1983.

______. *Nondestructive Testing Ultrasonic Testing* CT-6-4. 1967.

Gilmore, H., and Schwartz, H. *Integrated Product Testing & Evaluation.* Milwaukee, WI: ASQC Quality Press, 1986.

Grant, E., and Leavenworth, R. *Statistical Quality Control,* 5th ed. New York: McGraw-Hill, 1980.

Groocook, J. *Chain of Quality.* New York: John Wiley & Sons, 1986.

Halpin, J. *Zero Defects—A New Dimension in QA.* New York: McGraw-Hill, 1979.

Hansen, B., and Ghare, P. *Quality Control and Application.* Englewood Cliffs, NJ: Prentice-Hall, 1987.

Hayes, G. *Quality Assurance: Management and Technology,* 7th ed. Capistrano Beach, CA: Gallant/Charger Publications, Inc., 1985.

Hayes, G., and Romig, H. *Modern Quality Control.* Encino, CA: Glencoe Publishing Co., 1982.

Hoy, W., and Miskel, C. *Educational Administration,* 2d ed. New York: Random House, 1982.

Hughes, T. *Measurement and Control Basics.* Triangle Park, NC: Instrument Society of America, 1986.

Imai, M. *Kaizen, the Key to Japan's Competitive Success.* New York: Random House Business Division, 1986.

Ishikawa, K. *Guide to Quality Control.* White Plains, NY: Kraus International Publications, 1987.

______. *What Is Total Quality Control?* Englewood Cliffs, NJ: Prentice-Hall, 1985.

Johnson, H. *Manufacturing Processes,* 2d ed. Peoria, IL: Bennett Publishing Co., 1979.

Johnson, R. *Elementary Statistics,* 4th ed. Boston, MA: Duxbury Press, 1980.

Juran, J. *Juran on Planning for Quality.* New York: The Free Press, 1988.

______ . *Managerial Breakthrough.* New York: McGraw-Hill, 1964.

______ . *Quality Control Handbook,* 3d ed. New York: McGraw-Hill, 1974.

______ . *The Taylor System and Quality Control.* A Quality Progress Reprint. Milwaukee, WI: American Society for Quality Control, 1987.

Juran, J., and Gryna, F., Jr. *Quality Planning and Analysis,* 2d ed. New York: McGraw-Hill, 1980.

Kaplan, F. *The Quality System.* Radnor, PA: Clinton Book Co., 1980.

Kazanas, H.; Baker, G.; and Gregor, T. *Basic Manufacturing Processes.* New York: McGraw-Hill, 1981.

Lester, R.; Enrick, N.; and Mottley, H., Jr. *Quality Control for Profit,* 2d ed. New York: Marcel-Dekker, 1977.

Lu, D. *Kanban, Just-in-Time at Toyota.* Cambridge, MA: Productivity Press, 1986.

McClave, J., and Benson, G. *Statistics for Business & Economics.* San Francisco, CA: Dellen Publishing Co., 1979.

McGuiar, S., and Peabody, C. *Working Safely in Gamma Radiography.* Washington, DC: U.S. Nuclear Regulatory Commission, 1982.

Marlin, T.; Perkins, J.; Barton, G.; and Brisk, M. *Advanced Process Control Applications.* Triangle Park, NC: Instrument Society of America, 1988.

Mayer, R. *Production and Operations Management,* 4th ed. New York: McGraw-Hill, 1982.

Miles, R. *Macro Organizational Behavior.* Santa Monica, CA: Goodyear Publishing Co., 1980.

Miner, J., and Miner, M. *Personnel and Industrial Relations,* 3d ed. New York: Macmillan Publishing Co., 1977.

Murray, S. *Probability and Statistics.* New York: McGraw-Hill, 1975.

Naiman, A.; Rosenfeld, R.; and Zirkel, G. *Understanding Statistics,* 3d ed. New York: McGraw-Hill, 1983.

Niebel, B., and Draper, A. *Product Design and Process Engineering.* New York: McGraw-Hill, 1974.

Ott, E. *Process Quality Control.* New York: McGraw-Hill, 1975.

Pall, G. *Quality Process Management.* Englewood Cliffs, NJ: Prentice-Hall, 1987.

Person, R. *Essentials of Mathematics,* 4th ed. New York: John Wiley and Sons, 1979.

Peters, T. *Excellence in the Public Sector.* Washington, DC: WETA, 1989.

______. *Leadership Alliance.* Video Publishing House, Inc., 1988.

______. *Thriving on Chaos.* New York: Harper & Row, Publishers, 1987.

Platt, G. *Process Control.* Triangle Park, NC: Instrument Society of America, 1988.

Richards, L., and LaCava, J. *Business Statistics,* 2d ed. New York: McGraw-Hill, 1983.

Rieker, W. S. *Quality Control Circles,* 2d ed. California: Quality Control Circles, Inc., 1977.

Rosander, A. C. *Applications of Quality Control in the Service Industries.* New York: Marcel-Dekker, 1985.

Schollhammer, H., and Kuriloff, A. *Entrepreneurship and Small Business Management.* New York: John Wiley & Sons, 1979.

Shetty, Y., and Buehler, V. *Quality & Productivity Improvements.* Chicago, IL: Manufacturing Productivity Center, 1983.

Shewhart, W. A. *Economic Control of Quality of Manufactured Product.* New York: D. Van Nostrand Co., Inc., 1931.

Shores, R. *Survival of the Fittest.* Milwaukee, WI: ASQC Quality Press, 1988.

Sigl, C., and Quinn, R. *Radiography in Modern Industry,* 4th ed. Rochester, NY: Eastman Kodak Company, 1980.

Small, B. *Statistical Quality Control Handbook,* 2d ed. Indianapolis, IN: AT&T Technologies, 1985.

Spiegel, M. *Probability and Statistics.* New York: McGraw-Hill, 1975.

Stout, K. *Quality Control in Automation.* Englewood Cliffs, NJ: Prentice-Hall, 1985.

Suzaki, K. *Manufacturing Challenge.* New York: The Free Press, 1987.

Talley, D. *Management Audits for Excellence*, Milwaukee, WI: ASQC Press, 1988.

Unterweiser, P. *Failure Analysis.* Metals Park, OH: American Society for Metals, 1981.

U.S. Department of Defense. MIL-STD 105-D. *Sampling Procedures and Tables for Inspection by Attributes.* 29 April 1963.

______. MIL-STD 105-D. 1974.

______. MIL-STD 109. 1974.

______. MIL-STD 410-D. 1974.

Vendor-Vendee Technical Committee, American Society for Quality Control. *Procurement Quality Control,* 3d ed. Milwaukee, WI: American Society for Quality Control, 1985.

VonFlatern, J. *Dealing with Product Liability: One Manufacturer's New Approach.* A paper delivered at the A.B.A. National Institute Litigation in Aviation, 27–28 October 1988. Photocopied.

Wadsworth, H.; Stephens, K.; and Godfrey, A. *Modern Methods for Quality Control and Improvement.* New York: John Wiley and Sons, 1986.

Webber, R. *Management: Basic Elements of Managing Organizations,* 2d ed. Homewood, IL: Richard Irwin, Inc., 1979.

Weston, A. *Losing the Future.* NBC Special Report aired November 1988.

Winn, P., and Johnson, R. *Business Statistics.* New York: Macmillan Publishing Co., 1978.

Zenith Corp., nationwide television advertisement.

GLOSSARY

Absolute 1. Not relative, independent; e.g., absolute zero temperature, as distinct from zero on an arbitrary scale, as the celsius scale. 2. One way (as opposed to round trip), used with reference to distance measurement.

Absorption Changing part of a sound beam's mechanical energy into heat energy, evidenced by a slight increase in the temperature of specimen molecular particles. This is one cause of material loss attenuation.

Acceptable Quality Level (AQL) A value that is desirable for acceptance of a product.

Acceptance Sampling Making a decision about lot quality based on data from a sample taken from that lot.

Accuracy The amount a measurement differs from the true value.

Acoustic Impedance A measure of the work sound does to pass through a medium, equal in magnitude to the product of the sound velocity (C) multiplied by the medium's density (ρ).

Acoustic Interface The boundary between two media of different acoustic impedances.

Acoustic Zero For practical purposes, can be considered to be the point on the CRT display that represents the specimen entry surface.

Activate The act of bombarding a nucleus with neutrons until it becomes unstable.

Activity The number of rays given off per second by an isotope. Measured in curies.

Advisory Document A publication providing supplementary or informational material, which may be helpful but is not mandatory.

Agreement State A state that has signed an agreement with the NRC allowing the state to regulate certain activities using radioactive materials.

Alpha Particle A positively charged subatomic particle emitted from the nucleus of radioactive materials. Consists of two neutrons and two protons.

Amplifier An electronic device that increases signal strength fed into it by obtaining power from a source other than the input signal.

Amplitude Usually used in the combined term *echo amplitude,* signifying echo-pulse peak height, seen in linear form on the CRT screen above a reference line (usually the baseline).

Anomalies Deviation from the requirements or standard that may or may not cause rejection.

ANOVA A method of determining the equality of means of two distributions called Analysis of Variance.

Appraisal Costs associated with measuring, evaluating, or auditing products, components, and purchased materials to ensure conformance with quality standards and performance requirements.

Assembly Line Manufacturing A manufacturing method (usually linear) in which the assembly pieces are systematically arranged and the operations are highly repetitive.

Assignable Cause A variability process cause, not part of the random-cause system but inherent to that process.

Attenuation The loss of sound pressure in a traveling wave front caused by the reflection and/or absorption of some of the wave's sound pressure by the grain structure and/or porosity of the medium.

Attribute Quality type or description of a subject.

Attributes Data Data gathered from observing an elementary unit characteristic, which is either present or absent.

Attributes Sampling Plan An acceptance plan in which every item inspected is classified as good or defective, based on conformance or nonconformance to a standard. Accept/reject criteria depend on the number of sample defectives.

Audit An inspection of an organization's adherence to the established quality standard.

Back Echo or **Back-Wall Echo** The echo representing the specimen side opposite the side to which the transducer is coupled. This echo represents specimen thickness at this point.

Bank Account The formula that reveals the maximum amount of whole-body radiation that a person of a certain age may have received over his or her lifetime.

Bell Curve The plotted shape of a normal distribution, resembling that of a bell.

Beta Particle A negatively charged subatomic particle emitted from the nucleus of radioactive materials. Has the mass of an electron but comes from the nucleus.

Binomial Distribution The method used to find the probability that a sample selected from a population with a known percentage defective product will contain a given number of defects.

Blotting Agent A compound that draws the penetrant out of the defect.

Boundary The number that separates one class from the next.

Bremsstrahlung German for "breaking rays." The sudden reduction in speed of an electron as it nears the anode in an X-ray tube.

Calibration Comparison of a measurement standard or instrument with another of known higher-level accuracy to detect any variation in accuracy.

Calibration (Ultrasonics) The graduations (markings that indicate scale) of an instrument that allow measurements in definite units. The arbitrary 0–10 CRT scale may be calibrated in units of distance, converting the scale into time base information, and thus into distance information.

Capillary Action The action by which the surface of a liquid, where in contact with a solid, is elevated or depressed.

Cassette Film holder used during exposure.

Cathode Ray Tube (CRT) A vacuum tube that allows the direct observation of cathode ray behavior. It consists essentially of an electron gun producing an electron beam that, after passing between horizontal and vertical deflection plates, falls upon a luminescent screen; beam position can be observed by luminescence produced upon the screen. Electric potentials applied to the deflection plates are used to control beam position, and its movement across the screen, in any desired manner.

Cell Class.

Central Tendency The characteristic of data tending to group in a pattern around the center value.

Certification A written statement by an authorized party, which attests that specified equipment, process, or person complies with a specification, contract, or company policy.

Chronic Problem Problem that is long-standing and hard to solve.

Class A group of data within given parameters.

Class Midmark The number that occurs halfway between the upper limit and lower limit of the same class.

Class Width The distance between the upper limit of a section of grouped data and the lower limit.

Classical Approach Method used to find the probability when there is a finite number of possible outcomes, all outcomes are mutually exclusive, and all outcomes are equally likely to occur.

Clock Interval The clock period time elapsing between each clock pulse.

Coaxial Cable A cable consisting of a central conducting wire together with a concentric, cylindrical conductor, the space between the two being filled with a dielectric substance, e.g., polyethylene, air, etc. The outer conductor is normally connected to ground. A coaxial cable's main use is to transmit high-frequency power from one place to another with a minimum loss of energy.

Code Specification The document that prescribes approved procedures to be followed in a test.
Combination The grouping of data in which the order of the objects is not a consideration.
Conductive The ability of a material to carry an electrical current.
Conservative A worker whose desire for quality is superseded by his or her desire for security.
Continuous Manufacturing A manufacturing method that produces a long run of a single product. Examples of this type of operation are the production of plastic stretch film and the production of sewing yarn.
Contrast Relative difference in darkness of a radiograph.
Corrective Action Action taken to prevent recurrence of a discrepancy.
Couplant A material (usually a liquid) used between the transducer and the test specimen to eliminate air from this space and thus ensure sound-wave passage into and out of the specimen.
Critical Defect A defect that could result in hazardous or unsafe conditions for individuals using or maintaining the product.
Curie The activity of 37,000,000 disintegrations per second.
Decay The spontaneous act of an isotope undergoing change back to the stable state.
Decibel One tenth of a bel. A unit that compares levels of power. Two power levels, P1 and P2, are said to differ by n decibels when: $n = 10 \log 10\ P2/p1$. This unit is often used to express sound intensities. In this case, P2 is sound intensity under consideration and P1 is a reference level intensity. In the case of displayed voltages on a cathode ray tube screen, the relationship becomes: $n = 20 \log 10\ v2/v1$.
Defect (Discrepancy) Any deviation of an item from specified physical, metallurgical, or chemical requirements.
Defect Level The number of decibels of calibrated gain that must be added to the defect echo to bring its peak to the reference line on the CRT.
Defect Orientation The position of the suspected flaw relative to the angle of the ray.
Defective A product that does not meet specifications.
Delayed Effects Harmful effects from radiation exposure that occur many years after the fact.
Demagnetization The act of misaligning the domains in a ferromagnetic material.
Density The darkness of a radiograph.

Dependent Event An event in which the probability of occurrence depends on what events have previously occurred.
Detectability The ability to detect a given-size defect.
Differential Coil A coil arrangement using wire wrapped first in one direction and then in the opposite direction.
Discrimination The smallest increment a measuring instrument can accurately measure.
Dispersion The way in which data are spread out or distributed.
Distribution The manner in which individual pieces of data compare to each other when viewed as a group.
Dosimeter A personnel-monitoring device that reads total radiation dose.
Dual-Element Probe A probe containing two piezoelectric crystals, one of which transmits only, and one that receives only.
Dwell Time The amount of time the penetrant is allowed to remain on the surface of the part and be drawn into the flaw.
Elasticity A material's ability to return to original form and dimension when forces acting upon it are removed. If the forces are sufficiently large for deformation to cause a break in the material's molecular structure, elasticity is lost and the "elastic limit" is said to have been reached.
Electrical Zero 1. The point in time when a transmitter fires the initial pulse to the transducer and receiver. 2. The point on a cathode ray tube screen where an electron beam leaves the baseline, caused by an initial pulse signal coming from the transmitter.
Electromagnetic Magnetism produced by passing electrical current through a coil wrapped around a core material.
Electromagnetic Radiation Radiation consisting of photons of energy such as light.
Electron An elementary particle having a rest mass of 9.1091×10^{-31} kg, approximately 1/1836 that of a hydrogen atom, and bearing a negative electric charge of 1.06021×10^{-19} coulombs. The electron radius is 2.81777×10^{-15} meters. All atoms have electrons.
Electron Gun A cathode ray tube electron source, consisting of a cathode emitter, an anode with an aperture through which an electron beam can pass, and one or more focusing and control electrodes.
Element A section, component, or category describing the scope, intent, and requirements of an activity.
Encircling Coil A coil with its axis wound parallel to the surface of the specimen. Used in checking the outer surface of thin-wall pipe and tubing. This coil is sometimes called a through coil.

Equally Likely Requirement The condition in which each possible outcome has the same chance of occurring.

External Failures Costs generated by defective products being shipped to customers.

Faraday's Law Any time current passes through a conductor, a magnetic field is established in and around the conductor.

Fast Film Larger-grain films requiring less exposure to be developed.

Ferromagnetic Of or relating to substances with an abnormally high magnetic permeability, a definite saturation point, and appreciable residual magnetism.

Fill Factor An expression of the amount of space between a wall of tubing and an internal coil or an encircling coil; affects the efficiency of the test.

Film Contrast Differences in the darkness of a radiograph caused by film type.

Final Inspection Inspection of the final assembly.

First Article The first production unit scheduled for acceptance by inspector or customer.

Fitness for Use Selling price and the customer's end use.

Flux Lines of force surrounding a magnetic field.

Flux Density The number of flux lines per unit area.

Flux Leakage Characterized by the exit and entrance of flux lines in a ferromagnetic material.

Frequency Rate of occurrence.

Frequency (Ultrasonic) The number of cycles per second undergone or produced by an oscillating body.

Frequency Polygon A specialized graph in which the data are plotted with points marked at the frequency and connected.

Gain A term used in electronics with reference to an increase in signal power, usually expressed as the ratio of output power (for example, of an amplifier) to input power, in decibels.

Gamma Rays Electromagnetic radiation emitted from the nucleus of an atom as it decays.

Gauss The unit of measure of flux density.

Grouped Frequency Distribution The arrangement of data points compiled into groups and counted by the number of times the data are repeated.

Half-Life The amount of time required for the curie strength to be reduced by one-half.

Half-Value Layer The amount of material required to reduce the radiation level by one-half.

Hertz The derived frequency unit of a periodic phenomenon with a one-second period, equal to one cycle per second. Symbol: Hz. 1 Kilohertz (kHz) = 10^3 cycles per second; 1 megahertz (MHz) = 10^6 cycles per second. Named after Heinrich Hertz (1857 - 1894).

High Radiation Area An area where a person would receive in excess of 100 mrem per hour.

Histogram A specialized bar graph made up of a title, vertical scale, and horizontal scale.

Iceberg Effect A concept used in describing the total cost of poor quality; it implies that most (up to 75 percent) of the total cost of poor quality is hidden and not readily identifiable. Even though these costs are not seen they are present.

Impedance Total opposition to current flow in an ac circuit.

Implementation The execution of activities devised to meet requirements.

Incidence, Angle of The angle between a sound beam striking an acoustic interface, normal (perpendicular) to the surface at that point. Usually designated by the Greek symbol α (alpha).

Independent Event An event in which the probability of occurrence does not depend on previously occurring events.

Infant Mortality Failures occurring during the early portion of equipment life.

Inhibitor A worker who fears the quality effort will encroach on his or her rights.

Innovator A worker whose desire for quality outweighs everything else.

In-Process Inspection Inspection performed during manufacturing in an effort to prevent defects from occurring and to inspect characteristics.

Inspection A highly defined, close examination process.

Inspection Procedure A detailed progression outlining steps and resources used in the inspection.

Intensity The frequency (Hz) of electromagnetic radiation.

Interference (Specifically of Wave Motions) Vector addition or combination of waves (also, superposition).

Internal Coil A coil with its axis 90° or parallel to the inner surface of tubing or thin-wall pipe. Sometimes called a bobbin coil.

Internal Failures Costs associated with defective products, components, and materials that fail to meet quality requirements and cause manufacturing losses.

Inverse Square Law The mathematical expression of radiation dose versus distance from the source.

Ionizing Radiation Radiation that is capable of ripping electrons away from atoms.

IP Abbreviation of the term *initial pulse,* which is the electrical pulse sent out by the transmitter to the receiver and the transducer. The term *IP* generally refers to this trailing edge caused by the "ringing" (or continuing vibration) of the transducer crystal.

Isotope A material that has an imbalance in the nucleus of the atom.

Job Shop Manufacturing Type of manufacturing of (usually) a small number of parts being produced one at a time.

Longitudinal Wave Wave propagation characterized by particle movement parallel to the direction of wave propagation.

Lot Product manufactured to specific requirements using the same manufacturing methods.

Lower Limit The lowest possible number that could occur within a given class.

Magnetic Field The space around a magnet in which the lines of force act.

Major Defect One that may result in failure of the product.

Management by Disaster The tendency of mid-level managers to neglect long-range goals because of continuous crisis situations.

Mean The simple average of data.

Median Middle value of a sample when data are ranked in order, according to size.

Medium A substance through which a force acts or an effect is transmitted; surrounding or enveloping substance; environment.

Military Standards A group of documents issued and controlled by the federal government and used by both military and civilian industry to describe requirements for the manufacture of products.

Minor Defect One that will not reduce the usability of the product.

Mode Value that occurs most frequently.

Mode Conversion Changing a portion of a sound beam's energy into an opposite-mode wave, caused by reflection and/or refraction at incident angles other than 0°.

Molecule The smallest portion of a substance capable of existing independently and retaining properties of the original substance.

Mutually Exclusive The condition in which only one outcome is possible for a single trial.

Noise Signals picked up from irrelevant discontinuities.

Nonagreement State A state in which the NRC regulates the use of radioactive materials.

Nonconformance Any condition in which one or more product characteristics do not conform to requirements specified.

Normal Distribution The method used to find the probability that a selection of a given quantity will be made from a group of numbers that is normally distributed.

Normal Probe A transducer that sends sound into a test specimen perpendicular to the entry surface.

Objective Approach The method of determining probability through a "best guess" approach.

Ogive A specialized graph in which data are plotted as a cumulative (adding one point to the next) frequency or relative cumulative frequency.

Oscillator A device for producing sonic or ultrasonic pressure waves in a medium.

Outside Disturbance The external force acting on a system that is not inherent in the process, causing a deviation.

Parallel Extending in exactly the same direction so that there is neither divergence nor convergence; being an equal distance apart at all points.

Pareto Principle The proposition that 20 percent of the problems account for 80 percent of the effort.

Particulate Radiation Ionizing radiation consisting of subatomic particles alpha, beta, and positron.

Penetrameters Quality indicators used on radiographic films.

Penetration The ability of the test system to detect a given-size defect at a given distance.

Permeability The ease with which a material can be magnetized.

Permutation The grouping of data in which the order of the data is taken into account.

Perpendicular At right angles; a straight line making an angle of 90° with another line or plane.

Phase Points in the path of a wave motion are said to be points of equal phase if the displacements at those points, at any instant, are exactly similar; i.e., of the same magnitude and variation.

Philosophy The basic belief in what is considered the proper management style.

Phosphor A substance capable of "luminescence" (light emission from a body from any cause other than high temperature), storing energy (particularly from ionizing radiation) and later releasing it as light.

Piezoelectric Crystals A family of crystals that possess the characteristic ability to produce a) a voltage differential across their faces when deformed by an externally applied mechanical force, and b) a change in their own physical configuration (dimensions) when an external voltage is applied to them.

Poisson Distribution The method used to identify the probability of finding a product with a given number of defects if the average number of defects per product is known.

Population Also known as "universe," the entire set of observations of value in statistical investigation.

Positron A positively charged beta particle.

Precision The extent instruments repeat results with continued measurement.

Prevention Costs associated with personnel engaged in designing, implementing, and maintaining the quality system. The latter includes auditing the system.

Preventive Action Action taken to prevent occurrence of discrepancy.

Probability The numerical likelihood of an uncertain occurrence of an event.

Probe The transducer, or "search unit."

Process Inspection Inspections performed at specific intervals during the manufacturing process.

Prompt Effects Harmful effects from radiation exposure that occur within a few days or weeks of the exposure.

Prototype An initial unit or design produced prior to release for production.

Pulse A wave disturbance of short duration.

Pulse Repetition Rate The frequency with which a clock circuit sends out its trigger pulses to the sweep generator and transmitter, usually quoted in terms of pulses per second.

Quality The degree of conformance by an item to a governing criterion.

Quality Assurance All those planned and systematic actions necessary to provide adequate confidence that a system or product will perform satisfactorily in service.

Quality Audit An audit performed to determine overall quality program effectiveness, and to identify those areas requiring corrective action.

Quality Control All those actions, relating to the physical characteristics of the material, that provide a means to control quality to predetermined standards.

Quality Manual The residence point of procedures describing activities relative to the quality system.

Quantitative Describing the quantity in numerical form.

Radiation Area An area where a person would receive in excess of 5 mrem per hour.

Radio Frequency An oscillation frequency that falls within the range used in radio, i.e., 10 kH to 100,000 MH.

Random Sampling A process in which each element of a population has an equal probability of being sampled.

Range (Stastical) Difference between highest and lowest values in a group of data.

Range (Ultrasonic) The total distance (specimen depth) being displayed at any one time across the CRT screen.

Ray A line giving wave direction of advance at any point. This direction corresponds to that of the radius of curvature.

Reference Echo The echo from a reference reflector.

Reference Level The number of decibels of calibrated gain that must be added to the reference-echo signal to bring its peak to the reference line on the CRT.

Reference Reflector A known-size reflector at a known distance, such as a flat-bottomed hole.

Refraction Sound beam bending when passing through an acoustic interface at an incident angle other than 0°. The bending is caused by the difference in wave speed on either side of the interface, so refraction is accompanied by a wavelength change.

Refraction, Angle of The angle between a refracted sound beam and the perpendicular.

Registrar The person or organization responsible for the review and analysis of a program or activity in regard to compliance with stated requirements.

Relative Frequency A type of histogram indicating the percentage that each class composes of the entire data.

Relative Frequency Approach The method of determining probability by collecting sample data from a population and evaluating the number of times that the event being studied occurs in the sample.

Reliability The probability of a product performing without failure a specified function under given conditions for a specified period of time. (More simply, reliability is the chance that a product will work.)

Reliability Life Test A test used to determine whether a product meets specified requirements for the expected mean life.

REM Roentgen equivalent man. The unit of measure of radiation exposure to a human.

Residual Magnetism The magnetic field that remains after the magnetizing force has been removed.

Resolution Test system ability to distinguish defects at slightly different depths.

Restricted Area An area in which the licensee sets barricades to restrict public access.
Retentivity The ability of a material to retain a magnetic field after the magnetizing force has been removed.
Roentgen The amount of radiation required to give 1 cm^3 of air one electrostatic unit of charge at standard temperature and pressure.
Sample One or more units of product selected at random from the material or process represented.
Sample Size (n) Number of items in a sample.
Self-Actualization The state of mind where a person reaches out from a secure base to explore new areas.
Sensitivity The ability of the test system to detect a given-size defect at a given distance.
Sharpness The distance it takes to get from one area of contrast to another on a radiograph.
Shock Wave A particularly sudden and intense wave disturbance of short duration.
Single-Element Probe A probe containing only one piezoelectric crystal, which is used both to transmit only and to receive only.
Skin Effect The condition of having an extreme concentration of eddy currents near the surface with very little depth of penetration.
Sonic Of or relating to frequencies within human audible range, between 20 and 20,000 cycles/sec.
Sound Path Distance The distance from the transducer beam index to the reflector located in the specimen, measured along the actual path sound traveled.
Source Inspection Inspection performed at the vendor's facilities by the customer.
Specimen Contrast Difference in darkness of a radiograph caused by the geometry of the specimen.
Sporadic Problems Problems that are dramatic deviations from the status quo.
Standard 1. Test piece of known size, used to adjust the reading of a measuring instrument. 2. A specimen that is flawless. Used to compare reading to test specimen.
Standard Deviation The average numeric distance a point of data falls away from the mean of all the data.
Statistic A sample numerical value determined by counting or computation.
Statistics, Applied Mathematics that helps to describe or analyze data; also, any set of numerical data.

Subsonic Of or relating to frequencies below the human audible range; below 20 cycles/sec.

Supersonic Of or relating to movement through some medium at speeds greater than the speed of sound in that medium.

Surface Coil A coil with its axis oriented 90° to the plane of the surface. Usually hand held and sometimes called a probe.

Surface Wave Wave propagation characterized by an elliptical movement of particles (molecules) on a specimen surface penetrating the specimen to a depth of one wavelength.

Surveillance A loose inspection process.

Survey Meter Personnel-monitoring equipment that reads dose rate.

Taylor Method The management approach that breaks the manufacturing process into small tasks in a set pattern to be altered only by the management personnel.

Tolerance The degree of permissible variation from a predetermined dimension.

Total Quality Control The concept put forth by Armand Feigenbaum concerning quality throughout all levels of a company.

Trace The illuminated line on a cathode ray tube screen, caused by the luminescence of the phosphor layer hit by an electron beam.

Transverse Wave Wave propagation characterized by particle movement perpendicular to the direction of wave propagation.

Ultrasonic Of or relating to frequencies above human audible range, above 20,000 cycles/sec.

Ultrasonics The study of pressure waves, which are similar to sound waves but have frequencies above the human audible limit, above 20 kH.

Ungrouped Distribution The arrangement of data points in ascending order.

Ungrouped Frequency Distribution The arrangement of data points indicating the number of times the data are repeated.

Unrestricted Area An area where a person will not receive more than 2 mrem per hour.

Upper Limit The largest possible number that could occur within a given class.

Variable Counts or measurements of a subject.

Variance The standard deviation squared.

Variation Differences in a measured characteristic caused by random, chance disturbances and by identifiable disturbances.

Vendor Inspection Incoming inspection.

Vital Few and Trivial Many Phrase coined by Joseph Juran denoting the Pareto Principle.

Volt The derived unit of electrode potential defined as the difference in potential between two points on a conducting wire carrying a constant current of one ampere when power dissipated between these points is one watt. Named after Allesandro Volta (1745 - 1827).

Voltmeter An instrument for measuring potential difference between two points.

Wave A periodic disturbance in a medium (or in a vacuum, as in the case of electromagnetic waves), which may involve the elastic displacement of material particles or a periodic change in some physical quantity such as temperature, pressure, electric potential, or electromagnetic field strength.

Wave Form The shape of a wave, illustrated graphically by plotting the values of the periodic quantity against time.

Wave Front The locus of adjacent points, in the path of a wave motion, that possess the same phase.

Wave Motion The propagation of a periodic disturbance carrying energy. At any point along the path of a wave motion, a periodic displacement or vibration about the mean (average) position takes place. This may take the form of a displacement of electromagnetic vectors. The locus of these displacements at any instant is called a "wave." The wave motion moves forward a distance equal to its wavelength in the time taken for the displacement, at any point, to undergo a complete cycle about its mean position.

Wavelength The distance between like points on successive wave fronts, i.e., the distance between any two successive particles of an oscillating medium that are in the same phase. It is denoted by the Greek letter λ (lambda).

Wetting Agent A substance that lowers the surface tension of a liquid.

X Rays Electromagnetic radiation emitted from the interaction of electrons with atoms in electrical equipment.

Zero Defect The program implemented by Philip Crosby while at Martin Marietta Corporation.

APPENDIX A

Area Under the Normal Curve; continued p. 378. Proportion of the total area of the standard normal curve from $-\infty$ to z (z represents a normalized statistic). Source of Appendix A: Eugene L. Grant and Richard S. Leavenworth, *Statistical Quality Control,* 6th ed. (New York: McGraw-Hill, 1988), 666–667.

z	0.09	0.08	0.07	0.06	0.05	0.04	0.03	0.02	0.01	0.00
-3.5	0.00017	0.00017	0.00018	0.00019	0.00019	0.00020	0.00021	0.00022	0.00022	0.00023
-3.4	0.00024	0.00025	0.00026	0.00027	0.00028	0.00029	0.00030	0.00031	0.00033	0.00034
-3.3	0.00035	0.00036	0.00038	0.00039	0.00040	0.00042	0.00043	0.00045	0.00047	0.00048
-3.2	0.00050	0.00052	0.00054	0.00056	0.00058	0.00060	0.00062	0.00064	0.00066	0.00069
-3.1	0.00071	0.00074	0.00076	0.00079	0.00082	0.00085	0.00087	0.00090	0.00094	0.00097
-3.0	0.00100	0.00104	0.00107	0.00111	0.00114	0.00118	0.00122	0.00126	0.00131	0.00135
-2.9	0.0014	0.0014	0.0015	0.0015	0.0016	0.0016	0.0017	0.0017	0.0018	0.0019
-2.8	0.0019	0.0020	0.0021	0.0021	0.0022	0.0023	0.0023	0.0024	0.0025	0.0026
-2.7	0.0026	0.0027	0.0028	0.0029	0.0030	0.0031	0.0032	0.0033	0.0034	0.0035
-2.6	0.0036	0.0037	0.0038	0.0039	0.0040	0.0041	0.0043	0.0044	0.0045	0.0047
-2.5	0.0048	0.0049	0.0051	0.0052	0.0054	0.0055	0.0057	0.0059	0.0060	0.0062
-2.4	0.0064	0.0066	0.0068	0.0069	0.0071	0.0073	0.0075	0.0078	0.0080	0.0082
-2.3	0.0084	0.0087	0.0089	0.0091	0.0094	0.0096	0.0099	0.0102	0.0104	0.0107
-2.2	0.0110	0.0113	0.0116	0.0119	0.0122	0.0125	0.0129	0.0132	0.0136	0.0139
-2.1	0.0143	0.0146	0.0150	0.0154	0.0158	0.0162	0.0166	0.0170	0.0174	0.0179
-2.0	0.0183	0.0188	0.0192	0.0197	0.0202	0.0207	0.0212	0.0217	0.0222	0.0228
-1.9	0.0233	0.0239	0.0244	0.0250	0.0256	0.0262	0.0268	0.0274	0.0281	0.0287
-1.8	0.0294	0.0301	0.0307	0.0314	0.0322	0.0329	0.0336	0.0344	0.0351	0.0359
-1.7	0.0367	0.0375	0.0384	0.0392	0.0401	0.0409	0.0418	0.0427	0.0436	0.0446
-1.6	0.0455	0.0465	0.0475	0.0485	0.0495	0.0505	0.0516	0.0526	0.0537	0.0548
-1.5	0.0559	0.0571	0.0582	0.0594	0.0606	0.0618	0.0630	0.0643	0.0655	0.0668
-1.4	0.0681	0.0694	0.0708	0.0721	0.0735	0.0749	0.0764	0.0778	0.0793	0.0808
-1.3	0.0823	0.0838	0.0853	0.0869	0.0885	0.0901	0.0918	0.0934	0.0951	0.0968
-1.2	0.0985	0.1003	0.1020	0.1038	0.1057	0.1075	0.1093	0.1112	0.1131	0.1151
-1.1	0.1170	0.1190	0.1210	0.1230	0.1251	0.1271	0.1292	0.1314	0.1335	0.1357
-1.0	0.1379	0.1401	0.1423	0.1446	0.1469	0.1492	0.1515	0.1539	0.1562	0.1587
-0.9	0.1611	0.1635	0.1660	0.1685	0.1711	0.1736	0.1762	0.1788	0.1814	0.1841
-0.8	0.1867	0.1894	0.1922	0.1949	0.1977	0.2005	0.2033	0.2061	0.2090	0.2119
-0.7	0.2148	0.2177	0.2207	0.2236	0.2266	0.2297	0.2327	0.2358	0.2389	0.2420
-0.6	0.2451	0.2483	0.2514	0.2546	0.2578	0.2611	0.2643	0.2676	0.2709	0.2743
-0.5	0.2776	0.2810	0.2843	0.2877	0.2912	0.2946	0.2981	0.3015	0.3050	0.3085
-0.4	0.3121	0.3156	0.3192	0.3228	0.3264	0.3300	0.3336	0.3372	0.3409	0.3446
-0.3	0.3483	0.3520	0.3557	0.3594	0.3632	0.3669	0.3707	0.3745	0.3783	0.3821
-0.2	0.3859	0.3897	0.3936	0.3974	0.4013	0.4052	0.4090	0.4129	0.4168	0.4207
-0.1	0.4247	0.4286	0.4325	0.4364	0.4404	0.4443	0.4483	0.4522	0.4562	0.4602
-0.0	0.4641	0.4681	0.4721	0.4761	0.4801	0.4840	0.4880	0.4920	0.4960	0.5000

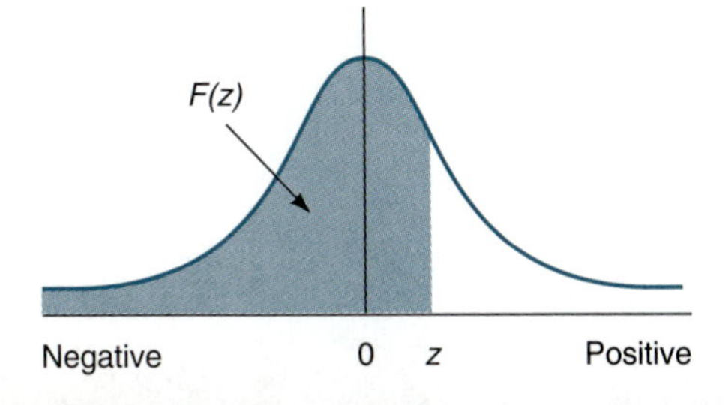

(continued)

z	0.00	0.01	0.02	0.03	0.04	0.05	0.06	0.07	0.08	0.09
+0.0	0.5000	0.5040	0.5080	0.5120	0.5160	0.5199	0.5239	0.5279	0.5319	0.5359
+0.1	0.5398	0.5438	0.5478	0.5517	0.5557	0.5596	0.5636	0.5675	0.5714	0.5753
+0.2	0.5793	0.5832	0.5871	0.5910	0.5948	0.5987	0.6026	0.6064	0.6103	0.6141
+0.3	0.6179	0.6217	0.6255	0.6293	0.6331	0.6368	0.6406	0.6443	0.6480	0.6517
+0.4	0.6554	0.6591	0.6628	0.6664	0.6700	0.6736	0.6772	0.6808	0.6844	0.6879
+0.5	0.6915	0.6950	0.6985	0.7019	0.7054	0.7088	0.7123	0.7157	0.7190	0.7224
+0.6	0.7257	0.7291	0.7324	0.7357	0.7389	0.7422	0.7454	0.7486	0.7517	0.7549
+0.7	0.7580	0.7611	0.7642	0.7673	0.7704	0.7734	0.7764	0.7794	0.7823	0.7852
+0.8	0.7881	0.7910	0.7939	0.7967	0.7995	0.8023	0.8051	0.8079	0.8106	0.8133
+0.9	0.8159	0.8186	0.8212	0.8238	0.8264	0.8289	0.8315	0.8340	0.8365	0.8389
+1.0	0.8413	0.8438	0.8461	0.8485	0.8508	0.8531	0.8554	0.8577	0.8599	0.8621
+1.1	0.8643	0.8665	0.8686	0.8708	0.8729	0.8749	0.8770	0.8790	0.8810	0.8830
+1.2	0.8849	0.8869	0.8888	0.8907	0.8925	0.8944	0.8962	0.8980	0.8997	0.9015
+1.3	0.9032	0.9049	0.9066	0.9082	0.9099	0.9115	0.9131	0.9147	0.9162	0.9177
+1.4	0.9192	0.9207	0.9222	0.9236	0.9251	0.9265	0.9279	0.9292	0.9306	0.9319
+1.5	0.9332	0.9345	0.9357	0.9370	0.9382	0.9394	0.9406	0.9418	0.9429	0.9441
+1.6	0.9452	0.9463	0.9474	0.9484	0.9495	0.9505	0.9515	0.9525	0.9535	0.9545
+1.7	0.9554	0.9564	0.9573	0.9582	0.9591	0.9599	0.9608	0.9616	0.9625	0.9633
+1.8	0.9641	0.9649	0.9656	0.9664	0.9671	0.9678	0.9686	0.9693	0.9699	0.9706
+1.9	0.9713	0.9719	0.9726	0.9732	0.9738	0.9744	0.9750	0.9756	0.9761	0.9767
+2.0	0.9773	0.9778	0.9783	0.9788	0.9793	0.9798	0.9803	0.9808	0.9812	0.9817
+2.1	0.9821	0.9826	0.9830	0.9834	0.9838	0.9842	0.9846	0.9850	0.9854	0.9857
+2.2	0.9861	0.9864	0.9868	0.9871	0.9875	0.9878	0.9881	0.9884	0.9887	0.9890
+2.3	0.9893	0.9896	0.9898	0.9901	0.9904	0.9906	0.9909	0.9911	0.9913	0.9916
+2.4	0.9918	0.9920	0.9922	0.9925	0.9927	0.9929	0.9931	0.9932	0.9934	0.9936
+2.5	0.9938	0.9940	0.9941	0.9943	0.9945	0.9946	0.9948	0.9949	0.9951	0.9952
+2.6	0.9953	0.9955	0.9956	0.9957	0.9959	0.9960	0.9961	0.9962	0.9963	0.9964
+2.7	0.9965	0.9966	0.9967	0.9968	0.9969	0.9970	0.9971	0.9972	0.9973	0.9974
+2.8	0.9974	0.9975	0.9976	0.9977	0.9977	0.9978	0.9979	0.9979	0.9980	0.9981
+2.9	0.9981	0.9982	0.9983	0.9983	0.9984	0.9984	0.9985	0.9985	0.9986	0.9986
+3.0	0.99865	0.99869	0.99874	0.99878	0.99882	0.99886	0.99889	0.99893	0.99896	0.99900
+3.1	0.99903	0.99906	0.99910	0.99913	0.99915	0.99918	0.99921	0.99924	0.99926	0.99929
+3.2	0.99931	0.99934	0.99936	0.99938	0.99940	0.99942	0.99944	0.99946	0.99948	0.99950
+3.3	0.99952	0.99953	0.99955	0.99957	0.99958	0.99960	0.99961	0.99962	0.99964	0.99965
+3.4	0.99966	0.99967	0.99969	0.99970	0.99971	0.99972	0.99973	0.99974	0.99975	0.99976
+3.5	0.99977	0.99978	0.99978	0.99979	0.99980	0.99981	0.99981	0.99982	0.99983	0.99983

Characteristic	Statistic	Normalized statistic z
Measurement	X	$\frac{X - \mu X}{\sigma}$
Subgroup average	$\bar{X}$	$\frac{\overline{X} - \mu_{\bar{X}}}{\sigma / \sqrt{n}}$
Binomial count	$np = c$	$\frac{c + 0.5 - n\mu_p}{\sqrt{n\mu_p(1 - \mu_p)}}$
Binomial fraction	$p = \frac{c}{n}$	$\frac{(c + 0.5)/n - \mu_p}{\sqrt{\mu_p(1 - \mu_p)/n}}$
Poisson count	c	$\frac{c + 0.5 - \mu_c}{\sqrt{\mu_c}}$

APPENDIX B

Right Tail Area of the X^2 Distribution†

υ	γ = **0.995**	**0.990**	**0.975**	**0.950**	**0.900**	**0.500**	**0.100**	**0.050**	**0.025**	**0.010**	**0.005**
1	0.00	0.00	0.00	0.00	0.02	0.45	2.71	3.84	5.02	6.63	7.88
2	0.01	0.02	0.05	0.10	0.21	1.39	4.61	5.99	7.38	9.21	10.60
3	0.07	0.11	0.22	0.35	0.58	2.37	6.25	7.81	9.35	11.34	12.84
4	0.21	0.30	0.48	0.71	1.06	3.36	7.78	9.49	11.14	13.28	14.86
5	0.41	0.55	0.83	1.15	1.61	4.35	9.24	11.07	12.83	15.09	16.75
6	0.68	0.87	1.24	1.64	2.20	5.35	10.65	12.59	14.45	16.81	18.55
7	0.99	1.24	1.69	2.17	2.83	6.35	12.02	14.07	16.01	18.48	20.28
8	1.34	1.65	2.18	2.73	3.49	7.34	13.36	15.51	17.53	20.09	21.96
9	1.73	2.09	2.70	3.33	4.17	8.34	14.68	16.92	19.02	21.67	23.59
10	2.16	2.56	3.25	3.94	4.87	9.34	15.99	18.31	20.48	23.21	25.19
11	2.60	3.05	3.82	4.57	5.58	10.34	17.28	19.68	21.92	24.72	26.76
12	3.07	3.57	4.40	5.23	6.30	11.34	18.55	21.03	23.34	26.22	28.30
13	3.57	4.11	5.01	5.89	7.04	12.34	19.81	22.36	24.74	27.69	29.82
14	4.07	4.66	5.63	6.57	7.79	13.34	21.06	23.68	26.12	29.14	31.32
15	4.60	5.23	6.26	7.26	8.55	14.34	22.31	25.00	27.49	30.58	32.80
16	5.14	5.81	6.91	7.96	9.31	15.34	23.54	26.30	28.85	32.00	34.27
17	5.70	6.41	7.56	8.67	10.09	16.34	24.77	27.59	30.19	33.41	35.72
18	6.26	7.01	8.23	9.39	10.87	17.34	25.99	28.87	31.53	34.81	37.16
19	6.84	7.63	8.91	10.12	11.65	18.34	27.20	30.14	32.85	36.19	38.58
20	7.43	8.26	9.59	10.85	12.44	19.34	28.41	31.41	34.17	37.57	40.00
21	8.03	8.90	10.28	11.59	13.24	20.34	29.62	32.67	35.48	38.93	41.40
22	8.64	9.54	10.98	12.34	14.04	21.34	30.81	33.92	36.78	40.29	42.80
23	9.26	10.20	11.69	13.09	14.85	22.34	32.01	35.17	38.08	41.64	44.18
24	9.89	10.86	12.40	13.85	15.66	23.34	33.20	36.42	39.36	42.98	45.56
25	10.52	11.52	13.12	14.61	16.47	24.34	34.38	37.65	40.65	44.31	46.93
26	11.16	12.20	13.84	15.38	17.29	25.34	35.56	38.89	41.92	45.64	48.29
27	11.81	12.88	14.57	16.15	18.11	26.34	36.74	40.11	43.19	46.96	49.65
28	12.46	13.57	15.31	16.93	18.94	27.34	37.92	41.34	44.46	48.28	50.99
29	13.12	14.26	16.05	17.71	19.77	28.34	39.09	42.56	45.72	49.59	52.34
30	13.79	14.95	16.79	18.49	20.60	29.34	40.26	43.77	46.98	50.89	53.67
40	20.71	22.16	24.43	26.51	29.05	39.34	51.80	55.76	59.34	63.69	66.77
50	27.99	29.71	32.36	34.76	37.69	49.33	63.17	67.50	71.42	76.15	79.49
70	43.28	45.44	48.76	51.74	55.33	69.33	85.53	90.53	95.02	100.42	104.22
100	67.33	70.06	74.22	77.93	82.36	99.33	118.50	124.34	129.56	135.81	140.17

†Taken by permission from Leland Blank, *Statistical Procedures for Engineering, Management, and Science,* McGraw-Hill Book Company, New York, 1980.

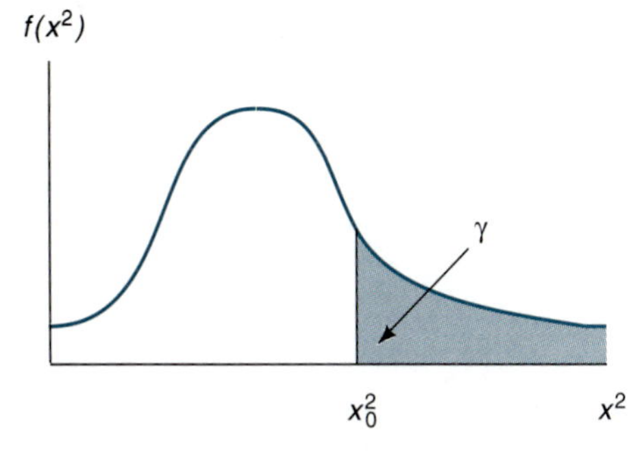

Given υ, the table gives the x_0^2 value with γ of the area above it; that is, $P(x^2 \geq x_0^2) = \gamma$

APPENDIX C

Factors for Estimating σ from $\bar{R}$, $\bar{s}$, or $\bar{\sigma}_{RMS}$ and σ_R from $\bar{R}$

Number of observations in subgroup, n	Factor d_2, $d_2 = \frac{\bar{R}}{\sigma}$	Factor d_3, $d_3 = \frac{\sigma_R}{\sigma}$	Factor c_2, $c_2 = \frac{\bar{\sigma}_{RMS}}{\sigma}$	Factor c_4, $c_4 = \frac{\bar{s}}{\sigma}$
2	1.128	0.8525	0.5642	0.7979
3	1.693	0.8884	0.7236	0.8862
4	2.059	0.8798	0.7979	0.9213
5	2.326	0.8641	0.8407	0.9400
6	2.534	0.8480	0.8686	0.9515
7	2.704	0.8332	0.8882	0.9594
8	2.847	0.8198	0.9027	0.9650
9	2.970	0.8078	0.9139	0.9693
10	3.078	0.7971	0.9227	0.9727
11	3.173	0.7873	0.9300	0.9754
12	3.258	0.7785	0.9359	0.9776
13	3.336	0.7704	0.9410	0.9794
14	3.407	0.7630	0.9453	0.9810
15	3.472	0.7562	0.9490	0.9823
16	3.532	0.7499	0.9523	0.9835
17	3.588	0.7441	0.9551	0.9845
18	3.640	0.7386	0.9576	0.9854
19	3.689	0.7335	0.9599	0.9862
20	3.735	0.7287	0.9619	0.9869
21	3.778	0.7242	0.9638	0.9876
22	3.819	0.7199	0.9655	0.9882
23	3.858	0.7159	0.9670	0.9887
24	3.895	0.7121	0.9684	0.9892
25	3.931	0.7084	0.9696	0.9896
30	4.086	0.6926	0.9748	0.9914
35	4.213	0.6799	0.9784	0.9927
40	4.322	0.6692	0.9811	0.9936
45	4.415	0.6601	0.9832	0.9943
50	4.498	0.6521	0.9849	0.9949
55	4.572	0.6452	0.9863	0.9954
60	4.639	0.6389	0.9874	0.9958
65	4.699	0.6337	0.9884	0.9961
70	4.755	0.6283	0.9892	0.9964
75	4.806	0.6236	0.9900	0.9966
80	4.854	0.6194	0.9906	0.9968
85	4.898	0.6154	0.9912	0.9970
90	4.939	0.6118	0.9916	0.9972
95	4.978	0.6084	0.9921	0.9973
100	5.015	0.6052	0.9925	0.9975

Estimate of $\sigma = \bar{R}/d_2$ or $\bar{s}/c_4$ or $\bar{\sigma}_{RMS}/c_2$; $\sigma_R = \bar{R}/d_3$. These factors assume sampling from a normal universe.

APPENDIX D

Factors for Determining from $\bar{R}$ the 3-sigma Control Limits for $\bar{X}$ and R Charts

Number of observations in subgroup, n	Factor for $\bar{X}$ chart, A_2	Factors for R chart: Lower control limit D_3	Factors for R chart: Upper control limit D_4
2	1.88	0	3.27
3	1.02	0	2.57
4	0.73	0	2.28
5	0.58	0	2.11
6	0.48	0	2.00
7	0.42	0.08	1.92
8	0.37	0.14	1.86
9	0.34	0.18	1.82
10	0.31	0.22	1.78
11	0.29	0.26	1.74
12	0.27	0.28	1.72
13	0.25	0.31	1.69
14	0.24	0.33	1.67
15	0.22	0.35	1.65
16	0.21	0.36	1.64
17	0.20	0.38	1.62
18	0.19	0.39	1.61
19	0.19	0.40	1.60
20	0.18	0.41	1.59

Upper control limit for $\bar{X} = UCL_{\bar{X}} = \bar{\bar{X}} + A_2\bar{R}$
Lower control limit for $\bar{X} = LCL_{\bar{X}} = \bar{\bar{X}} - A_2\bar{R}$

(If aimed-at or standard value $\bar{X}_0$ is used rather than $\bar{\bar{X}}$ as the central line on the control chart, $\bar{X}_0$ should be substituted for $\bar{\bar{X}}$ in the preceding formulas.)

Upper control limit for $R = UCL_R = D_4\bar{R}$
Lower control limit for $R = LCL_R = D_3\bar{R}$

All factors are based on the normal distribution.

APPENDIX E

Factors for Determining from $\bar{s}$ and $\bar{\sigma}_{RMS}$ the 3-sigma Control Limits for $\bar{X}$ and s or σ_{RMS} Charts

Number of observations in subgroup, n	Factor for $\bar{X}$ chart using $\bar{\sigma}_{RMS}$, A_1	Factor for $\bar{X}$ chart using $\bar{s}$, A_3	Factors for s or σ_{RMS} charts	
			Lower control limit B_3	Upper control limit B_4
2	3.76	2.66	0	3.27
3	2.39	1.95	0	2.57
4	1.88	1.63	0	2.27
5	1.60	1.43	0	2.09
6	1.41	1.29	0.03	1.97
7	1.28	1.18	0.12	1.88
8	1.17	1.10	0.19	1.81
9	1.09	1.03	0.24	1.76
10	1.03	0.98	0.28	1.72
11	0.97	0.93	0.32	1.68
12	0.93	0.89	0.35	1.65
13	0.88	0.85	0.38	1.62
14	0.85	0.82	0.41	1.59
15	0.82	0.79	0.43	1.57
16	0.79	0.76	0.45	1.55
17	0.76	0.74	0.47	1.53
18	0.74	0.72	0.48	1.52
19	0.72	0.70	0.50	1.50
20	0.70	0.68	0.51	1.49
21	0.68	0.66	0.52	1.48
22	0.66	0.65	0.53	1.47
23	0.65	0.63	0.54	1.46
24	0.63	0.62	0.55	1.45
25	0.62	0.61	0.56	1.44
30	0.56	0.55	0.60	1.40
35	0.52	0.51	0.63	1.37
40	0.48	0.48	0.66	1.34
45	0.45	0.45	0.68	1.32
50	0.43	0.43	0.70	1.30
55	0.41	0.41	0.71	1.29
60	0.39	0.39	0.72	1.28
65	0.38	0.37	0.73	1.27
70	0.36	0.36	0.74	1.26
75	0.35	0.35	0.75	1.25
80	0.34	0.34	0.76	1.24
85	0.33	0.33	0.77	1.23
90	0.32	0.32	0.77	1.23
95	0.31	0.31	0.78	1.22
100	0.30	0.30	0.79	1.21

Upper control limit for $\bar{X} = UCL_{\bar{X}} = \bar{\bar{X}} + A_3\bar{s} = \bar{\bar{X}} + A_1\bar{\sigma}_{RMS}$. Lower control limit for $\bar{X} = LCL_{\bar{X}} = \bar{\bar{X}} - A_3\bar{s} = \bar{\bar{X}} - A_1\bar{\sigma}_{RMS}$. (If aimed-at or standard value $\bar{\bar{X}}_0$ is used rather than $\bar{\bar{X}}$ as the central line on the control chart, $\bar{X}_0$ should be substituted for $\bar{\bar{X}}$ in the preceding formulas.) Upper control limit for s or $\sigma_{RMS} = UCL = B_4\bar{s} = B_4\bar{\sigma}_{RMS}$. Lower control limit for s or $\sigma_{RMS} = LCL = B_3\bar{s} = B_3\bar{\sigma}_{RMS}$. All factors are based on the normal distribution.

APPENDIX F

Factors for Determining from σ the 3-sigma Control Limits for $\bar{X}$, R, and s or σ_{RMS} Charts

Number of observations in subgroup, n	Factors for $\bar{X}$ chart, A	Factors for R chart: Lower control limit D_1	Factors for R chart: Upper control limit D_2	Factors for σ_{RMS} chart: Lower control limit B_1	Factors for σ_{RMS} chart: Upper control limit B_2	Factors for s chart: Lower control limit B_5	Factors for s chart: Upper control limit B_6
2	2.12	0	3.69	0	1.84	0	2.61
3	1.73	0	4.36	0	1.86	0	2.28
4	1.50	0	4.70	0	1.81	0	2.09
5	1.34	0	4.92	0	1.76	0	1.96
6	1.22	0	5.08	0.03	1.71	0.03	1.87
7	1.13	0.20	5.20	0.10	1.67	0.11	1.81
8	1.06	0.39	5.31	0.17	1.64	0.18	1.75
9	1.00	0.55	5.39	0.22	1.61	0.23	1.71
10	0.95	0.69	5.47	0.26	1.58	0.28	1.67
11	0.90	0.81	5.53	0.30	1.56	0.31	1.64
12	0.87	0.92	5.59	0.33	1.54	0.35	1.61
13	0.83	1.03	5.65	0.36	1.52	0.37	1.59
14	0.80	1.12	5.69	0.38	1.51	0.40	1.56
15	0.77	1.21	5.74	0.41	1.49	0.42	1.54
16	0.75	1.28	5.78	0.43	1.48	0.44	1.53
17	0.73	1.36	5.82	0.44	1.47	0.46	1.51
18	0.71	1.43	5.85	0.46	1.45	0.48	1.50
19	0.69	1.49	5.89	0.48	1.44	0.49	1.48
20	0.67	1.55	5.92	0.49	1.43	0.50	1.47
21	0.65			0.50	1.42	0.52	1.46
22	0.64			0.52	1.41	0.53	1.45
23	0.63			0.53	1.41	0.54	1.44
24	0.61			0.54	1.40	0.55	1.43
25	0.60			0.55	1.39	0.56	1.42
30	0.55			0.59	1.36	0.60	1.38
35	0.51			0.62	1.33	0.63	1.36
40	0.47			0.65	1.31	0.66	1.33
45	0.45			0.67	1.30	0.68	1.31
50	0.42			0.68	1.28	0.69	1.30
55	0.40			0.70	1.27	0.71	1.28
60	0.39			0.71	1.26	0.72	1.27
65	0.37			0.72	1.25	0.73	1.26
70	0.36			0.74	1.24	0.74	1.25
75	0.35			0.75	1.23	0.75	1.24
80	0.34			0.75	1.23	0.76	1.24
85	0.33			0.76	1.22	0.77	1.23
90	0.32			0.77	1.22	0.77	1.22
95	0.31			0.77	1.21	0.78	1.22
100	0.30			0.78	1.20	0.78	1.21

$UCL_{\bar{X}} = \mu + A\sigma$ $LCL_{\bar{X}} = \mu - A\sigma$ (If actual average is to be used rather than standard or aimed-at average, $\bar{\bar{X}}$ should be substituted for μ in the preceding formulas.)

$UCL_R = D_2\sigma$ $UCL_s = B_6\sigma$ $UCL_{\sigma RMS} = B_2\sigma$

Central line$_R = d_2\sigma$ Central line$_s = c_4\sigma$ Central line$_{\sigma RMS} = c_2\sigma$

$LCL_R = D_1\sigma$ $LCL_s = B_5\sigma$ $LCL_{\sigma RMS} = B_1\sigma$

APPENDIX G

Random Numbers

07	28	68	61	81	38	11	98	34	74	64	03	48	09	18	10	15	25	98	80
29	24	86	11	41	21	16	12	96	17	56	61	49	32	48	35	43	29	34	12
76	05	58	54	35	55	35	59	07	19	00	92	65	95	34	88	26	32	61	36
95	01	20	28	66	31	15	92	14	33	39	98	55	85	71	35	82	04	51	64
73	89	25	53	83	33	75	79	98	20	09	06	76	92	43	42	55	86	41	67
41	58	46	41	68	72	73	78	34	65	87	08	10	93	46	00	32	48	29	68
53	46	33	57	86	99	47	87	14	55	98	93	72	15	77	23	13	26	37	20
39	46	65	77	16	92	33	65	57	49	18	41	87	68	05	23	73	33	55	49
40	98	58	06	54	13	55	31	86	06	34	94	43	59	08	54	86	44	59	84
06	45	65	80	97	46	95	38	82	01	88	12	28	75	93	39	33	60	00	48
84	72	36	35	94	11	36	23	17	09	95	90	26	46	90	70	81	40	77	38
61	14	68	60	77	44	75	28	56	67	36	58	03	82	16	76	39	12	73	70
07	47	15	19	64	62	17	97	36	08	22	55	58	81	17	77	83	65	75	05
70	43	84	46	41	98	44	54	23	72	39	79	53	16	88	04	66	00	66	43
57	10	02	26	17	12	56	48	43	97	65	06	21	97	65	97	95	77	93	01
95	01	58	34	51	77	89	80	79	72	60	94	43	05	89	83	88	15	09	58
53	00	18	66	58	39	02	95	62	79	35	52	01	06	50	18	98	88	87	81
51	86	20	34	89	54	54	61	15	00	96	89	11	34	05	18	26	77	17	23
38	63	42	41	87	99	37	18	91	08	55	42	27	51	69	48	94	14	70	96
47	77	39	28	14	56	98	96	73	22	31	67	20	90	85	04	01	87	42	17
26	20	46	66	36	28	98	66	97	56	78	29	19	53	46	08	20	30	55	61
58	58	28	68	36	45	83	66	12	05	17	37	74	90	81	86	99	04	17	90
80	83	75	20	32	63	09	41	69	12	43	82	63	40	08	89	71	89	68	44
40	90	05	68	85	00	90	91	49	16	23	00	26	56	52	66	71	22	63	40
77	38	50	26	29	57	56	31	37	52	88	88	37	72	14	52	73	79	23	79
51	62	77	67	70	21	17	88	22	26	66	77	78	55	87	14	39	07	31	67
66	81	52	18	87	47	01	60	71	73	90	72	90	39	37	64	44	26	82	07
67	72	78	24	07	12	61	67	78	85	92	68	95	24	69	57	74	13	28	64
14	29	00	91	50	43	64	63	85	17	54	46	92	58	58	52	97	54	84	09
30	89	99	07	56	26	49	27	83	67	52	35	36	93	63	60	15	71	16	34
26	42	43	27	81	79	67	35	84	28	64	59	79	16	11	54	85	34	01	49
98	05	34	47	71	14	87	98	70	21	53	51	01	46	60	71	19	33	62	43
02	82	10	42	11	62	87	83	16	96	34	46	04	25	33	69	55	37	82	29
99	88	34	85	46	77	12	00	89	17	04	48	85	62	32	77	08	24	88	65
83	59	57	38	84	22	08	75	21	10	58	75	87	70	19	07	94	83	09	37
76	27	52	23	67	14	39	88	57	00	72	71	21	68	81	49	24	94	19	37
03	80	24	56	17	64	66	90	80	09	62	03	65	61	66	39	83	87	41	95
40	86	98	74	63	72	14	00	08	38	25	25	37	93	89	96	74	66	36	06
38	02	78	20	39	15	04	67	68	27	46	22	43	79	26	45	45	17	66	13
19	51	85	12	56	95	63	15	44	74	88	26	02	10	68	09	84	86	26	81

Source: Eugene L. Grant and Richard S. Leavenworth, *Statistical Quality Control,* 6th ed. (New York: McGraw-Hill, 1988), 701. Reproduced with permission of McGraw-Hill.

APPENDIX H

Binomial Probabilities; continued on p. 386–393

$n = 2$

	p	.05	.10	.20	.30	.40	.50	.60	.70	.80	.90	.95
	0	.9025	.8100	.6400	.4900	.3600	.2500	.1600	.0900	.0400	.0100	.0025
s	1	.0950	.1800	.3200	.4200	.4800	.5000	.4800	.4200	.3200	.1800	.0950
	2	.0025	.0100	.0400	.0900	.1600	.2500	.3600	.4900	.6400	.8100	.9025

$n = 3$

	p	.05	.10	.20	.30	.40	.50	.60	.70	.80	.90	.95
	0	.8574	.7290	.5120	.3430	.2160	.1250	.0640	.0270	.0080	.0010	.0001
	1	.1354	.2430	.3840	.4410	.4320	.3750	.2880	.1890	.0960	.0270	.0071
s	2	.0071	.0270	.0960	.1890	.2880	.3750	.4320	.4410	.3840	.2430	.1354
	3	.0001	.0010	.0080	.0270	.0640	.1250	.2160	.3430	.5120	.7290	.8574

$n = 4$

	p	.05	.10	.20	.30	.40	.50	.60	.70	.80	.90	.95
	0	.8145	.6561	.4096	.2401	.1296	.0625	.0256	.0081	.0016	.0001	.0000
	1	.1715	.2916	.4096	.4116	.3456	.2500	.1536	.0756	.0256	.0036	.0005
s	2	.0135	.0486	.1536	.2646	.3456	.3750	.3456	.2646	.1536	.0486	.0135
	3	.0005	.0036	.0256	.0756	.1536	.2500	.3456	.4116	.4096	.2916	.1715
	4	.0000	.0001	.0016	.0081	.0256	.0625	.1296	.2401	.4096	.6561	.8145

$n = 5$

	p	.05	.10	.20	.30	.40	.50	.60	.70	.80	.90	.95
	0	.7738	.5905	.3277	.1681	.0778	.0313	.0102	.0024	.0003	.0000	.0000
	1	.2036	.3281	.4096	.3602	.2592	.1563	.0768	.0284	.0064	.0005	.0000
	2	.0214	.0729	.2048	.3087	.3456	.3125	.2304	.1323	.0512	.0081	.0011
s	3	.0011	.0081	.0512	.1323	.2304	.3125	.3456	.3087	.2048	.0729	.0214
	4	.0000	.0005	.0064	.0284	.0768	.1563	.2592	.3602	.4096	.3281	.2036
	5	.0000	.0000	.0003	.0024	.0102	.0313	.0778	.1681	.3277	.5905	.7738

Source of Appendix H: Arnold Naiman, Robert Rosenfeld, and Gene Zirkel, *Understanding Statistics,* 3d ed. (New York: McGraw-Hill, 1983), 304–309. Reproduced with permission of McGraw-Hill.

$n = 6$

s \ p	.05	.10	.20	.30	.40	.50	.60	.70	.80	.90	.95
0	.7351	.5314	.2621	.1176	.0467	.0156	.0041	.0007	.0001	.0000	.0000
1	.2321	.3543	.3932	.3025	.1866	.0938	.0369	.0102	.0015	.0001	.0000
2	.0305	.0984	.2458	.3241	.3110	.2344	.1382	.0595	.0154	.0012	.0001
3	.0021	.0146	.0819	.1852	.2765	.3125	.2765	.1852	.0819	.0146	.0021
4	.0001	.0012	.0154	.0595	.1382	.2344	.3110	.3241	.2458	.0984	.0305
5	.0000	.0001	.0015	.0102	.0369	.0938	.1866	.3025	.3932	.3543	.2321
6	.0000	.0000	.0001	.0007	.0041	.0156	.0467	.1176	.2621	.5314	.7351

$n = 7$

s \ p	.05	.10	.20	.30	.40	.50	.60	.70	.80	.90	.95
0	.6983	.4783	.2097	.0824	.0280	.0078	.0016	.0002	.0000	.0000	.0000
1	.2573	.3720	.3670	.2471	.1306	.0547	.0172	.0036	.0004	.0000	.0000
2	.0406	.1240	.2753	.3177	.2613	.1641	.0774	.0250	.0043	.0002	.0000
3	.0036	.0230	.1147	.2269	.2903	.2734	.1935	.0972	.0287	.0026	.0002
4	.0002	.0026	.0287	.0972	.1935	.2734	.2903	.2269	.1147	.0230	.0036
5	.0000	.0002	.0043	.0250	.0774	.1641	.2613	.3177	.2753	.1240	.0406
6	.0000	.0000	.0004	.0036	.0172	.0547	.1306	.2471	.3670	.3720	.2573
7	.0000	.0000	.0000	.0002	.0016	.0078	.0280	.0824	.2097	.4783	.6983

$n = 8$

s \ p	.05	.10	.20	.30	.40	.50	.60	.70	.80	.90	.95
0	.6634	.4305	.1678	.0576	.0168	.0039	.0007	.0001	.0000	.0000	.0000
1	.2793	.3826	.3355	.1977	.0896	.0313	.0079	.0012	.0001	.0000	.0000
2	.0515	.1488	.2936	.2965	.2090	.1094	.0413	.0100	.0011	.0000	.0000
3	.0054	.0331	.1468	.2541	.2787	.2188	.1239	.0467	.0092	.0004	.0000
4	.0004	.0046	.0459	.1361	.2322	.2734	.2322	.1361	.0459	.0046	.0004
5	.0000	.0004	.0092	.0467	.1239	.2188	.2787	.2541	.1468	.0331	.0054
6	.0000	.0000	.0011	.0100	.0413	.1094	.2090	.2965	.2936	.1488	.0515
7	.0000	.0000	.0001	.0012	.0079	.0313	.0896	.1977	.3355	.3826	.2793
8	.0000	.0000	.0000	.0001	.0007	.0039	.0168	.0576	.1678	.4305	.6634

$n = 9$

s / p	.05	.10	.20	.30	.40	.50	.60	.70	.80	.90	.95
0	.6302	.3874	.1342	.0404	.0101	.0020	.0003	.0000	.0000	.0000	.0000
1	.2985	.3874	.3020	.1556	.0605	.0176	.0035	.0004	.0000	.0000	.0000
2	.0629	.1722	.3020	.2668	.1612	.0703	.0212	.0039	.0003	.0000	.0000
3	.0077	.0446	.1762	.2668	.2508	.1641	.0743	.0210	.0028	.0001	.0000
4	.0006	.0074	.0661	.1715	.2508	.2461	.1672	.0735	.0165	.0008	.0000
5	.0000	.0008	.0165	.0735	.1672	.2461	.2508	.1715	.0661	.0074	.0006
6	.0000	.0001	.0028	.0210	.0743	.1641	.2508	.2668	.1762	.0446	.0077
7	.0000	.0000	.0003	.0039	.0212	.0703	.1612	.2668	.3020	.1722	.0629
8	.0000	.0000	.0000	.0004	.0035	.0176	.0605	.1556	.3020	.3874	.2985
9	.0000	.0000	.0000	.0000	.0003	.0020	.0101	.0404	.1342	.3874	.6302

$n = 10$

s / p	.05	.10	.20	.30	.40	.50	.60	.70	.80	.90	.95
0	.5987	.3487	.1074	.0282	.0060	.0010	.0001	.0000	.0000	.0000	.0000
1	.3151	.3874	.2684	.1211	.0403	.0098	.0016	.0001	.0000	.0000	.0000
2	.0746	.1937	.3020	.2335	.1209	.0439	.0106	.0014	.0001	.0000	.0000
3	.0105	.0574	.2013	.2668	.2150	.1172	.0425	.0090	.0008	.0000	.0000
4	.0010	.0112	.0881	.2001	.2508	.2051	.1115	.0368	.0055	.0001	.0000
5	.0001	.0015	.0264	.1029	.2007	.2461	.2007	.1029	.0264	.0015	.0001
6	.0000	.0001	.0055	.0368	.1115	.2051	.2508	.2001	.0881	.0112	.0010
7	.0000	.0000	.0008	.0090	.0425	.1172	.2150	.2668	.2013	.0574	.0105
8	.0000	.0000	.0001	.0014	.0106	.0439	.1209	.2335	.3020	.1937	.0746
9	.0000	.0000	.0000	.0001	.0016	.0098	.0403	.1211	.2684	.3874	.3151
10	.0000	.0000	.0000	.0000	.0001	.0010	.0060	.0282	.1074	.3487	.5987

$n = 11$

s / p	.05	.10	.20	.30	.40	.50	.60	.70	.80	.90	.95
0	.5688	.3138	.0859	.0198	.0036	.0005	.0000	.0000	.0000	.0000	.0000
1	.3293	.3835	.2362	.0932	.0266	.0054	.0007	.0000	.0000	.0000	.0000
2	.0867	.2131	.2953	.1998	.0887	.0269	.0052	.0005	.0000	.0000	.0000
3	.0137	.0710	.2215	.2568	.1774	.0806	.0234	.0037	.0002	.0000	.0000
4	.0014	.0158	.1107	.2201	.2365	.1611	.0701	.0173	.0017	.0000	.0000
5	.0001	.0025	.0388	.1321	.2207	.2256	.1471	.0566	.0097	.0003	.0000
6	.0000	.0003	.0097	.0566	.1471	.2256	.2207	.1321	.0388	.0025	.0001
7	.0000	.0000	.0017	.0173	.0701	.1611	.2365	.2201	.1107	.0158	.0014
8	.0000	.0000	.0002	.0037	.0234	.0806	.1774	.2568	.2215	.0710	.0137
9	.0000	.0000	.0000	.0005	.0052	.0269	.0887	.1998	.2953	.2131	.0867
10	.0000	.0000	.0000	.0000	.0007	.0054	.0266	.0932	.2362	.3835	.3293
11	.0000	.0000	.0000	.0000	.0000	.0005	.0036	.0198	.0859	.3138	.5688

$n = 12$

	p	.05	.10	.20	.30	.40	.50	.60	.70	.80	.90	.95
	0	.5404	.2824	.0687	.0138	.0022	.0002	.0000	.0000	.0000	.0000	.0000
	1	.3413	.3766	.2062	.0712	.0174	.0029	.0003	.0000	.0000	.0000	.0000
	2	.0988	.2301	.2835	.1678	.0639	.0161	.0025	.0002	.0000	.0000	.0000
	3	.0173	.0852	.2362	.2397	.1419	.0537	.0125	.0015	.0001	.0000	.0000
	4	.0021	.0213	.1329	.2311	.2128	.1208	.0420	.0078	.0005	.0000	.0000
	5	.0002	.0038	.0532	.1585	.2270	.1934	.1009	.0291	.0033	.0000	.0000
s	6	.0000	.0005	.0155	.0792	.1766	.2256	.1766	.0792	.0155	.0005	.0000
	7	.0000	.0000	.0033	.0291	.1009	.1934	.2270	.1585	.0532	.0038	.0002
	8	.0000	.0000	.0005	.0078	.0420	.1208	.2128	.2311	.1329	.0213	.0021
	9	.0000	.0000	.0001	.0015	.0125	.0537	.1419	.2397	.2362	.0852	.0173
	10	.0000	.0000	.0000	.0002	.0025	.0161	.0639	.1678	.2835	.2301	.0988
	11	.0000	.0000	.0000	.0000	.0003	.0029	.0174	.0712	.2062	.3766	.3413
	12	.0000	.0000	.0000	.0000	.0000	.0002	.0022	.0138	.0687	.2824	.5404

$n = 13$

	p	.05	.10	.20	.30	.40	.50	.60	.70	.80	.90	.95
	0	.5133	.2542	.0550	.0097	.0013	.0001	.0000	.0000	.0000	.0000	.0000
	1	.3512	.3672	.1787	.0540	.0113	.0016	.0001	.0000	.0000	.0000	.0000
	2	.1109	.2448	.2680	.1388	.0453	.0095	.0012	.0001	.0000	.0000	.0000
	3	.0214	.0997	.2457	.2181	.1107	.0349	.0065	.0006	.0000	.0000	.0000
	4	.0028	.0277	.1535	.2337	.1845	.0873	.0243	.0034	.0001	.0000	.0000
	5	.0003	.0055	.0691	.1803	.2214	.1571	.0656	.0142	.0011	.0000	.0000
	6	.0000	.0008	.0230	.1030	.1968	.2095	.1312	.0442	.0058	.0001	.0000
s	7	.0000	.0001	.0058	.0442	.1312	.2095	.1968	.1030	.0230	.0008	.0000
	8	.0000	.0000	.0011	.0142	.0656	.1571	.2214	.1803	.0691	.0055	.0003
	9	.0000	.0000	.0001	.0034	.0243	.0873	.1845	.2337	.1535	.0277	.0028
	10	.0000	.0000	.0000	.0006	.0065	.0349	.1107	.2181	.2457	.0997	.0214
	11	.0000	.0000	.0000	.0001	.0012	.0095	.0453	.1388	.2680	.2448	.1109
	12	.0000	.0000	.0000	.0000	.0001	.0016	.0113	.0540	.1787	.3672	.3512
	13	.0000	.0000	.0000	.0000	.0000	.0001	.0013	.0097	.0550	.2542	.5133

$n = 14$

s \ p	.05	.10	.20	.30	.40	.50	.60	.70	.80	.90	.95
0	.4877	.2288	.0440	.0068	.0008	.0001	.0000	.0000	.0000	.0000	.0000
1	.3593	.3559	.1539	.0407	.0073	.0009	.0001	.0000	.0000	.0000	.0000
2	.1229	.2570	.2501	.1134	.0317	.0056	.0005	.0000	.0000	.0000	.0000
3	.0259	.1142	.2501	.1943	.0845	.0222	.0033	.0002	.0000	.0000	.0000
4	.0037	.0349	.1720	.2290	.1549	.0611	.0136	.0014	.0000	.0000	.0000
5	.0004	.0078	.0860	.1963	.2066	.1222	.0408	.0066	.0003	.0000	.0000
6	.0000	.0013	.0322	.1262	.2066	.1833	.0918	.0232	.0020	.0000	.0000
7	.0000	.0002	.0092	.0618	.1574	.2095	.1574	.0618	.0092	.0002	.0000
8	.0000	.0000	.0020	.0232	.0918	.1833	.2066	.1262	.0322	.0013	.0000
9	.0000	.0000	.0003	.0066	.0408	.1222	.2066	.1963	.0860	.0078	.0004
10	.0000	.0000	.0000	.0014	.0136	.0611	.1549	.2290	.1720	.0349	.0037
11	.0000	.0000	.0000	.0002	.0033	.0222	.0845	.1943	.2501	.1142	.0259
12	.0000	.0000	.0000	.0000	.0005	.0056	.0317	.1134	.2501	.2570	.1229
13	.0000	.0000	.0000	.0000	.0001	.0009	.0073	.0407	.1539	.3559	.3593
14	.0000	.0000	.0000	.0000	.0000	.0001	.0008	.0068	.0440	.2288	.4877

$n = 15$

s \ p	.05	.10	.20	.30	.40	.50	.60	.70	.80	.90	.95
0	.4633	.2059	.0352	.0047	.0005	.0000	.0000	.0000	.0000	.0000	.0000
1	.3658	.3432	.1319	.0305	.0047	.0005	.0000	.0000	.0000	.0000	.0000
2	.1348	.2669	.2309	.0916	.0219	.0032	.0003	.0000	.0000	.0000	.0000
3	.0307	.1285	.2501	.1700	.0634	.0139	.0016	.0001	.0000	.0000	.0000
4	.0049	.0428	.1876	.2186	.1268	.0417	.0074	.0006	.0000	.0000	.0000
5	.0006	.0105	.1032	.2061	.1859	.0916	.0245	.0030	.0001	.0000	.0000
6	.0000	.0019	.0430	.1472	.2066	.1527	.0612	.0116	.0007	.0000	.0000
7	.0000	.0003	.0138	.0811	.1771	.1964	.1181	.0348	.0035	.0000	.0000
8	.0000	.0000	.0035	.0348	.1181	.1964	.1771	.0811	.0138	.0003	.0000
9	.0000	.0000	.0007	.0116	.0612	.1527	.2066	.1472	.0430	.0019	.0000
10	.0000	.0000	.0001	.0030	.0245	.0916	.1859	.2061	.1032	.0105	.0006
11	.0000	.0000	.0000	.0006	.0074	.0417	.1268	.2186	.1876	.0428	.0049
12	.0000	.0000	.0000	.0001	.0016	.0139	.0634	.1700	.2501	.1285	.0307
13	.0000	.0000	.0000	.0000	.0003	.0032	.0219	.0916	.2309	.2669	.1348
14	.0000	.0000	.0000	.0000	.0000	.0005	.0047	.0305	.1319	.3432	.3658
15	.0000	.0000	.0000	.0000	.0000	.0000	.0005	.0047	.0352	.2059	.4633

$n = 16$

s \ p	.05	.10	.20	.30	.40	.50	.60	.70	.80	.90	.95
0	.4401	.1853	.0281	.0033	.0003	.0000	.0000	.0000	.0000	.0000	.0000
1	.3706	.3294	.1126	.0228	.0030	.0002	.0000	.0000	.0000	.0000	.0000
2	.1463	.2745	.2111	.0732	.0150	.0018	.0001	.0000	.0000	.0000	.0000
3	.0359	.1423	.2463	.1465	.0468	.0085	.0008	.0000	.0000	.0000	.0000
4	.0061	.0514	.2001	.2040	.1014	.0278	.0040	.0002	.0000	.0000	.0000
5	.0008	.0137	.1201	.2099	.1623	.0667	.0142	.0013	.0000	.0000	.0000
6	.0001	.0028	.0550	.1649	.1983	.1222	.0392	.0056	.0002	.0000	.0000
7	.0000	.0004	.0197	.1010	.1889	.1746	.0840	.0185	.0012	.0000	.0000
8	.0000	.0001	.0055	.0487	.1417	.1964	.1417	.0487	.0055	.0001	.0000
9	.0000	.0000	.0012	.0185	.0840	.1746	.1889	.1010	.0197	.0004	.0000
10	.0000	.0000	.0002	.0056	.0392	.1222	.1983	.1649	.0550	.0028	.0001
11	.0000	.0000	.0000	.0013	.0142	.0667	.1623	.2099	.1201	.0137	.0008
12	.0000	.0000	.0000	.0002	.0040	.0278	.1014	.2040	.2001	.0514	.0061
13	.0000	.0000	.0000	.0000	.0008	.0085	.0468	.1465	.2463	.1423	.0359
14	.0000	.0000	.0000	.0000	.0001	.0018	.0150	.0732	.2111	.2745	.1463
15	.0000	.0000	.0000	.0000	.0000	.0002	.0030	.0228	.1126	.3294	.3706
16	.0000	.0000	.0000	.0000	.0000	.0000	.0003	.0033	.0281	.1853	.4401

$n = 17$

s \ p	.05	.10	.20	.30	.40	.50	.60	.70	.80	.90	.95
0	.4181	.1668	.0225	.0023	.0002	.0000	.0000	.0000	.0000	.0000	.0000
1	.3741	.3150	.0957	.0169	.0019	.0001	.0000	.0000	.0000	.0000	.0000
2	.1575	.2800	.1914	.0581	.0102	.0010	.0001	.0000	.0000	.0000	.0000
3	.0415	.1556	.2393	.1245	.0341	.0052	.0004	.0000	.0000	.0000	.0000
4	.0076	.0605	.2093	.1868	.0796	.0182	.0021	.0001	.0000	.0000	.0000
5	.0010	.0175	.1361	.2081	.1379	.0472	.0081	.0006	.0000	.0000	.0000
6	.0001	.0039	.0680	.1784	.1839	.0944	.0242	.0026	.0001	.0000	.0000
7	.0000	.0007	.0267	.1201	.1927	.1484	.0571	.0095	.0004	.0000	.0000
8	.0000	.0001	.0084	.0644	.1606	.1855	.1070	.0276	.0021	.0000	.0000
9	.0000	.0000	.0021	.0276	.1070	.1855	.1606	.0644	.0084	.0001	.0000
10	.0000	.0000	.0004	.0095	.0571	.1484	.1927	.1201	.0267	.0007	.0000
11	.0000	.0000	.0001	.0026	.0242	.0944	.1839	.1784	.0680	.0039	.0001
12	.0000	.0000	.0000	.0006	.0081	.0472	.1379	.2081	.1361	.0175	.0010
13	.0000	.0000	.0000	.0001	.0021	.0182	.0796	.1868	.2093	.0605	.0076
14	.0000	.0000	.0000	.0000	.0004	.0052	.0341	.1245	.2393	.1556	.0415
15	.0000	.0000	.0000	.0000	.0001	.0010	.0102	.0581	.1914	.2800	.1575
16	.0000	.0000	.0000	.0000	.0000	.0001	.0019	.0169	.0957	.3150	.3741
17	.0000	.0000	.0000	.0000	.0000	.0000	.0002	.0023	.0225	.1668	.4181

$n = 18$

	p	.05	.10	.20	.30	.40	.50	.60	.70	.80	.90	.95
	0	.3972	.1501	.0180	.0016	.0001	.0000	.0000	.0000	.0000	.0000	.0000
	1	.3763	.3002	.0811	.0126	.0012	.0001	.0000	.0000	.0000	.0000	.0000
	2	.1683	.2835	.1723	.0458	.0069	.0006	.0000	.0000	.0000	.0000	.0000
	3	.0473	.1680	.2297	.1046	.0246	.0031	.0002	.0000	.0000	.0000	.0000
	4	.0093	.0700	.2153	.1681	.0614	.0117	.0011	.0000	.0000	.0000	.0000
	5	.0014	.0218	.1507	.2017	.1146	.0327	.0045	.0002	.0000	.0000	.0000
	6	.0002	.0052	.0816	.1873	.1655	.0708	.0145	.0012	.0000	.0000	.0000
	7	.0000	.0010	.0350	.1376	.1892	.1214	.0374	.0046	.0001	.0000	.0000
	8	.0000	.0002	.0120	.0811	.1734	.1669	.0771	.0149	.0008	.0000	.0000
s	9	.0000	.0000	.0033	.0386	.1284	.1855	.1284	.0386	.0033	.0000	.0000
	10	.0000	.0000	.0008	.0149	.0771	.1669	.1734	.0811	.0120	.0002	.0000
	11	.0000	.0000	.0001	.0046	.0374	.1214	.1892	.1376	.0350	.0010	.0000
	12	.0000	.0000	.0000	.0012	.0145	.0708	.1655	.1873	.0816	.0052	.0002
	13	.0000	.0000	.0000	.0002	.0045	.0327	.1146	.2017	.1507	.0218	.0014
	14	.0000	.0000	.0000	.0000	.0011	.0117	.0614	.1681	.2153	.0700	.0093
	15	.0000	.0000	.0000	.0000	.0002	.0031	.0246	.1046	.2297	.1680	.0473
	16	.0000	.0000	.0000	.0000	.0000	.0006	.0069	.0458	.1723	.2835	.1683
	17	.0000	.0000	.0000	.0000	.0000	.0001	.0012	.0126	.0811	.3002	.3763
	18	.0000	.0000	.0000	.0000	.0000	.0000	.0001	.0016	.0180	.1501	.3972

$n = 19$

s \ p	.05	.10	.20	.30	.40	.50	.60	.70	.80	.90	.95
0	.3774	.1351	.0144	.0011	.0001	.0000	.0000	.0000	.0000	.0000	.0000
1	.3774	.2852	.0685	.0093	.0008	.0000	.0000	.0000	.0000	.0000	.0000
2	.1787	.2852	.1540	.0358	.0046	.0003	.0000	.0000	.0000	.0000	.0000
3	.0533	.1796	.2182	.0869	.0175	.0018	.0001	.0000	.0000	.0000	.0000
4	.0112	.0798	.2182	.1491	.0467	.0074	.0005	.0000	.0000	.0000	.0000
5	.0018	.0266	.1636	.1916	.0933	.0222	.0024	.0001	.0000	.0000	.0000
6	.0002	.0069	.0955	.1916	.1451	.0518	.0085	.0005	.0000	.0000	.0000
7	.0000	.0014	.0443	.1525	.1797	.0961	.0237	.0022	.0000	.0000	.0000
8	.0000	.0002	.0166	.0981	.1797	.1442	.0532	.0077	.0003	.0000	.0000
9	.0000	.0000	.0051	.0514	.1464	.1762	.0976	.0220	.0013	.0000	.0000
10	.0000	.0000	.0013	.0220	.0976	.1762	.1464	.0514	.0051	.0000	.0000
11	.0000	.0000	.0003	.0077	.0532	.1442	.1797	.0981	.0166	.0002	.0000
12	.0000	.0000	.0000	.0022	.0237	.0961	.1797	.1525	.0443	.0014	.0000
13	.0000	.0000	.0000	.0005	.0085	.0518	.1451	.1916	.0955	.0069	.0002
14	.0000	.0000	.0000	.0001	.0024	.0222	.0933	.1916	.1636	.0266	.0018
15	.0000	.0000	.0000	.0000	.0005	.0074	.0467	.1491	.2182	.0798	.0112
16	.0000	.0000	.0000	.0000	.0001	.0018	.0175	.0869	.2182	.1796	.0533
17	.0000	.0000	.0000	.0000	.0000	.0003	.0046	.0358	.1540	.2852	.1787
18	.0000	.0000	.0000	.0000	.0000	.0000	.0008	.0093	.0685	.2852	.3774
19	.0000	.0000	.0000	.0000	.0000	.0000	.0001	.0011	.0144	.1351	.3774

$n = 20$

s \ p	.05	.10	.20	.30	.40	.50	.60	.70	.80	.90	.95
0	.3585	.1216	.0115	.0008	.0000	.0000	.0000	.0000	.0000	.0000	.0000
1	.3774	.2702	.0576	.0068	.0005	.0000	.0000	.0000	.0000	.0000	.0000
2	.1887	.2852	.1369	.0278	.0031	.0002	.0000	.0000	.0000	.0000	.0000
3	.0596	.1901	.2054	.0716	.0123	.0011	.0000	.0000	.0000	.0000	.0000
4	.0133	.0898	.2182	.1304	.0350	.0046	.0003	.0000	.0000	.0000	.0000
5	.0022	.0319	.1746	.1789	.0746	.0148	.0013	.0000	.0000	.0000	.0000
6	.0003	.0089	.1091	.1916	.1244	.0370	.0049	.0002	.0000	.0000	.0000
7	.0000	.0020	.0545	.1643	.1659	.0739	.0146	.0010	.0000	.0000	.0000
8	.0000	.0004	.0222	.1144	.1797	.1201	.0355	.0039	.0001	.0000	.0000
9	.0000	.0001	.0074	.0654	.1597	.1602	.0710	.0120	.0005	.0000	.0000
10	.0000	.0000	.0020	.0308	.1171	.1762	.1171	.0308	.0020	.0000	.0000
11	.0000	.0000	.0005	.0120	.0710	.1602	.1597	.0654	.0074	.0001	.0000
12	.0000	.0000	.0001	.0039	.0355	.1201	.1797	.1144	.0222	.0004	.0000
13	.0000	.0000	.0000	.0010	.0146	.0739	.1659	.1643	.0545	.0020	.0000
14.	.0000	.0000	.0000	.0002	.0049	.0370	.1244	.1916	.1091	.0089	.0003
15	.0000	.0000	.0000	.0000	.0013	.0148	.0746	.1789	.1746	.0319	.0022
16	.0000	.0000	.0000	.0000	.0003	.0046	.0350	.1304	.2182	.0898	.0133
17	.0000	.0000	.0000	.0000	.0000	.0011	.0123	.0716	.2054	.1901	.0596
18	.0000	.0000	.0000	.0000	.0000	.0002	.0031	.0278	.1369	.2852	.1887
19	.0000	.0000	.0000	.0000	.0000	.0000	.0005	.0068	.0576	.2702	.3774
20	.0000	.0000	.0000	.0000	.0000	.0000	.0000	.0008	.0115	.1216	.3585

APPENDIX I

Summation of Terms of Poisson's Exponential Binomial Limit (1000 × probability of c or less occurrences of event that has average number of occurrences equal to μ_c or μ_{np})

	c									
μ_c or μ_{np}	0	1	2	3	4	5	6	7	8	9
0.02	980	1000								
0.04	961	999	1000							
0.06	942	998	1000							
0.08	923	997	1000							
0.10	905	995	1000							
0.15	861	990	999	1000						
0.20	819	982	999	1000						
0.25	779	974	998	1000						
0.30	741	963	996	1000						
0.35	705	951	994	1000						
0.40	670	938	992	999	1000					
0.45	638	925	989	999	1000					
0.50	607	910	986	998	1000					
0.55	577	894	982	998	1000					
0.60	549	878	977	997	1000					
0.65	522	861	972	996	999	1000				
0.70	497	844	966	994	999	1000				
0.75	472	827	959	993	999	1000				
0.80	449	809	953	991	999	1000				
0.85	427	791	945	989	998	1000				
0.90	407	772	937	987	998	1000				
0.95	387	754	929	984	997	1000				
1.00	368	736	920	981	996	999	1000			
1.1	333	699	900	974	995	999	1000			
1.2	301	663	879	966	992	998	1000			
1.3	273	627	857	957	989	998	1000			
1.4	247	592	833	946	986	997	999	1000		
1.5	223	558	809	934	981	996	999	1000		
1.6	202	525	783	921	976	994	999	1000		
1.7	183	493	757	907	970	992	998	1000		
1.8	165	463	731	891	964	990	997	999	1000	
1.9	150	434	704	875	956	987	997	999	1000	
2.0	135	406	677	857	947	983	995	999	1000	

(continued)

μ_c or μ_{np}	c 0	1	2	3	4	5	6	7	8	9
2.2	111	355	623	819	928	975	993	998	1000	
2.4	091	308	570	779	904	964	988	997	999	1000
2.6	074	267	518	736	877	951	983	995	999	1000
2.8	061	231	469	692	848	935	976	992	998	999
3.0	050	199	423	647	815	916	966	988	996	999
3.2	041	171	380	603	781	895	955	983	994	998
3.4	033	147	340	558	744	871	942	977	992	997
3.6	027	126	303	515	706	844	927	969	988	996
3.8	022	107	269	473	668	816	909	960	984	994
4.0	018	092	238	433	629	785	889	949	979	992
4.2	015	078	210	395	590	753	867	936	972	989
4.4	012	066	185	359	551	720	844	921	964	985
4.6	010	056	163	326	513	686	818	905	955	980
4.8	008	048	143	294	476	651	791	887	944	975
5.0	007	040	125	265	440	616	762	867	932	968
5.2	006	034	109	238	406	581	732	845	918	960
5.4	005	029	095	213	373	546	702	822	903	951
5.6	004	024	082	191	342	512	670	797	886	941
5.8	003	021	072	170	313	478	638	771	867	929
6.0	002	017	062	151	285	446	606	744	847	916

	10	11	12	13	14	15	16
2.8	1000						
3.0	1000						
3.2	1000						
3.4	999	1000					
3.6	999	1000					
3.8	998	999	1000				
4.0	997	999	1000				
4.2	996	999	1000				
4.4	994	998	999	1000			
4.6	992	997	999	1000			
4.8	990	996	999	1000			
5.0	986	995	998	999	1000		
5.2	982	993	997	999	1000		
5.4	977	990	996	999	1000		
5.6	972	988	995	998	999	1000	
5.8	965	984	993	997	999	1000	
6.0	957	980	991	996	999	999	1000

(continued)

μ_c or μ_{np}	c = 0	1	2	3	4	5	6	7	8	9
6.2	002	015	054	134	259	414	574	716	826	902
6.4	002	012	046	119	235	384	542	687	803	886
6.6	001	010	040	105	213	355	511	658	780	869
6.8	001	009	034	093	192	327	480	628	755	850
7.0	001	007	030	082	173	301	450	599	729	830
7.2	001	006	025	072	156	276	420	569	703	810
7.4	001	005	022	063	140	253	392	539	676	788
7.6	001	004	019	055	125	231	365	510	648	765
7.8	000	004	016	048	112	210	338	481	620	741
8.0	000	003	014	042	100	191	313	453	593	717
8.5	000	002	009	030	074	150	256	386	523	653
9.0	000	001	006	021	055	116	207	324	456	587
9.5	000	001	004	015	040	089	165	269	392	522
10.0	000	000	003	010	029	067	130	220	333	458

μ_c or μ_{np}	10	11	12	13	14	15	16	17	18	19
6.2	949	975	989	995	998	999	1000			
6.4	939	969	986	994	997	999	1000			
6.6	927	963	982	992	997	999	999	1000		
6.8	915	955	978	990	996	998	999	1000		
7.0	901	947	973	987	994	998	999	1000		
7.2	887	937	967	984	993	997	999	999	1000	
7.4	871	926	961	980	991	996	998	999	1000	
7.6	854	915	954	976	989	995	998	999	1000	
7.8	835	902	945	971	986	993	997	999	1000	
8.0	816	888	936	966	983	992	996	998	999	1000
8.5	763	849	909	949	973	986	993	997	999	999
9.0	706	803	876	926	959	978	989	995	998	999
9.5	645	752	836	898	940	967	982	991	996	998
10.0	583	697	792	864	917	951	973	986	993	997

μ_c or μ_{np}	20	21	22
8.5	1000		
9.0	1000		
9.5	999	1000	
10.0	998	999	1000

(continued)

μ_c or μ_{np}	c: 0	1	2	3	4	5	6	7	8	9
10.5	000	000	002	007	021	050	102	179	279	397
11.0	000	000	001	005	015	038	079	143	232	341
11.5	000	000	001	003	011	028	060	114	191	289
12.0	000	000	001	002	008	020	046	090	155	242
12.5	000	000	000	002	005	015	035	070	125	201
13.0	000	000	000	001	004	011	026	054	100	166
13.5	000	000	000	001	003	008	019	041	079	135
14.0	000	000	000	000	002	006	014	032	062	109
14.5	000	000	000	000	001	004	010	024	048	088
15.0	000	000	000	000	001	003	008	018	037	070

μ_c or μ_{np}	10	11	12	13	14	15	16	17	18	19
10.5	521	639	742	825	888	932	960	978	988	994
11.0	460	579	689	781	854	907	944	968	982	991
11.5	402	520	633	733	815	878	924	954	974	986
12.0	347	462	576	682	772	844	899	937	963	979
12.5	297	406	519	628	725	806	869	916	948	969
13.0	252	353	463	573	675	764	835	890	930	957
13.5	211	304	409	518	623	718	798	861	908	942
14.0	176	260	358	464	570	669	756	827	883	923
14.5	145	220	311	413	518	619	711	790	853	901
15.0	118	185	268	363	466	568	664	749	819	875

μ_c or μ_{np}	20	21	22	23	24	25	26	27	28	29
10.5	997	999	999	1000						
11.0	995	998	999	1000						
11.5	992	996	998	999	1000					
12.0	988	994	997	999	999	1000				
12.5	983	991	995	998	999	999	1000			
13.0	975	986	992	996	998	999	1000			
13.5	965	980	989	994	997	998	999	1000		
14.0	952	971	983	991	995	997	999	999	1000	
14.5	936	960	976	986	992	996	998	999	999	1000
15.0	917	947	967	981	989	994	997	998	999	1000

(continued)

μ_c or μ_{np}	c									
	4	**5**	**6**	**7**	**8**	**9**	**10**	**11**	**12**	**13**
16	000	001	004	010	022	043	077	127	193	275
17	000	001	002	005	013	026	049	085	135	201
18	000	000	001	003	007	015	030	055	092	143
19	000	000	001	002	004	009	018	035	061	098
20	000	000	000	001	002	005	011	021	039	066
21	000	000	000	000	001	003	006	013	025	043
22	000	000	000	000	001	002	004	008	015	028
23	000	000	000	000	000	001	002	004	009	017
24	000	000	000	000	000	000	001	003	005	011
25	000	000	000	000	000	000	001	001	003	006
	14	**15**	**16**	**17**	**18**	**19**	**20**	**21**	**22**	**23**
16	368	467	566	659	742	812	868	911	942	963
17	281	371	468	564	655	736	805	861	905	937
18	208	287	375	469	562	651	731	799	855	899
19	150	215	292	378	469	561	647	725	793	849
20	105	157	221	297	381	470	559	644	721	787
21	072	111	163	227	302	384	471	558	640	716
22	048	077	117	169	232	306	387	472	556	637
23	031	052	082	123	175	238	310	389	472	555
24	020	034	056	087	128	180	243	314	392	473
25	012	022	038	060	092	134	185	247	318	394
	24	**25**	**26**	**27**	**28**	**29**	**30**	**31**	**32**	**33**
16	978	987	993	996	998	999	999	1000		
17	959	975	985	991	995	997	999	999	1000	
18	932	955	972	983	990	994	997	998	999	1000
19	893	927	951	969	980	988	993	996	998	999
20	843	888	922	948	966	978	987	992	995	997
21	782	838	883	917	944	963	976	985	991	994
22	712	777	832	877	913	940	959	973	983	989
23	635	708	772	827	873	908	936	956	971	981
24	554	632	704	768	823	868	904	932	953	969
25	473	553	629	700	763	818	863	900	929	950

μ_c or μ_{np}	c 34	35	36	37	38	39	40	41	42	43
19	999	1000								
20	999	999	1000							
21	997	998	999	999	1000					
22	994	996	998	999	999	1000				
23	988	993	996	997	999	999	1000			
24	979	987	992	995	997	998	999	999	1000	
25	966	978	985	991	994	997	998	999	999	1000

INDEX